Green Corrosion Inhibition
Fundamentals, Design, Synthesis and Applications

Green Chemistry Series

Titles in the series:
1: The Future of Glycerol: New Uses of a Versatile Raw Material
2: Alternative Solvents for Green Chemistry
3: Eco-Friendly Synthesis of Fine Chemicals
4: Sustainable Solutions for Modern Economies
5: Chemical Reactions and Processes under Flow Conditions
6: Radical Reactions in Aqueous Media
7: Aqueous Microwave Chemistry
8: The Future of Glycerol: 2nd Edition
9: Transportation Biofuels: Novel Pathways for the Production of Ethanol, Biogas and Biodiesel
10: Alternatives to Conventional Food Processing
11: Green Trends in Insect Control
12: A Handbook of Applied Biopolymer Technology: Synthesis, Degradation and Applications
13: Challenges in Green Analytical Chemistry
14: Advanced Oil Crop Biorefineries
15: Enantioselective Homogeneous Supported Catalysis
16: Natural Polymers Volume 1: Composites
17: Natural Polymers Volume 2: Nanocomposites
18: Integrated Forest Biorefineries
19: Sustainable Preparation of Metal Nanoparticles: Methods and Applications
20: Alternative Solvents for Green Chemistry: 2nd Edition
21: Natural Product Extraction: Principles and Applications
22: Element Recovery and Sustainability
23: Green Materials for Sustainable Water Remediation and Treatment
24: The Economic Utilisation of Food Co-Products
25: Biomass for Sustainable Applications: Pollution Remediation and Energy
26: From C–H to C–C Bonds: Cross-Dehydrogenative-Coupling
27: Renewable Resources for Biorefineries
28: Transition Metal Catalysis in Aerobic Alcohol Oxidation
29: Green Materials from Plant Oils

30: Polyhydroxyalkanoates (PHAs) Based Blends, Composites and Nanocomposites
31: Ball Milling Towards Green Synthesis: Applications, Projects, Challenges
32: Porous Carbon Materials from Sustainable Precursors
33: Heterogeneous Catalysis for Today's Challenges: Synthesis, Characterization and Applications
34: Chemical Biotechnology and Bioengineering
35: Microwave-Assisted Polymerization
36: Ionic Liquids in the Biorefinery Concept: Challenges and Perspectives
37: Starch-based Blends, Composites and Nanocomposites
38: Sustainable Catalysis: With Non-endangered Metals, Part 1
39: Sustainable Catalysis: With Non-endangered Metals, Part 2
40: Sustainable Catalysis: Without Metals or Other Endangered Elements, Part 1
41: Sustainable Catalysis: Without Metals or Other Endangered Elements, Part 2
42: Green Photo-active Nanomaterials
43: Commercializing Biobased Products: Opportunities, Challenges, Benefits, and Risks
44: Biomass Sugars for Non-Fuel Applications
45: White Biotechnology for Sustainable Chemistry
46: Green and Sustainable Medicinal Chemistry: Methods, Tools and Strategies for the 21st Century Pharmaceutical Industry
47: Alternative Energy Sources for Green Chemistry
48: High Pressure Technologies in Biomass Conversion
49: Sustainable Solvents: Perspectives from Research, Business and International Policy
50: Fast Pyrolysis of Biomass: Advances in Science and Technology
51: Catalyst-free Organic Synthesis
52: Hazardous Reagent Substitution: A Pharmaceutical Perspective
53: Alternatives to Conventional Food Processing: 2nd Edition
54: Sustainable Synthesis of Pharmaceuticals: Using Transition Metal Complexes as Catalysts
55: Intensification of Biobased Processes
56: Sustainable Catalysis for Biorefineries
57: Supercritical and Other High-pressure Solvent Systems: For Extraction, Reaction and Material Processing
58: Biobased Aerogels: Polysaccharide and Protein-based Materials
59: Rubber Recycling: Challenges and Developments
60: Green Chemistry for Surface Coatings, Inks and Adhesives: Sustainable Applications
61: Green Synthetic Processes and Procedures
62: Resource Recovery from Wastes: Towards a Circular Economy
63: Flow Chemistry: Integrated Approaches for Practical Applications
64: Transition Towards a Sustainable Biobased Economy

65: Transportation Biofuels: Pathways for Production: 2nd Edition
66: Challenges in Green Analytical Chemistry: 2nd Edition
67: CO_2-switchable Materials: Solvents, Surfactants, Solutes and Solids
68: Green Toxicology: Making Chemicals Benign by Design
69: Sustainable and Functional Redox Chemistry
70: Waste Valorisation Using Ionic Liquids
71: Natural Product Extraction: Principles and Applications: 2nd Edition
72: Renewable Resources for Surface Coatings, Inks and Adhesives
73: Green Corrosion Inhibition: Fundamentals, Design, Synthesis and Applications

How to obtain future titles on publication:
A standing order plan is available for this series. A standing order will bring delivery of each new volume immediately on publication.

For further information please contact:
Book Sales Department, Royal Society of Chemistry, Thomas Graham House, Science Park, Milton Road, Cambridge, CB4 0WF, UK
Telephone: +44 (0)1223 420066, Fax: +44 (0)1223 420247
Email: booksales@rsc.org
Visit our website at www.rsc.org/books

Green Corrosion Inhibition
Fundamentals, Design, Synthesis and Applications

By

Chandrabhan Verma
King Fahd University of Petroleum and Minerals, Saudi Arabia
Email: Chandraverma.rs.apc@itbhu.ac.in

Mumtaz A. Quraishi
King Fahd University of Petroleum and Minerals, Saudi Arabia
Email: mumtaz.quraishi@kfupm.edu.sa

and

Dheeraj Singh Chauhan
King Fahd University of Petroleum and Minerals, Saudi Arabia
Email: dschauhan.bhu@gmail.com

ROYAL SOCIETY OF CHEMISTRY

Green Chemistry Series No. 73

Print ISBN: 978-1-83916-702-7
PDF ISBN: 978-1-83916-746-1
EPUB ISBN: 978-1-83916-747-8
Print ISSN: 1757-7039
Electronic ISSN: 1757-7047

A catalogue record for this book is available from the British Library

The Royal Society of Chemistry is a charity, registered in England and Wales, Number 207890, and a company incorporated in England by Royal Charter (Registered No. RC000524), registered office: Burlington House, Piccadilly, London W1J 0BA, UK, Telephone: +44 (0) 20 7437 8656.

For further information see our website at www.rsc.org

Printed in the United Kingdom by CPI Group (UK) Ltd, Croydon, CR0 4YY, UK

Preface

Development and understanding of corrosion inhibition action and mechanisms is one of the foremost areas of surface science, electrochemistry, and materials science and engineering. This is related to a wide field of applications, and a great deal of experimental and computational resources are devoted to research and development of novel inhibitor molecules. In the past few years, a considerable amount of attention has been paid to the development of environmentally benign corrosion inhibitors. This is because of the strict guidelines issued by the various government agencies safeguarding and protecting the environment from harmful and toxic discharge of industrial effluents. This book aims to provide a comprehensive overview of the various types of green corrosion inhibitors, their design protocols, and performance.

Chapter 1 presents some of the fundamental aspects of corrosion, and corrosion inhibitors. Chapter 2 describes some of the principles of green chemistry and certain metrics relevant to development of environmentally benign inhibitors. Some of the major experimental (gravimetric, electrochemical and surface analyses) and computational approaches (*e.g.* DFT, and atomistic simulations) for analysis and estimation of inhibition performance are given in Chapter 3. To avoid unnecessary confusion and to facilitate easy understanding we have refrained from emphasizing mathematical aspects, and provided only the most important theories, and equations.

Chapters 4–11 respectively focus attention on the various types of green and sustainable corrosion inhibitors developed from ionic liquids, synthetic organics, natural extracts, biopolymers, amino acids, surfactants, oleochemicals, and pharmaceutical products. In each of these chapters, the significance of the inhibitor type, greenness, and importance in corrosion inhibition is outlined. In each of the chapters, some of the key issues on the

Green Chemistry Series No. 73
Green Corrosion Inhibition: Fundamentals, Design, Synthesis and Applications
By Chandrabhan Verma, Mumtaz A. Quraishi and Dheeraj Singh Chauhan

Published by the Royal Society of Chemistry, www.rsc.org

prospects of commercialization of these corrosion inhibitors are also discussed. This is followed by a brief overview of literature on some examples of each kind of corrosion inhibitors.

The book attempts to not only cover the various inhibitor types, but also provides a glimpse of their applicability in various industries, such as oil-well stimulation, oil-gas transport, desalination, *etc.* Accordingly, some of the structures of inhibitors, methods of chemical modifications, and inhibition performance are also discussed.

The major target readers of this book include scientists, researchers, as well as students working in the disciplines of chemistry, materials science, chemical engineering, mechanical engineering, and computational studies. We hope that this book will provide a handy reference tool for students and researchers working in the area of green corrosion inhibitors. Special thanks to Dr Helen Armes, commissioning editor, for her dedicated support and help during this project. In the end all thanks to the Royal Society Chemistry for publishing the book.

Chandrabhan Verma
M. A. Quraishi
Dheeraj Singh Chauhan

Author Biography

Chandrabhan Verma
Interdisciplinary Research Center for Advanced Materials, King Fahd University of Petroleum and Minerals, Dhahran 31261, Saudi Arabia.

Chandrabhan Verma works at the Interdisciplinary Center for Research in Advanced Materials, King Fahd University of Petroleum and Minerals (KFUPM), Saudi Arabia. He obtained his PhD in Material Science/Chemistry from the Indian Institute of Technology (Banaras Hindu University), Varanasi, India. He is a member of the American Chemical Society (ACS) and serves as a reviewer and editorial board member for the internationally recognized ACS, RSC, Elsevier, Wiley and Springer platforms. Dr Verma is Associate Editor-in-Chief of the Organic Chemistry Plus Journal. He has published more than 150 papers in peer-reviewed journals, having a *h*-index of 44 and more than 6200 citations. Currently, Dr Verma has edited 8 books for the ACS, Elsevier, RSC, Springer and Wiley. Dr Verma has received several awards for his academic achievements.

Green Chemistry Series No. 73
Green Corrosion Inhibition: Fundamentals, Design, Synthesis and Applications
By Chandrabhan Verma, Mumtaz A. Quraishi and Dheeraj Singh Chauhan

Published by the Royal Society of Chemistry, www.rsc.org

Mumtaz A. Quraishi
Interdisciplinary Research Center for Advanced Materials, King Fahd University of Petroleum and Minerals, Dhahran 31261, Saudi Arabia.

Prof. Mumtaz A. Quraishi is a Chair Professor at the Interdisciplinary Center for Research in Advanced Materials, King Fahd University of Petroleum and Minerals (KFUPM), Saudi Arabia. He obtained his PhD in synthetic organic chemistry in 1986 from Kurukshetra University and was awarded a DSc in 2004 from Aligarh Muslim University, Aligarh, in the field of Corrosion Inhibition of Industrial Metals and Alloys. Before joining KFUPM, he was a full Professor at IIT BHU, Varanasi, India. He also served as Head (Chairman) of the Department of Chemistry at IIT BHU. He has teaching experience of more than 35 years. He has received several national and international awards. Dr Quraishi is an associate editor of Current Material Science, Bentham, and a member of the editorial board of more than 30 international journals. Dr Quraishi is a fellow of the Royal Society of Chemistry, UK, and a member of the American Chemical Society. He has published more than 400 papers in peer-reviewed journals, having a h-index of 91 and more than 26 500 citations. His global status is one in terms of a very high h-index in the field of Corrosion Inhibitors. He authored a book entitled *Heterocyclic Organic Corrosion Inhibitors Principles and Applications*, Elsevier 2020.

Dr Dheeraj Singh Chauhan
Modern National Chemicals, Second Industrial City, Dammam 31421, Saudi Arabia.

Dr Dheeraj Singh Chauhan is working as a Corrosion Specialist at Modern National Chemicals, Dammam, Saudi Arabia. He received his BSc (Hons) (2002) and MSc (2004) degrees from the Banaras Hindu University (BHU), India, and PhD (2012) from the Indian Institute of Technology (IIT, BHU). Soon after, he joined Nanjing University of Science and Technology, China, as a Post-Doctoral Research Fellow (2013–2015). After that, Dr Chauhan worked as a Post-Doctoral Research Fellow at the Center of Research Excellence in Corrosion, King Fahd University of Petroleum and Minerals, Saudi Arabia (2018–2020). Currently, he is working on the development of corrosion inhibitors for the oil and gas industry. He has over 90 publications with more than 2700 citations and an h-index of 32.

Contents

Green Chemistry Series No. 73
Green Corrosion Inhibition: Fundamentals, Design, Synthesis and Applications
By Chandrabhan Verma, Mumtaz A. Quraishi and Dheeraj Singh Chauhan

Published by the Royal Society of Chemistry, www.rsc.org

CHAPTER 1

Fundamentals of Corrosion and Corrosion Inhibition

1.1 Corrosion: Basics and Adverse Effects

Corrosion is derived from the term "corrode", meaning to be eaten away gradually or to eat away, mainly by chemical reactions. According to the National Association of Chemical Engineers (NACE), corrosion is well-defined as the "weakening of materials (usually metals) or their properties due to their reaction through the environment".[1] In nature, most metals are present in their combined or ore forms, and the processing and manufacturing of pure metals require the input of energy. In their purified forms, most metals experience very high chemical reactivity and therefore undergo disintegration reactions with the ingredients of the atmosphere to form corrosion products (rusts and scales). Obviously, in the absence of moisture, corrosion proceeds through chemical reactions and is called dry corrosion or chemical corrosion, and in the presence of moisture, corrosion occurs through electrochemical methods and is known as wet corrosion or electrochemical corrosion.

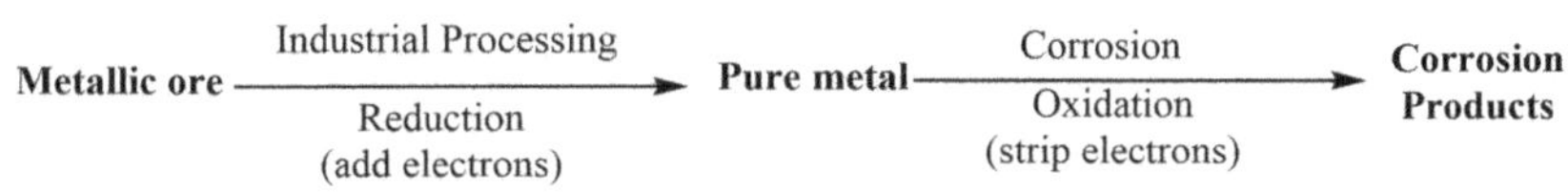

Corrosion causes extremely high safety and economic losses. As per the NACE estimation, the current cost of corrosion is about US $2.5 trillion (≈3.4% of the world's GDP).[2] Because of the increasing demands, consumption and utilization of metallic structures, especially in the building and construction sectors, the price of corrosion is anticipated to increase.[3] Besides economic losses, there are numerous accidents have been reported due to corrosion failure, which causes huge safety and life losses.

Green Chemistry Series No. 73
Green Corrosion Inhibition: Fundamentals, Design, Synthesis and Applications
By Chandrabhan Verma, Mumtaz A. Quraishi and Dheeraj Singh Chauhan

Published by the Royal Society of Chemistry, www.rsc.org

1.2 Corrosion Inhibition Using Organic Compounds

Due to very high security and monetary damage, numerous approaches to corrosion mitigation have been industrialized, depending upon the nature, material under investigation and the environment. Noticeably, annual savings of US \$375 (15%) to \$875 (35%) billion can be made by applying the formerly prevailing procedures of corrosion anticipation.[2,4] Some of the common methods of corrosion mitigation are alloying (galvanization), painting, coating, dealloying, cathodic protection and the use of corrosion inhibitors (Figure 1.1). In aqueous electrolytes, metallic corrosion proceeds through electrochemical mechanisms where the employment of organic compounds, particularly heterocyclic compounds, is the most common, frequently utilized and economical method of corrosion mitigation.[5–7] Literature search advises that the organic compounds operate by adsorbing on the metal surface.

These compounds adsorb on the metal surface using their electron dense sites, including π-electrons of triple and double bonds such as >C=C<, –C≡N, >C=O, >C=S, >C=N–, –N=N– –C≡C–, and –N=O *etc.* and functional groups, for example –OH, –CN, –OMe, $-CONH_2$, –O–, $-COOC_2H_5$, –SH, –COOH, $-NO_2$ *etc.*[6,8] After getting adsorbed, these compounds retard both anodic and cathodic Tafel reactions. Adsorption of an organic corrosion inhibitor (INH)

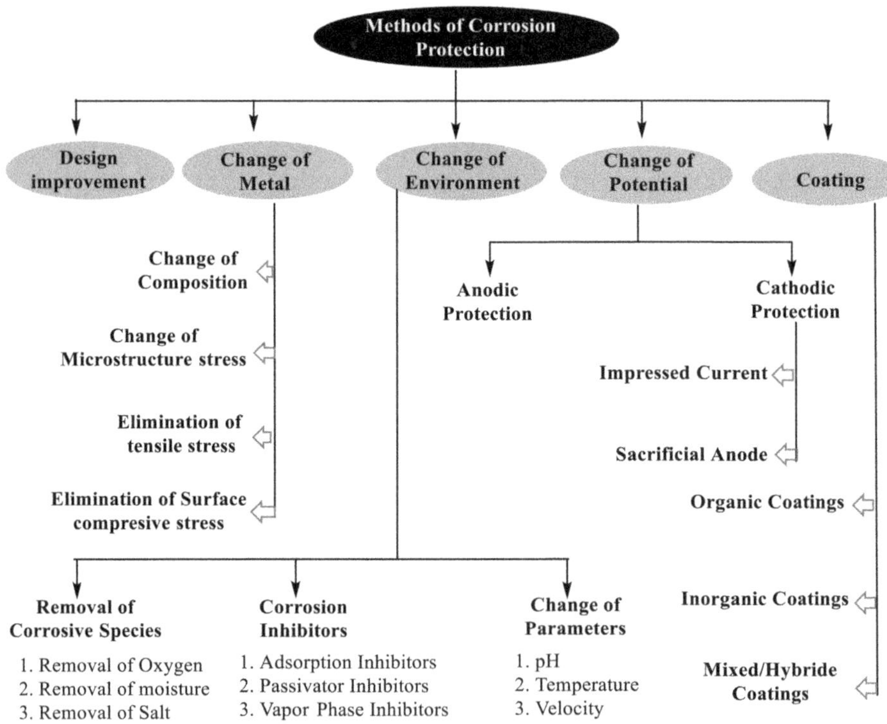

Figure 1.1 Methods of corrosion prevention.

and the retardation of corrosion (of iron) in an aqueous electrolyte are illustrated as follows:[9,10]

(i) Tafel reactions without corrosion inhibitor

$$Fe + H_2O \leftrightarrow FeOH_{ads} + H^+ + e^- \tag{1.1}$$

$$FeOH_{ads} \leftrightarrow FeOH^+ + e^- \text{ (slowest and rate determining step)} \tag{1.2}$$

$$FeOH^+ \leftrightarrow Fe^{2+} + H_2O \tag{1.3}$$

$$Fe + H^+ \leftrightarrow (FeH^+)_{ads} \tag{1.4}$$

$$(FeH^+)_{ads} + e^- \leftrightarrow (FeH)_{ads} \tag{1.5}$$

$$(FeH)_{ads} + H^+ + e^- \leftrightarrow Fe + H_2 \tag{1.6}$$

Eqn (1.1)–(1.3) represent the dissolution of iron at the anode, and eqn (1.4)–(1.6) denote the dissolution of iron at the cathode. Noticeably, at the anode, iron discharges in the ferrous form or as ferric ions, and at the cathode, it discharges in its neutral form. Adsorption of INH on the anodic site and retardation of the corrosion rate are presented below:

(ii) Tafel reactions with corrosion inhibitor (INH)

$$Fe + H_2O \leftrightarrow (FeH_2O)_{ads} \tag{1.7}$$

$$(FeH_2O)_{ads} + INH \leftrightarrow (Fe\text{-}INH^-)_{ads} + H^+ + INH \tag{1.8}$$

$$(FeH_2O)_{ads} + INH \leftrightarrow (Fe\text{-}INH)_{ads} + H_2O \tag{1.9}$$

$$(Fe\text{-}INH^-)_{ads} \leftrightarrow (Fe\text{-}INH)_{ads} + e^- \text{ (rate determining and slowest step)} \tag{1.10}$$

$$(Fe\text{-}INH^-)_{ads} \leftrightarrow (Fe\text{-}INH^+)_{ads} + e^- \tag{1.11}$$

$$(FeOH)_{ads} + (Fe\text{-}INH^+)_{ads} \leftrightarrow (FeOH^+)_{ads} + (Fe\text{-}INH)_{ads} \tag{1.12}$$

$$(FeOH^+)_{ads} + H^+ \leftrightarrow Fe^{2+} + H_2O \tag{1.13}$$

From the above schemes it can be evidently realized that INH molecules replace the previously adsorbed water (H_2O) molecules. This results in the

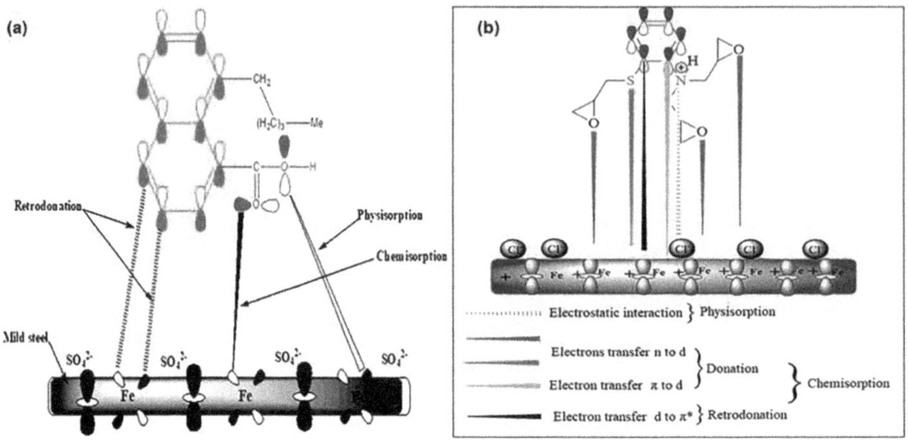

Figure 1.2 Diagrammatic illustration of physisorption and chemisorption of (a) a phytochemical.[12] Reproduced from ref. 12 with permission from the Royal Society of Chemistry; and (b) an epoxy resin on an iron surface.[13] Reproduced from ref. 13 with permission from the Royal Society of Chemistry.

formation of $(Fe\text{-}INH)_{ads}$ and a reduction in the amount of $(Fe\text{-}INH^-)_{ads}$, which is the rate decisive phase. It is well documented in the literature that, in an aqueous acidic solution, heteroatoms get protonated and therefore such organic corrosion inhibitors occur in the cationic form. On the other hand, metal surfaces acquire a negative charge due to the adsorption of counter ions of the electrolyte. Therefore, these positively charged moieties interact *via* electrostatic force of attraction, *i.e.* physisorption. For physisorption, Gibbs free energy (ΔG_{ads}) values are -20 kJ mol^{-1} or less negative. However, protonated/cationic heteroatoms return to their neutral form after gaining the electrons. The free electron pair(s) of heteroatoms interact with the metal surface and adsorb chemically, *i.e.* chemisorption. Unlike physisorption, chemisorption has Gibbs free energy (ΔG_{ads}) values of -40 kJ mol^{-1} or more negative. Nevertheless, these compounds adsorb on metallic surfaces in aqueous solutions using physiochemisorption mechanisms for which Gibbs free energy (ΔG_{ads}) values lie between -40 kJ mol^{-1} and -20 kJ mol^{-1}.[11] Many authors reported the physiochemisorption mechanism of organic corrosion inhibitors as adsorption on the metallic surface. Illustrative diagrams for the physiochemisorption of some organic corrosion inhibitors are presented in Figure 1.2.

1.3 Different Families of Heterocyclic Corrosion Inhibitors

Literature study suggests that various series of organic compounds are utilized as effective corrosion inhibitors. A detailed description of various types of organic corrosion inhibitors, especially green corrosion inhibitors, is described in various chapters of this book. In the present chapter, the corrosion inhibition potential of some selected series of heterocyclic compounds has been described.

1.3.1 Quinoline and its Derivatives as Corrosion Inhibitors

Quinoline is a bicyclic nitrogen-based heterocyclic that has the molecular formula C_9H_7N in which the pyridine ring is fused with the benzene.[14,15] Quinoline and its derivatives, mainly 8-hydroxyquinoline, exhibit very high biological and anticorrosive activities and the quinoline moiety is present in various drugs. Quinoline has five double bonds and one nitrogen atom, *i.e.* 10π and two non-bonding electrons, through which it easily gets adsorbed on the metallic surface and offers strong bonding. Suitably substituted quinoline, especially 8-hydroxyquinoline, 8-aminoquinoline and 8-nitroquinoline, can form chelating complexes with the metal atoms/ions. Quinoline and 8-hydroxyquinoline are widely investigated as corrosion inhibitors for different metals and alloys. Quinoline is slightly soluble in cold water but can be fully dissolved in hot water.[16–18] Obviously, quinoline derivatives exhibit superior anticorrosive activity compared to purified quinoline. This might be attributed to the combined effect of the chelating tendency of quinoline derivatives and enhanced solubility in the presence of polar substitutes. 8-Hydroquinoline and its derivatives are broadly used as corrosion inhibitors and a summary of the anticorrosive properties of some major reports are illustrated in Table 1.1.

These compounds are widely used in various experimental systems.[19] Similarly to the majority of organic corrosion inhibitors, quinoline based corrosion inhibitors act by forming the corrosion protective film. Using their π- and non-bonding electrons, quinoline based compounds easily get adsorbed onto metal surfaces. Their adsorption most likely depicts the Langmuir adsorption isotherm. After getting adsorbed, quinoline based corrosion inhibitors retard the Tafel reactions and perform as mixed-type inhibitors. Obviously, the presence of quinoline based compounds in the corrosive electrolyte diminishes the magnitude of current density (i_{corr}), which is attributed to the adsorption of these molecules at the active sites responsible for the corrosion. Generally, the presence of either type of substituents upsurges the %IE of quinoline, however electron donating groups have relatively more prominent effects compared to the electron donating substituents. Our research team, when studying the corrosion inhibition potential of 2-amino-4-arylquinoline-3-carbonitriles (AACs) for mild steel in 1 M HCl, displayed that the electron withdrawing nitro group has negative and electron donating hydroxyl group has positive effects compared to the non-substituented AAC molecule.[20] Using electrochemical analyses, it was observed that AACs behave as mixed- and interface-type corrosion inhibitors and become effective by blocking the active sites and adsorbing at the interface. DFT study indicates that the nitro group decreases the FMOs (HOMO

Table 1.1 A summary of some main reports on quinoline derivatives as corrosion inhibitors for different systems.

Chemical structure/abbreviation	Electrolyte	Metal	Conc./%IE	Ref.
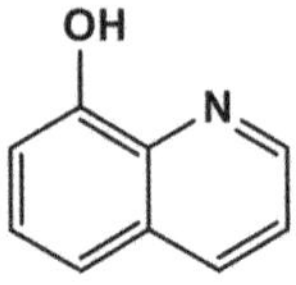 8-HQ	15% HCl	X60 steel	0.4 wt% + KI/83.27%	24
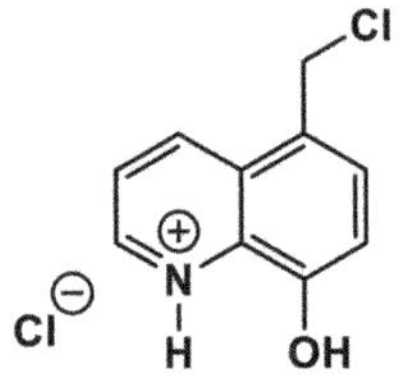	1 M HCl	XC38 steel	1.0 mM/97%	27
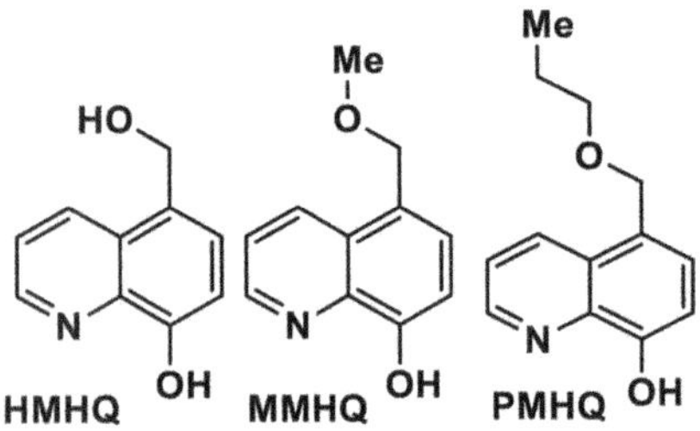 HMHQ MMHQ PMHQ	1 M HCl	Carbon steel	10^{-3} M (PMHQ)/94%	29
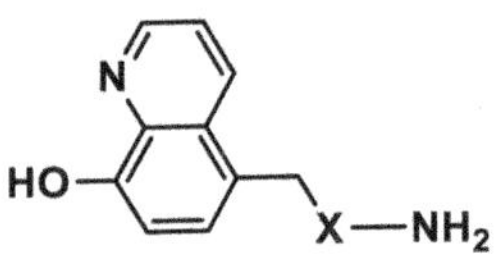 HMQN: X= NH AMQN: X= O	1 M HCl	Carbon steel	10^{-3} M (HMQN)/97.0%	31
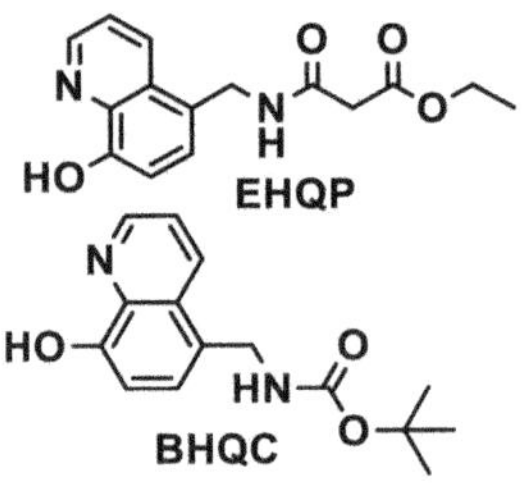 EHQP BHQC	1 M HCl	Carbon steel	10^{-3} M (BHQC)/95.2%	33
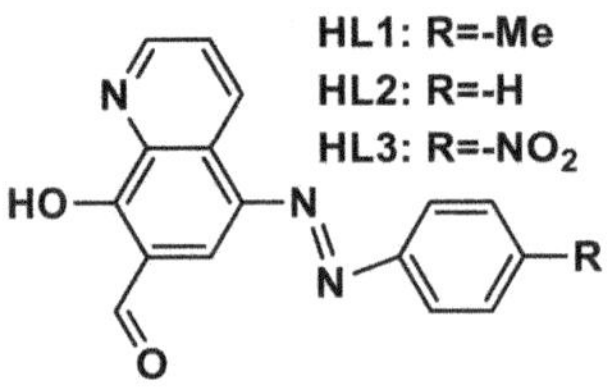 HL1: R=-Me HL2: R=-H HL3: R=-NO_2	2 M HCl	Carbon steel	11×10^{-6} M/87.6%	35

Chemical structure/abbreviation	Electrolyte	Metal	Conc./%IE	Ref.
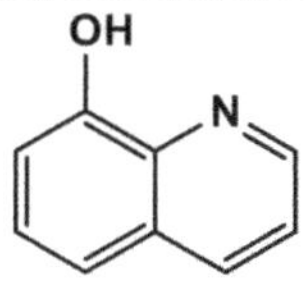 8-HQ	3% NaCl	Mg-AZ91D alloy	NA	25 and 26
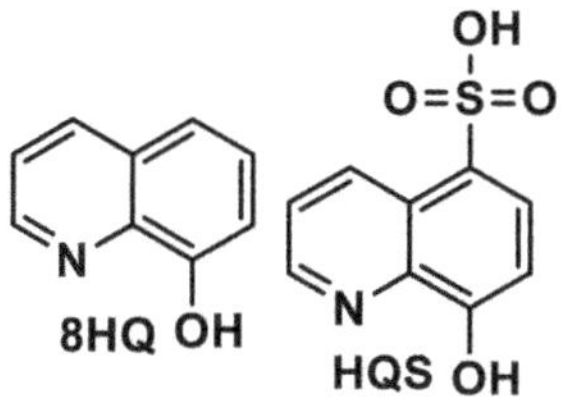 8HQ, HQS	3.5% NaCl	Al 2024-T3	NA	28
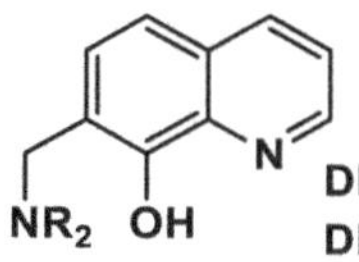 DMQ: R=-Me DPQ: R=-C_3H_7	1 M HCl	C40E steel	10^{-3} M (DPQ)/96.1%	30
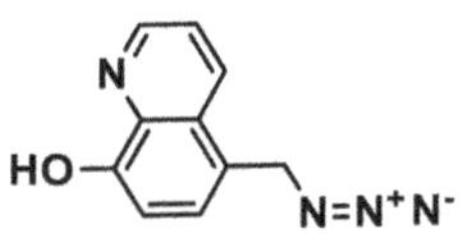 copper (II) and zinc (II).	1 M HCl	Mild steel	10^{-3} M $[Zn(Q)_2]$/90.95%	32
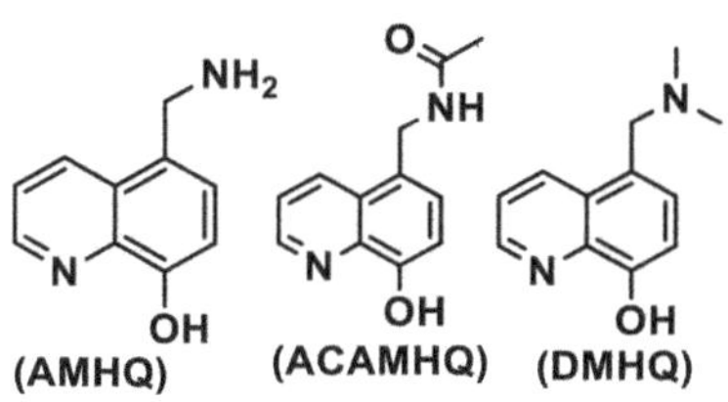 (AMHQ) (ACAMHQ) (DMHQ)	1 M HCl	Carbon steel	10^{-3} M (DMHQ)/92.8%	34
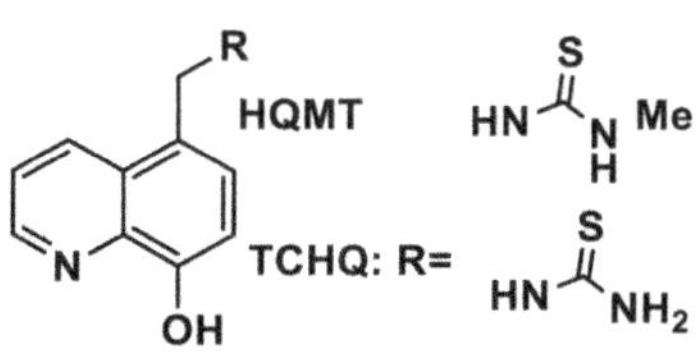 HQMT, TCHQ	1 M HCl	Carbon steel	10^{-3} M (HQMT)/95.6%	36

(*continued*)

Table 1.1 (*continued*)

Chemical structure/abbreviation	Electrolyte	Metal	Conc./%IE	Ref.
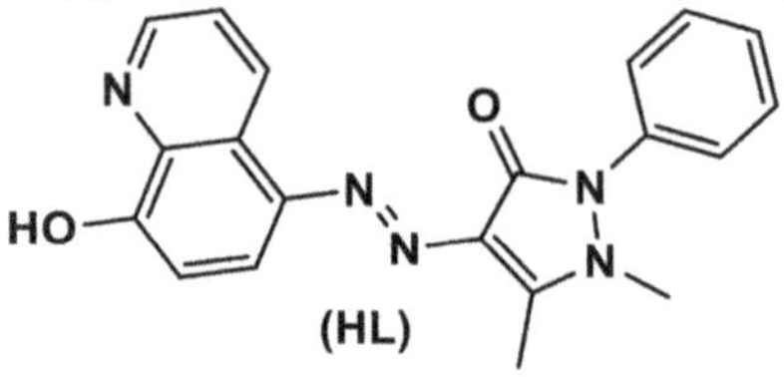 (HL)	2 M HCl	Carbon steel	11×10^{-6} M (HL)/94.6%	37
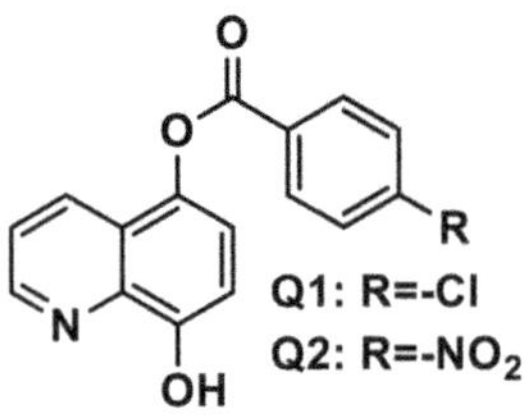 Q1: R=-Cl; Q2: R=-NO_2	1 M HCl	Mild steel	10^{-3} M (Q1)/95.2%	39
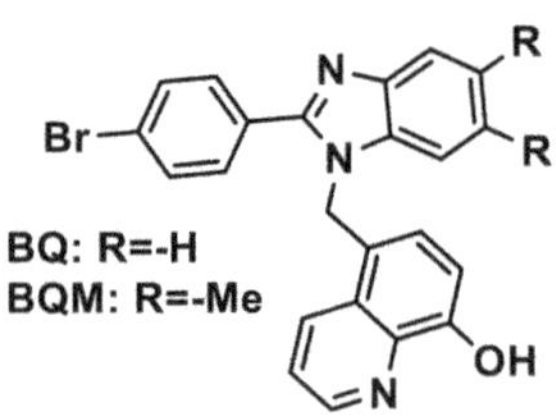 BQ: R=-H; BQM: R=-Me	1 M HCl	Mild steel	10^{-3} M (BQM)/98.0%	41
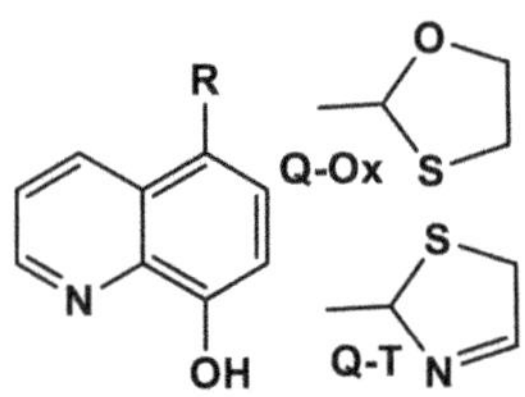 Q-Ox; Q-T	1.0 M HCl	Mild steel	10^{-3} M (Q-T)/97.6%	43
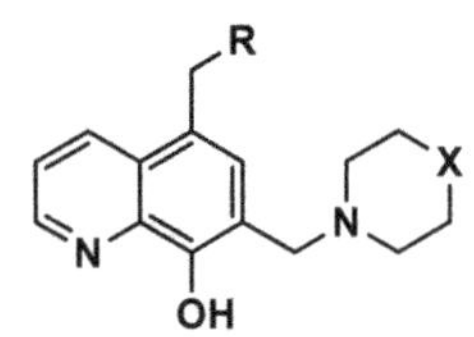 HM1: R=-N_3; X=O; HM2: R=-CN; X=O; HM3: R=-N_3; X=CH_2	1 M HCl	Mild steel	10^{-3} M (HM-3)/90%	45
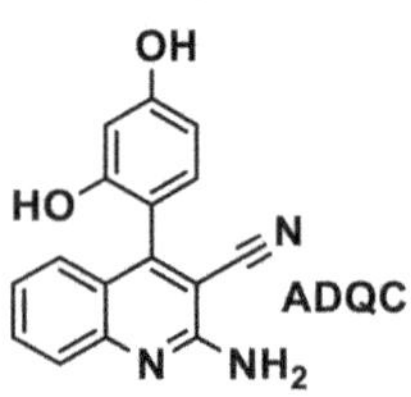 ADQC	1 M HCl	SAE 1006 steel	25 mg L^{-1} (ADQC)/96.12%	47

Chemical structure/abbreviation	Electrolyte	Metal	Conc./%IE	Ref.
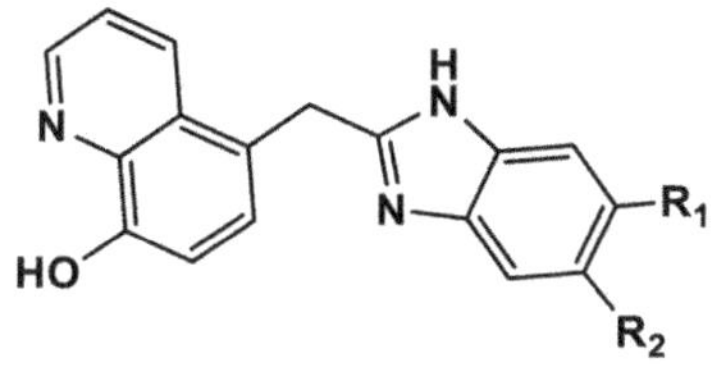 **BIMQ: R_1=R_2=-H; MBMQ; R_1= -H; R_2=-CH_3; CBMQ: R_1=-H; R_2=-Cl; DCBMQ: R_1= R_2=-Cl**	2 M H_3PO_4	Carbon steel	10^{-3} M (DCBMQ)/96.4%	38
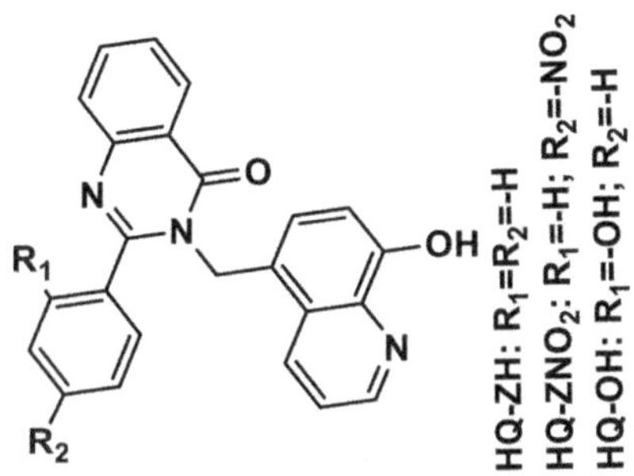	1 M HCl	Mild steel	10^{-3} M ($HQ-ZNO_2$)/95.9%	40
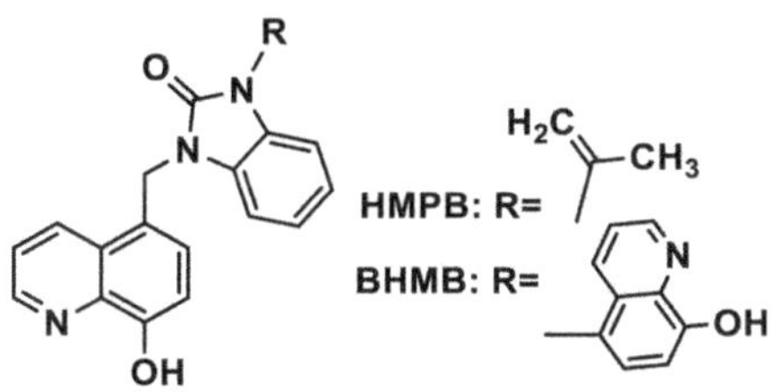	1 M HCl	Mild steel	10^{-3} M (BHMB)/91.2%	42
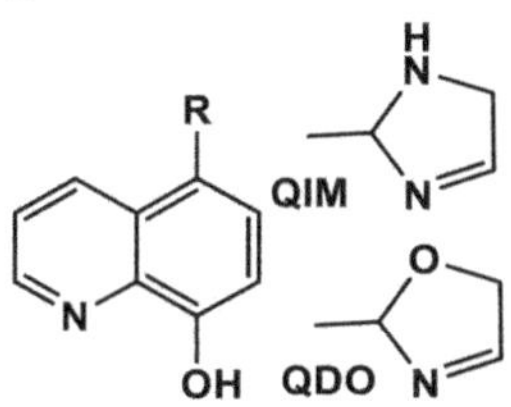	1 M HCl	Mild steel	10^{-3} M (QIM)/92.5%	44
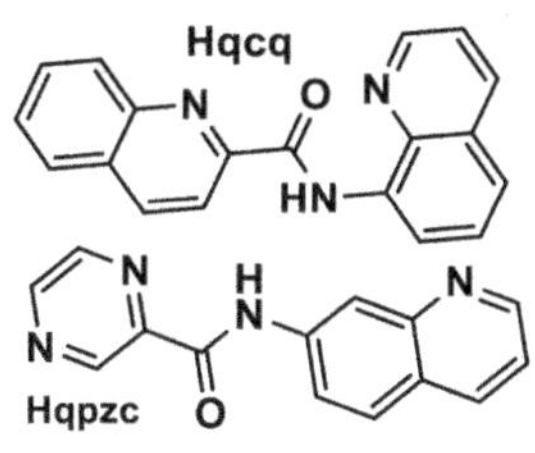	1 M HCl	Mild steel	100 ppm (Hqcq)/94%	46
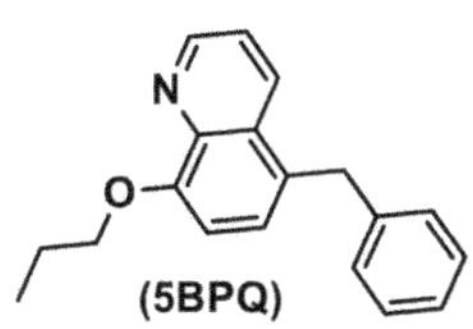	0.5 M H_2SO_4	Q235 steel	8.0 mM (5BPQ)/97.5%	48

(continued)

Table 1.1 (*continued*)

Chemical structure/abbreviation	Electrolyte	Metal	Conc./%IE	Ref.
AAC-1: R=-NO_2 AAC-2: R=-H AAC-3: R=-OH 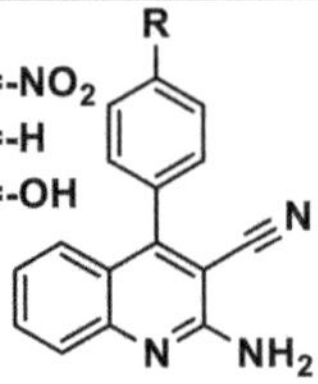	1 M HCl	Mild steel	50 ppm (AAC-3)/96.9%	20
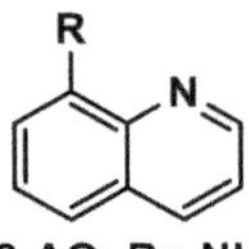 8-AQ: R=-NH_2 8-NQ: R=-NO_2	3% NaCl	Al-AA5052 alloy	0.02 mM (8-NQ)/89.5%	49
Q-1: R=-H Q-2: R=-Me Q-3: R=-OMe Q-4: R=-NMe_2 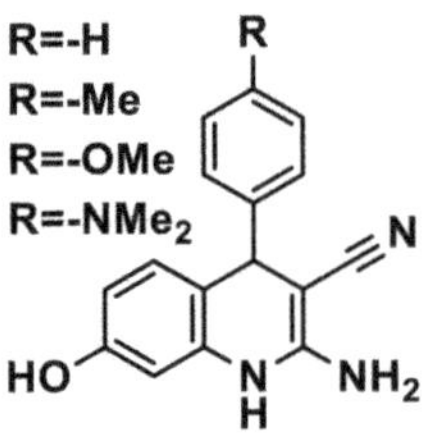	1 M HCl	Mild steel	150 ppm (Q-4)/98.1%	22
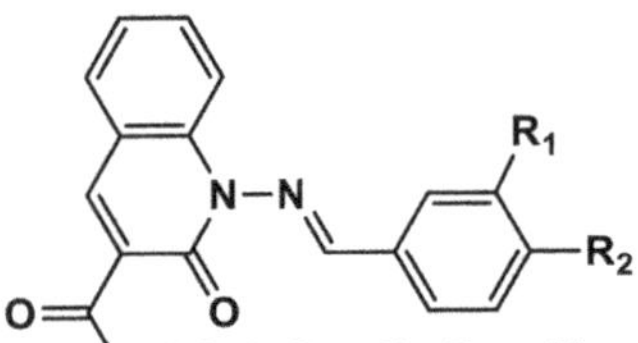 AQ-1:-R_1=-H; -R_2=--Me AQ-2:R_1=-H; -R_2=--OH AQ-3: R_1=-NO_2; -R_2=--H	15% HCl	N80 steel	0.1 M (AQ-3)/94.2%	52
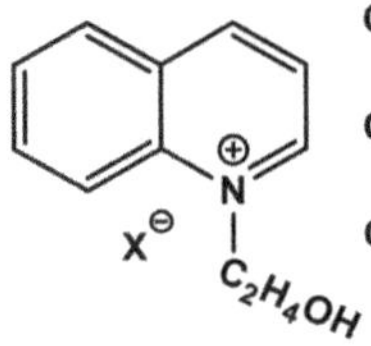 Compound 1: X= $\overset{\ominus}{O}$-Ph Compound 2: X= $\overset{\ominus}{O}$-Ph-p-Me Compound 3: X= $\overset{\ominus}{O}$-Ph-p-NO_2	26% NH_4Cl	Zinc–Mg batteries	500 (Comp-1)/97.9%	54 and 55
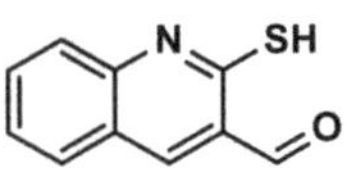 MQC 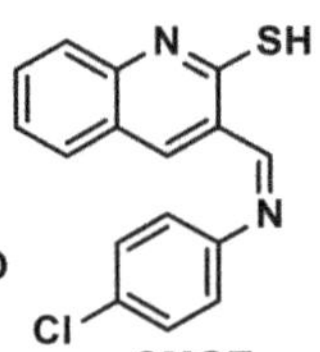 CMQT	1 M HCl	Mild steel	25 ppm (MQC)/94.26%	56

Chemical structure/abbreviation	Electrolyte	Metal	Conc./%IE	Ref.
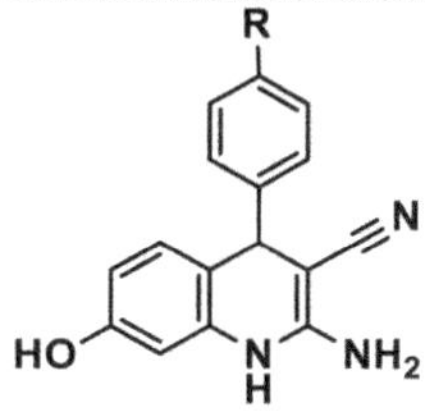	NA	Fe (110) surface	NA	21
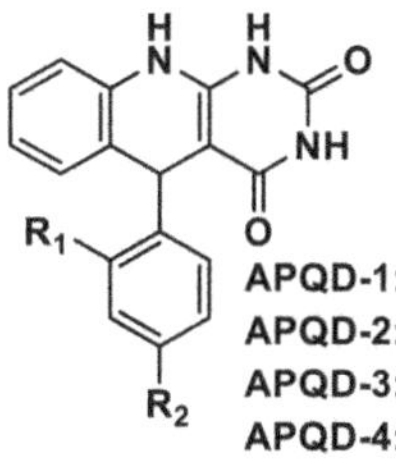	1 M HCl	Mild steel	20 ppm (APQD-4)/96.50%	50
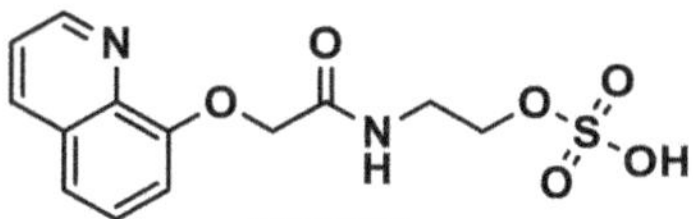	1.0 M $HClO_4$	Al	10^{-2} M (8HQD)/98.2%	51
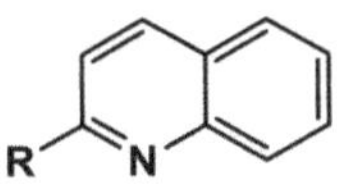	0.5 M HCl	Mild steel	0.1 M (QLD)/94.2%	53
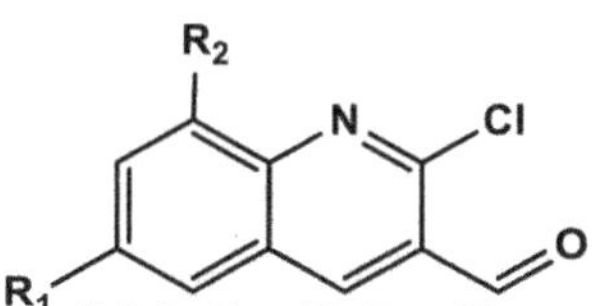	1 M HCl	Mild steel	5×10^{-3} (QA-3)/95%	23
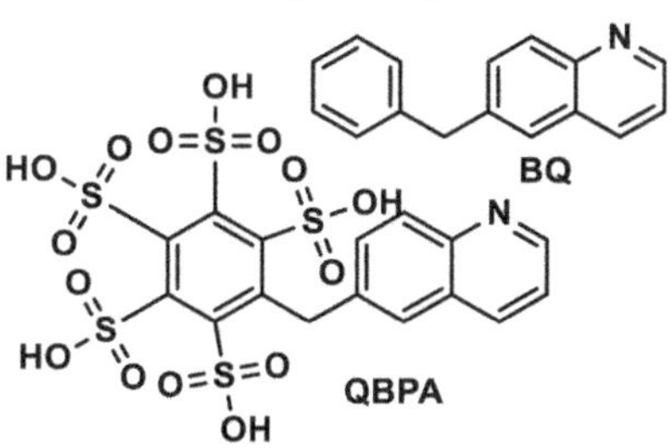	Mild steel/1 M HCl	Mild steel	1 mM (QBPA)/97.2%	57

(*continued*)

Table 1.1 (*continued*)

Chemical structure/abbreviation	Electrolyte	Metal	Conc./%IE	Ref.
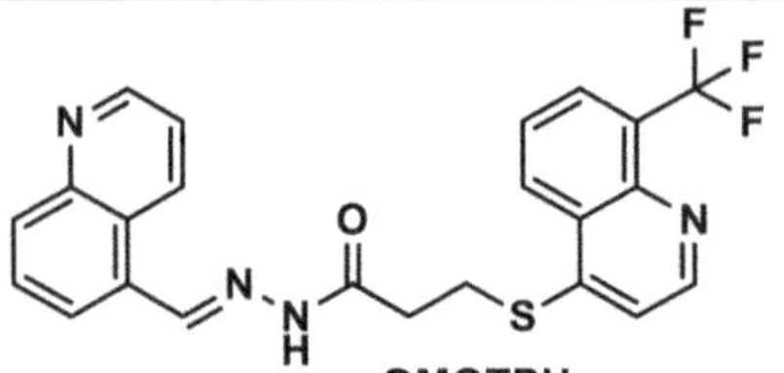 QMQTPH	1 & 2 M HCl	Mild steel	500 ppm (QMQTPH)/90.8%	58
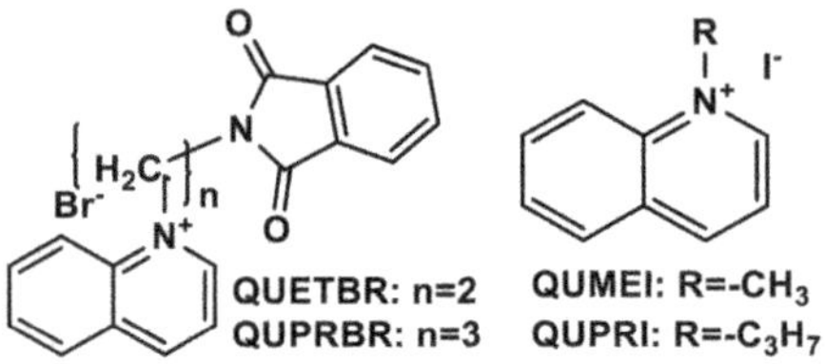 QUETBR: n=2 QUPRBR: n=3 QUMEI: R=-CH_3 QUPRI: R=-C_3H_7	0.5 M H_2SO_4	Mild steel	200 ppm (QUPRBR)/98.6%	61
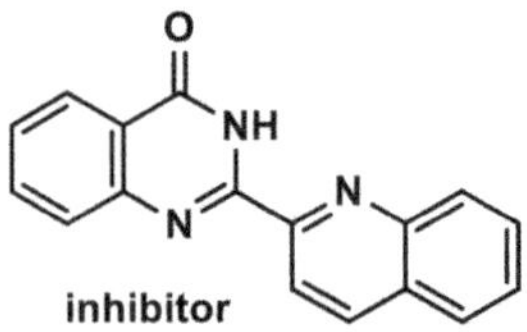 inhibitor	1 M HCl	Mild steel	0.185 mM/92.4%	63

and LUMO) contributions and the hydroxyl group increases FMOs contributions (Figure 1.3).

Similar remarks are also reported for %IE of substituted quinoline elsewhere (Table 1.1). Singh and coworkers[22] reported the corrosion inhibition effects of four novel quinoline derivatives containing different groups (–H, –Me, –OMe and –NMe_2) for mild steel/1 M HCl using experimental methods. Weight loss investigation suggests that inhibition effectiveness of the studied quinoline derivatives followed the order: Q-4 (–NMe_2) > Q-3 (–OMe) > Q-2 (–Me) > Q-1 (–H). This order of inhibition efficiency can be explained with the help of the electron donating abilities of the substituents. The greater the electron donating tendency, the greater the corrosion inhibition effectiveness.

Potentiodynamic polarization investigation suggests that these compounds retard either Tafel reactions or act as mainly mixed-type corrosion inhibitors, though Q-4 shows a little cathodic predominance. Anodic and cathodic Tafel curves with and without Q-1–4 at different concentrations are presented in Figure 1.4. It can be seen that both anodic and cathodic Tafel curves are greatly changed in the presence of the investigated quinoline derivatives. This finding suggests that Q-1–4 has a marked effect on anodic and cathodic Tafel curves. It can also be seen that corrosion current density (i_{corr}) values greatly decrease in the presence of quinoline derivatives. This observation indicates that investigated compounds

Chemical structure/abbreviation	Electrolyte	Metal	Conc./%IE	Ref.
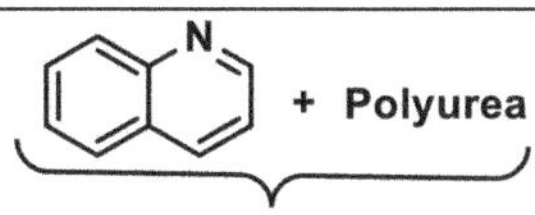 PU anticorrosive coating	5 wt% HCl	Mild steel	360 h time/67.7%	59 and 60
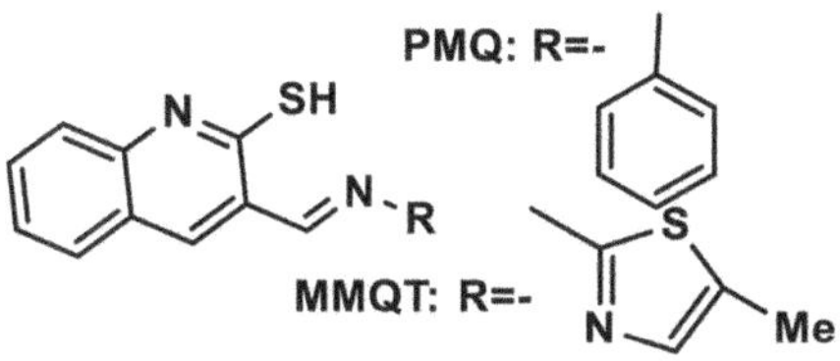	NA	Mild steel surface	NA	62
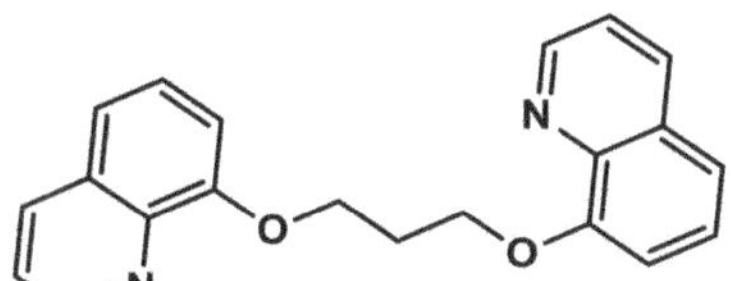 BQYP	2 M H_2SO_4	Mild steel	1.0 mM (BQYP)/91.7%	64

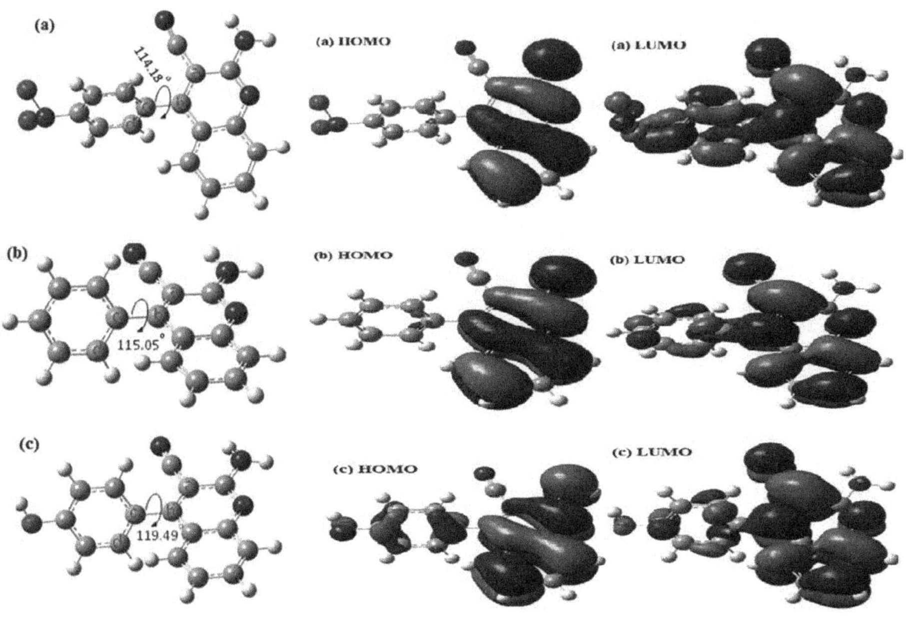

Figure 1.3 Frontier molecular orbital (optimized, HOMO & LUMO) pictures of (a) AAC-1, (b) AAC-2 and (c) AAC-3.[20] Reproduced from ref. 20 with permission from the Royal Society of Chemistry.

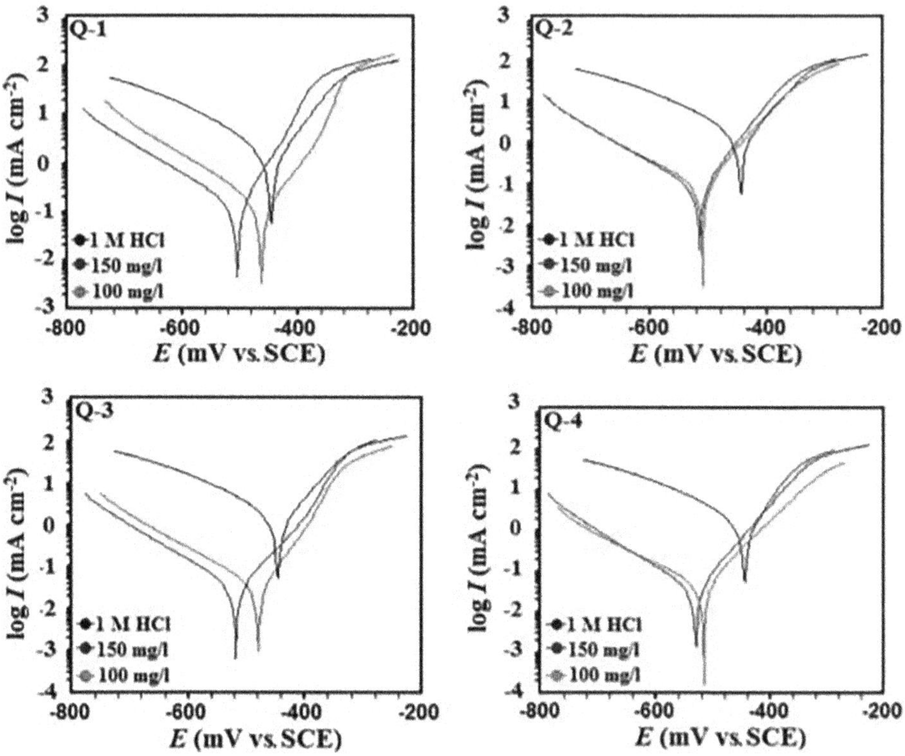

Figure 1.4 Anodic and cathodic Tafel polarization curves for mild steel corrosion in the absence and presence of different concentrations of Q-1–4.[22] Reproduced from ref. 22 with permission from Elsevier, Copyright 2016.

become effective by adsorbing and blocking the active sites responsible for corrosion. All investigated quinoline derivatives adsorb on the metallic surface and diminish corrosion. The adsorption mode of the studied ionic liquids on the metallic surface was reinforced by SEM, AFM and XPS methods.

Recently, in another report, Lgaz *et al.*[23] reported the corrosion inhibition power of nitro and chloro substituted quinoline derivatives. The inhibition efficiencies of the studied quinoline derivatives follow the sequence: QA3 ($-NO_2$ + –Cl) > QA2 ($-NO_2$) > QA1 (–Cl). The highest inhibition efficiency of QA3 can be explained on the basis of the presence of both nitro and chloro substituents that can act as adsorption centres during the interaction of QA3 with the metallic surface. Tafel polarization curves and Nyquist plots for mild steel corrosion in 1 M HCl with and without QA1–3 are illustrated in Figure 1.5. It can be clearly seen that in the presence of quinoline derivatives, the magnitude of the Nyquist curves increases significantly, indicating that QA1–3 adsorbs at the interface of mild steel and 1 M HCl and builds a protective film which creates a barrier for charge transfer processes. It can also be

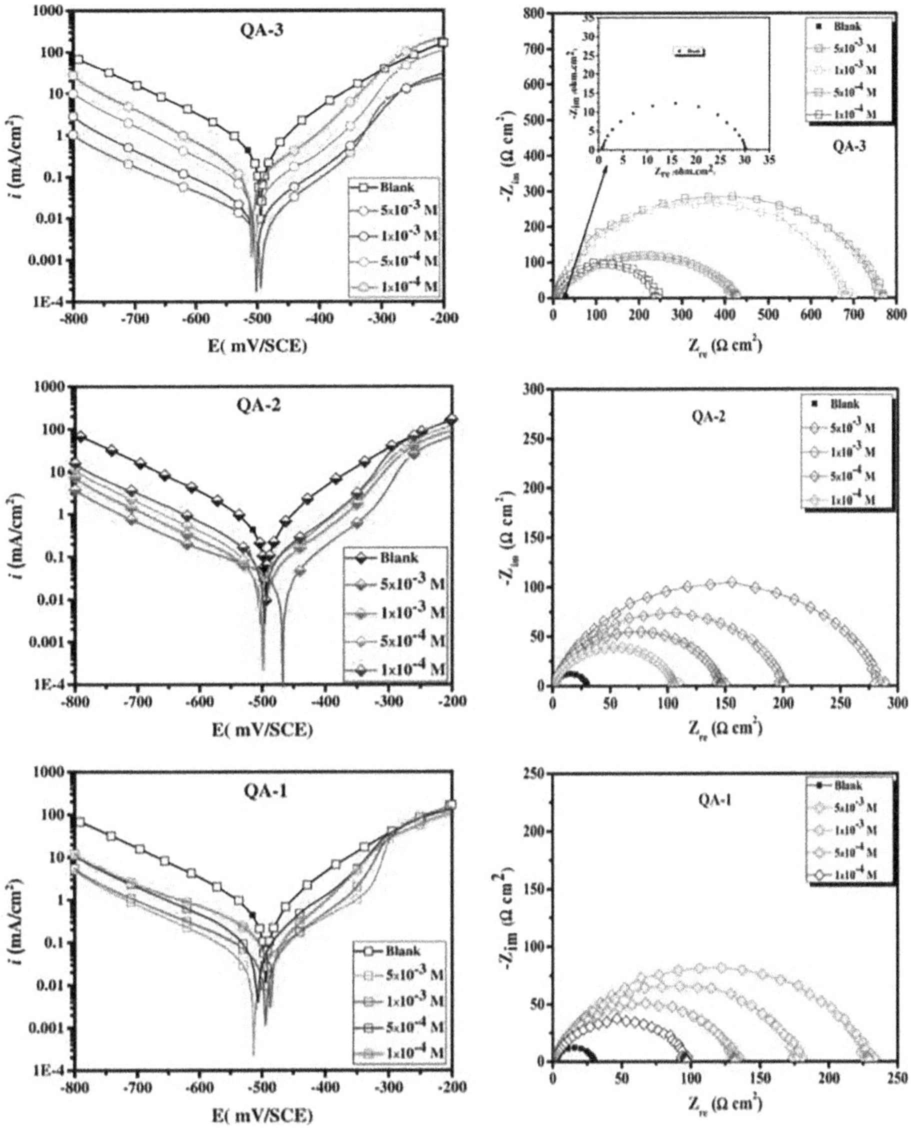

Figure 1.5 Tafel polarization and Nyquist curves for mild steel corrosion in 1 M HCl in the absence and presence of QAs.[23] Reproduced from ref. 23 with permission from Elsevier, Copyright 2017.

seen that both anodic and cathodic curves were affected by QA1–3 without causing any appreciable shift in the magnitude of corrosion potential (E_{corr}). This observation indicates that QA1–3 act as mixed-type corrosion inhibitors, and they become effective by blocking the active sites responsible for the corrosion.

1.3.2 Pyridine and its Derivatives as Corrosion Inhibitors

Pyridine is another significant N-based heterocyclic compound (C_5H_5N) that possesses high corrosion inhibition potential.[65] Pyridine can be regarded as an analogue of benzene in which one of the carbons is replaced by a nitrogen atom. Pyridine and pyridine derivatives exhibit remarkable biological activities.[66,67] Pyridine derivatives, specifically substituted at the 2nd and 6th positions, show extremely high anticorrosive activities.[68] Suitable substituted pyridine derivatives can act as mono-, bi- or polydentate ligands and can form chelating complexes with metal atoms/ions. Pyridine and pyridine derivatives constrain metal deterioration through the adsorption mechanism. The modes of interaction of pyridine and pyridine derivatives appropriately substituted at the 2nd and 6th positions are illustrated in Figure 1.6.

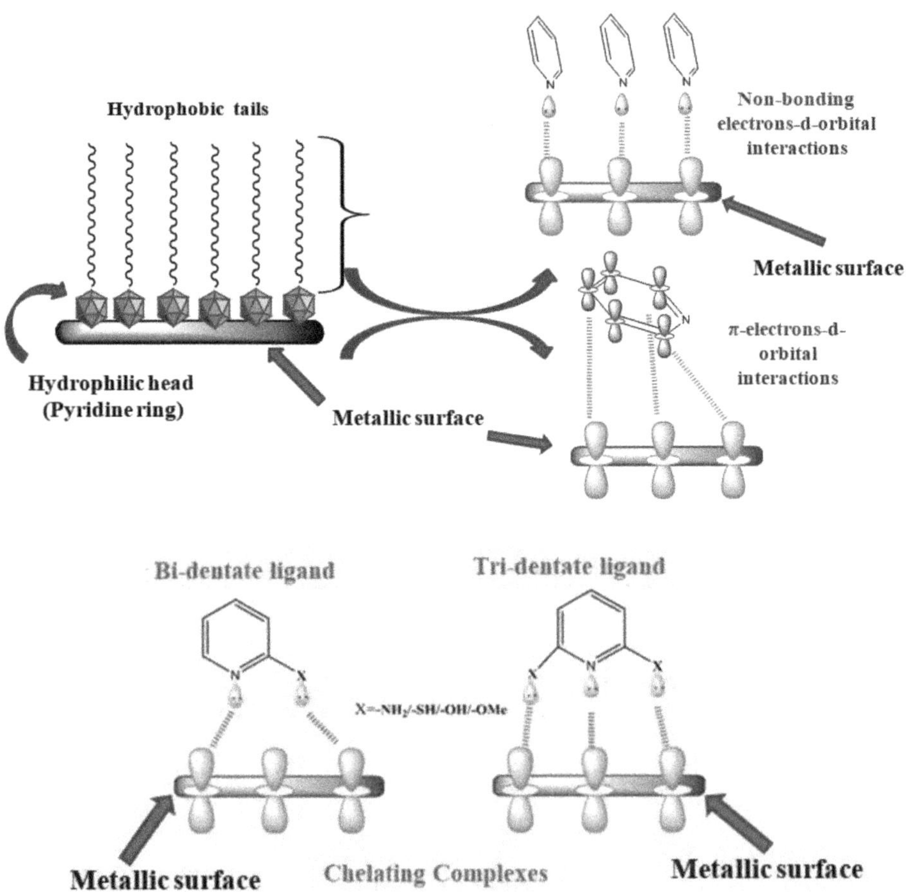

Figure 1.6 Adsorption behaviour of pyridine and pyridine derivatives suitably substituted at the 2nd and 6th positions.[68] Reproduced from ref. 68 with permission from Elsevier, Copyright 2020.

Similar to the quinoline based corrosion inhibitors, the effect of functional groups on the corrosion inhibition potential of pyridine derivatives has been tested and described extensively.[68] They become meaningful by adsorbing on the metal surface, mainly following through Langmuir adsorption isotherm model. Generally, electron donating groups exert positive effects on the corrosion inhibition performance of pyridine derivatives. However, this is not true in all cases. Ansari *et al.*,[69] studying the corrosion inhibition effectiveness of substituted pyridine derivatives, observed that their effectiveness decreases on introducing the electron donating –Me and –OMe substituents. Their inhibition effectiveness followed the sequence: PC-1 (–H) > PC-2 (–Me) > PC-3 (–OMe). Various experimental and computational methods were used to demonstrate the corrosion inhibition potential of PCs, and results derived from all methods were consistent with each other. Surface investigations show that PCs adsorb on the metal surface and build a corrosion protective film. The increase in the surface smoothness in SEM analysis and change in the elemental composition in EDX analysis in the presence of PCs validate this conclusion (Figure 1.7).

It can be seen from the SEM images that the mild steel surface was badly corroded and damaged in the absence of PCs; however, in the presence of PCs, especially in the presence of PC-1, the surface morphology of mild steel is greatly improved. This is attributed to the adsorption and formation of a protective film by PCs. A DFT study was also conducted to reinforce the experimental results and a good agreement was derived between the results of DFT and experimental analyses. DFT results suggest that PCs interact with the metal surface using donor–acceptor interactions and the pyridine ring greatly participates in the charge sharing. Corrosion inhibition effects of pyridine derivatives have been widely reported and a summary of some major reports is presented in Table 1.2.

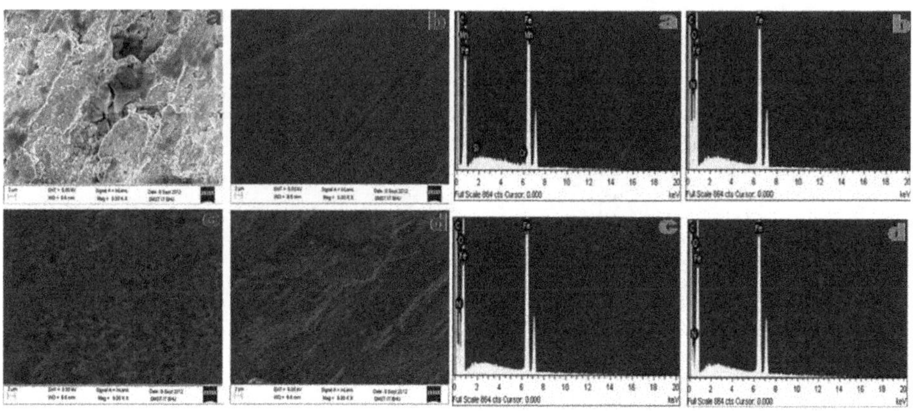

Figure 1.7 SEM images and EDX spectra of mild steel surfaces corroded in 1 M HCl (a) without and with (b) PC-1, (c) PC-2 and (d) PC-3.[69] Reproduced from ref. 69 with permission from Elsevier, Copyright 2015.

Table 1.2 A summary of pyridine derivatives as corrosion inhibitors.

Chem. Str. and abbreviation	Electrolytes	Metal	Conc./%IE	Ref.
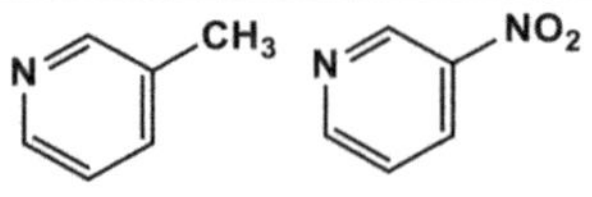 3-MP 3-NM	1 M KOH	Al	5×10^{-4} M (MP)/53.35%	70
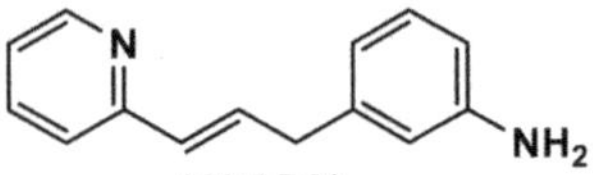 (4AAPA)	1 M HCl	Mild steel	60 ppm/93.44%	72
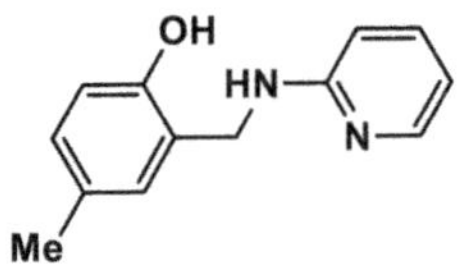	MS/1 M HCl	Mild steel	0.005 M/96.2%	74
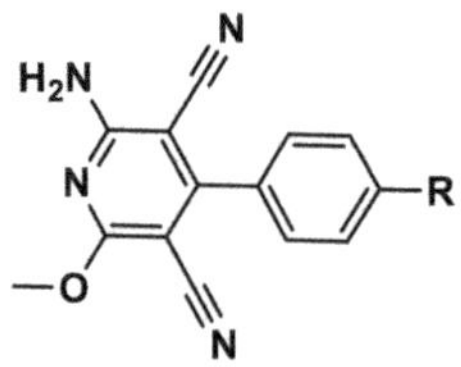 PC-1: R=-H; PC-2:R=-Me PC-3: R=-OMe	MS/1 M HCl	Mild steel	400 ppm (PC-1)/97.4%)	69
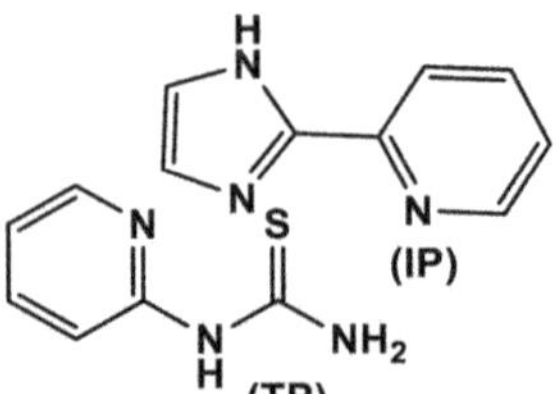 (IP) (TP)	MS/1 M HCl	Mild steel	4×10^{-4} M (TP)/96.66 at%	77
AMP: R_1=R_2=-H ADP: R_1=R_2=-OH 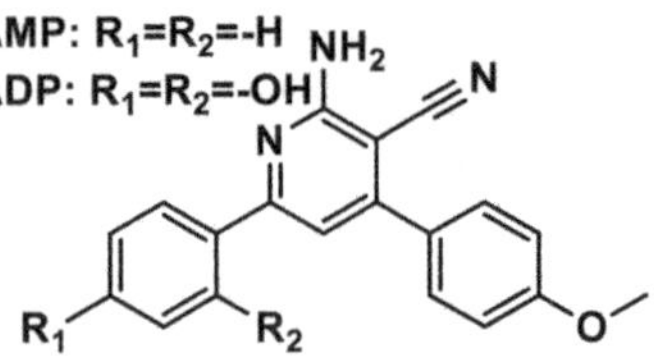	15% HCl	N80 steel	200 ppm (ADP)/92.31%	79
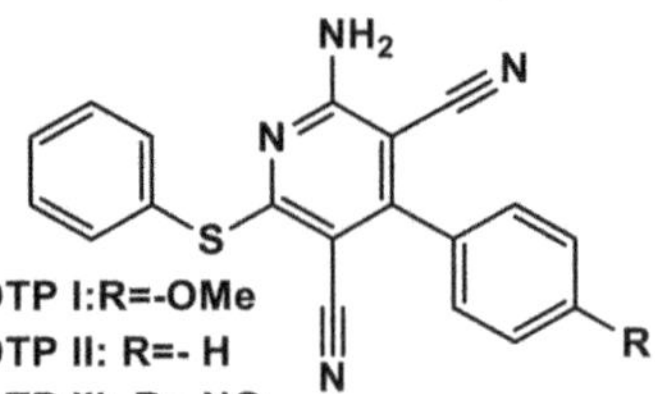 ADTP I:R=-OMe ADTP II: R=- H ADTP III: R=-NO_2	1 M HCl	Mild steel	1.52 mM (I)/97.6%	82

Chem. Str. and abbreviation	Electrolytes	Metal	%IE/Conc.	Ref.
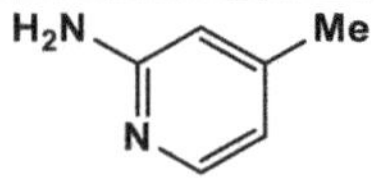 (AMP)	0.5 M HCl	Mild steel	50 mM/94.4%	71
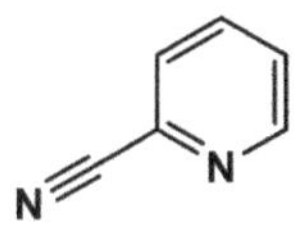 2-PCN	0.1 M HCl	Mild steel	10 mM/94.7%	73
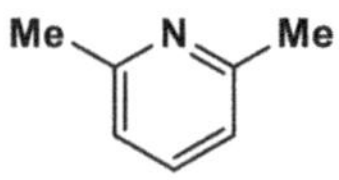 (DMP)	Water	Al	0.0187 M/86%	75
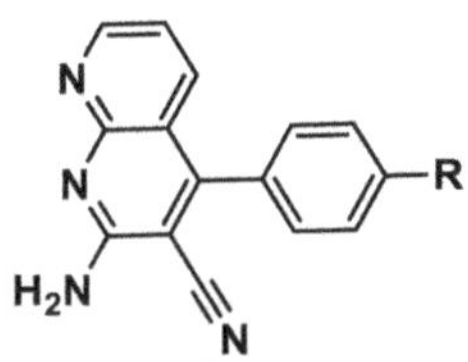 ANC-1: R=-OMe; ANC-2: R=-Me ANC-3: R=-NO_2	15% HCl	N80 steel	200 ppm (ANC-1)/93.9%	76
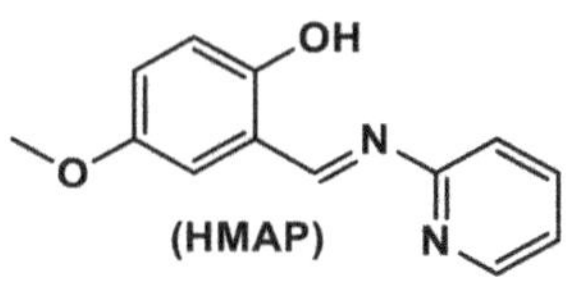 (HMAP)	1 M HCl	Mild steel	0.5 g L^{-1}/90.0%	78
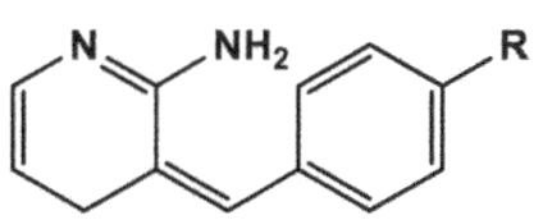 A:R=-H; B:R=-Me C:R=-Cl	Acidic media	Carbon steel/	5×10^{-3} M (A)/98.48%	80 and 81
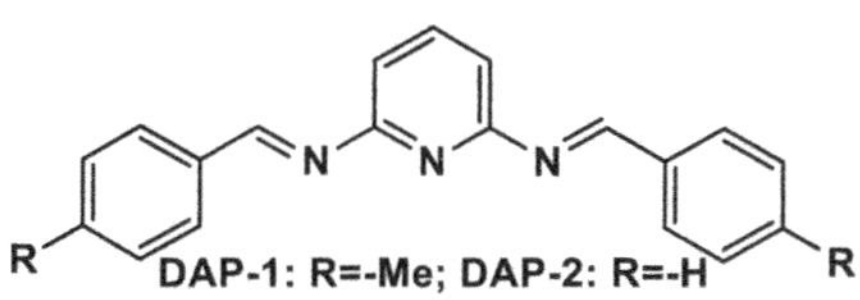 DAP-1: R=-Me; DAP-2: R=-H DAP-3: R=-NO_2	1 M HCl	Mild steel	50 ppm (DAP-1)/98.5%	83

(*continued*)

Table 1.2 (*continued*)

Chem. Str. and abbreviation	Electrolytes	Metal	Conc./%IE	Ref.
2PCOX 3PCOX 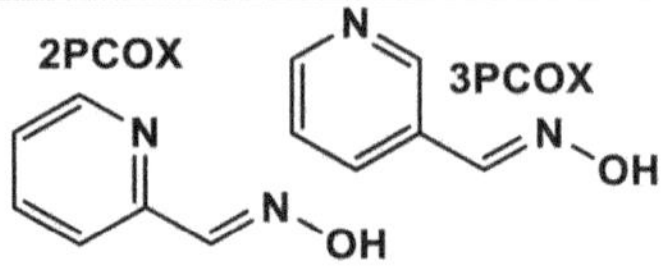	1 M HCl	Mild steel	1.0 mM/85.2 at%	84
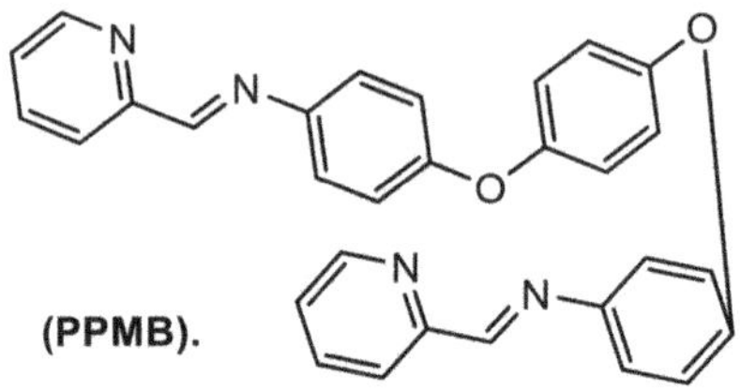 (PPMB).	1 M HCl	Mild steel	5 mM (PPMB)/93 at%	86
APMA 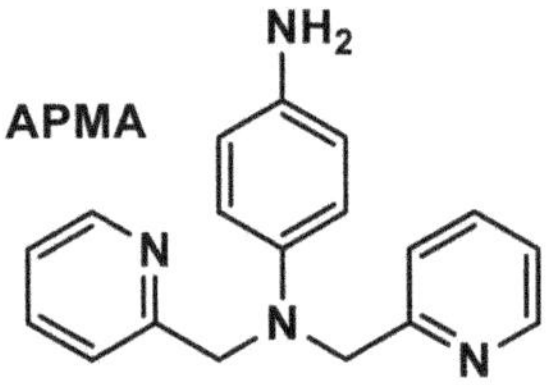	1 M HCl	Mild steel	2 mM/81%	88
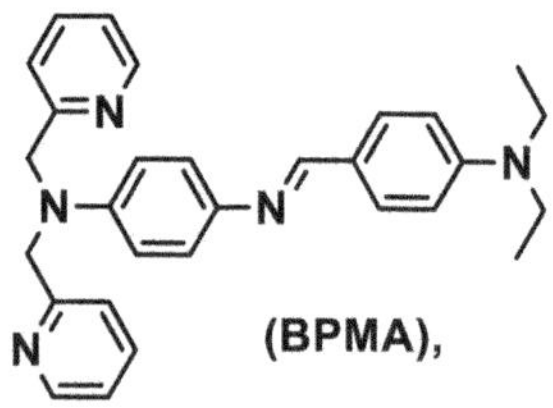 (BPMA),	1 M HCl	Q235 mild steel	1 mM/91.9%	90
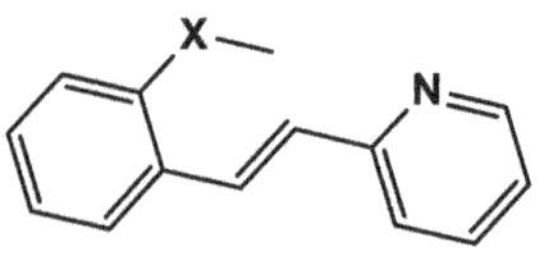 PMAM: X=NH; PMTM: X=S; PMPM: X=O	1 M HCl	Mild steel	10 mM (PMAP)/95%	92
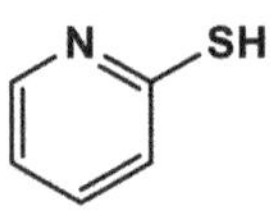 (P2T) 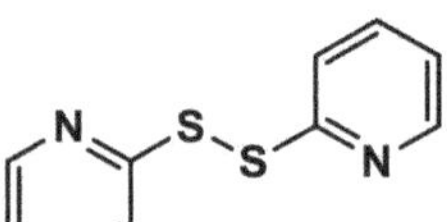 (2PD)	0.1 M HCl	Mild steel	200 ppm (P2T)/98.3 at%	94

Chem. Str. and abbreviation	Electrolytes	Metal	%IE/Conc.	Ref.
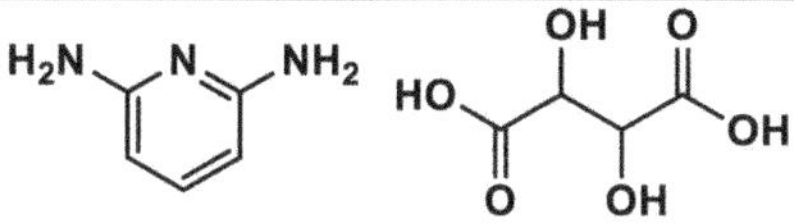 DAP TTA	0.5 M HCl	Mild steel	10 mM (DAP)/69.7%	85
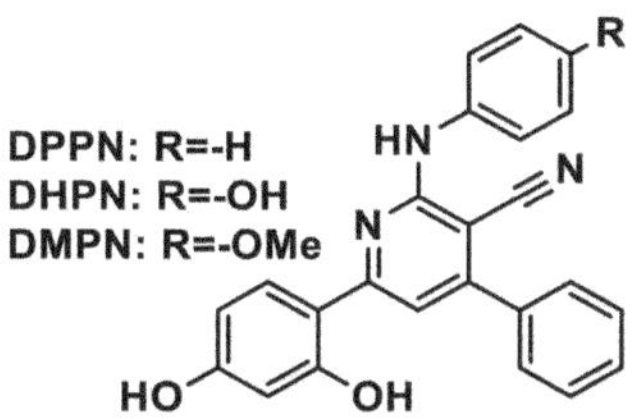 DPPN: R=-H DHPN: R=-OH DMPN: R=-OMe	1 M HCl	Mild steel	20.2×10^{-5} M/95.8%	87
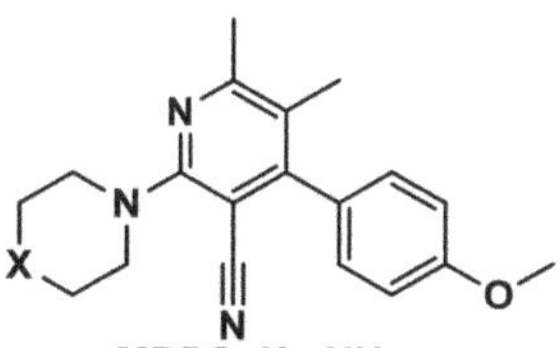 MDPC: X=-NH MDMC: X=O	15% HCl	N80 steel	100 ppm (MDPC)/95.5%	89
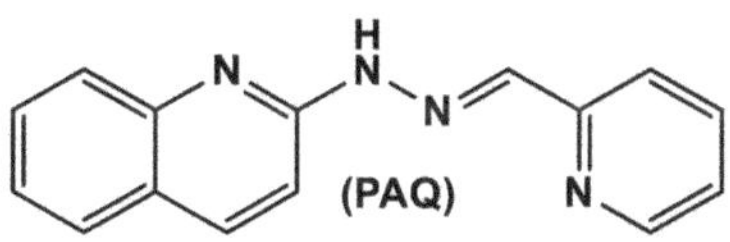 (PAQ)	1 M HCl	Q235 steel	3.24 mM/95.41%	91
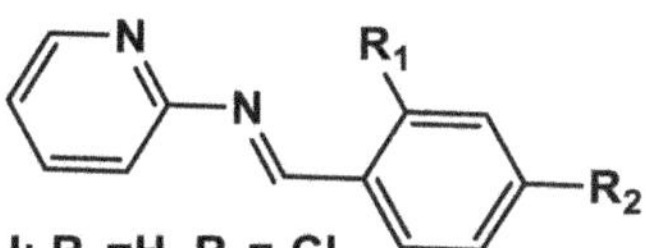 Comp I: R_1=H, R_2=-Cl Comp II: R_1=R_2=-H Comp III: R_1=OH, R_2=-H Comp VI: R_1=-H, R_2=-OMe Comp V: R_1=H, R_2=-NMe_2	1 M HCl	Carbon steel	10^{-3} M (V)/89.9%	93
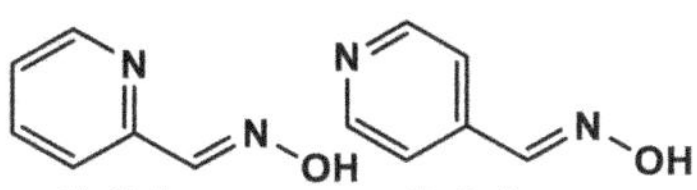 Hp2ylm Hp3ylm	0.1 M HNO_3	Copper	1.0 mM (Hp3ylm)/90.14%	95

(continued)

Table 1.2 (*continued*)

Chem. Str. and abbreviation	Electrolytes	Metal	Conc./%IE	Ref.
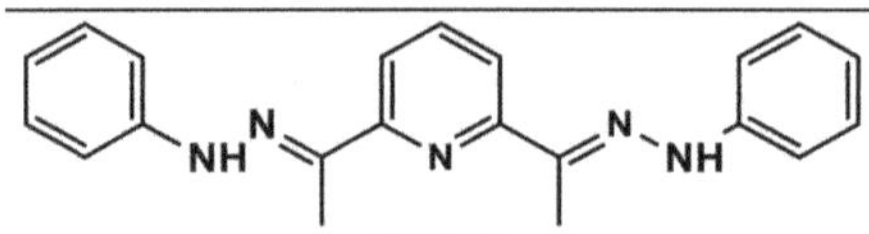 (BPEP)	1 M HCl	Zn	50 ppm/97.92%	96
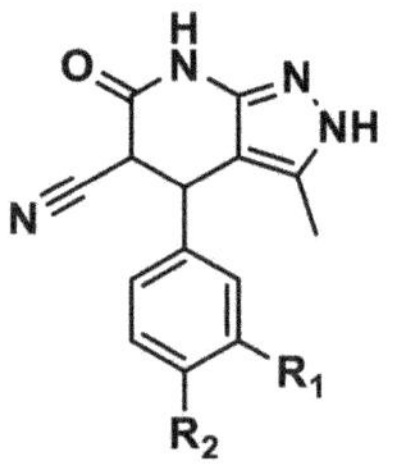 PP1: R_1=-H; R_2=-Br PP2: R_1=-H; R_2=-Cl PP3: R_1=-OPh; R_2=-H	1 M HCl	Mild steel	100 ppm (PP1)/94.92%	98
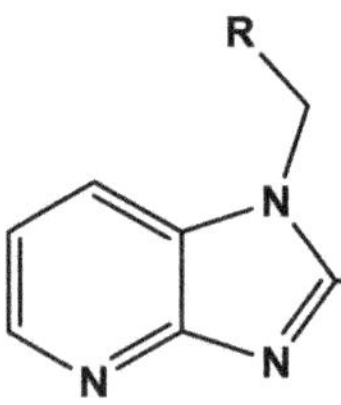 KA2: R= -CH=CH_2 KA3: R=	1 M HCl	Mild steel	10^{-3} M (KA2)/90%	100
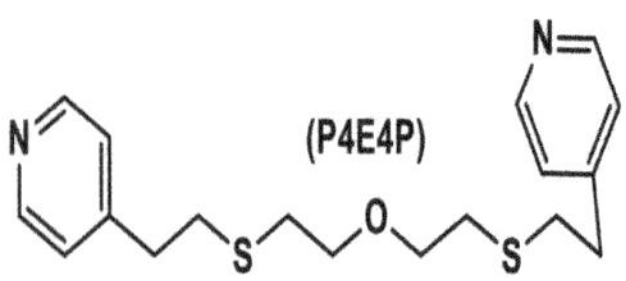 (P4E4P)	1 M HCl	Mild steel	1 mM/97%	102
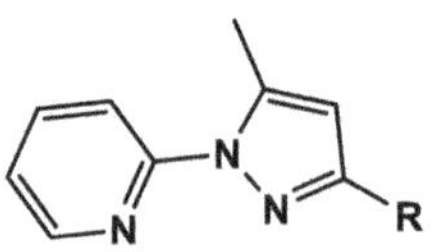 P1: R=-CH_2OH P2: R: -$COOC_2H_5$	1 M HCl	Mild steel	10^{-3} M (P1)/94%	103

Chem. Str. and abbreviation	Electrolytes	Metal	%IE/Conc.	Ref.
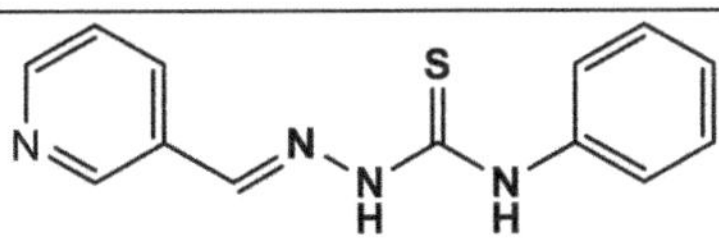 (3-PCPTC) 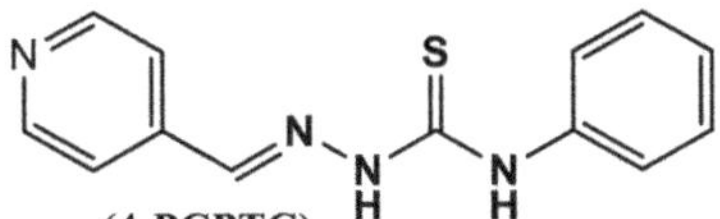 (4-PCPTC)	1 M HCl	Mild steel	1.0 mM (3-PTPTC)/97%	97
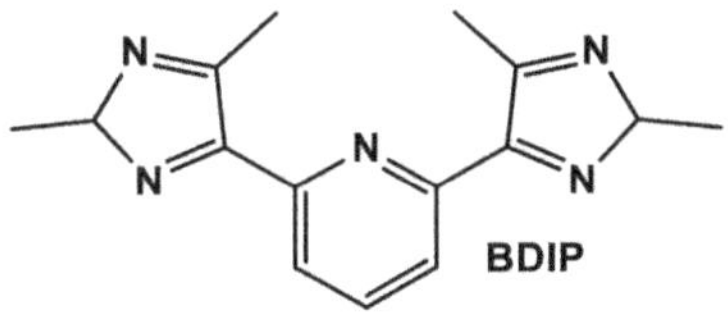 BDIP	1 M HCl	Mild steel	10^{-3} M/97.4%	99
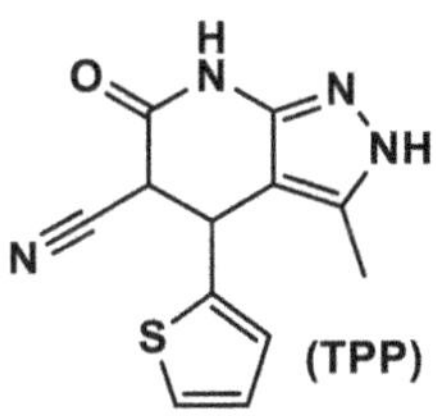 (TPP)	0.5 M HCl	Mild steel	200 ppm/96.8%	101
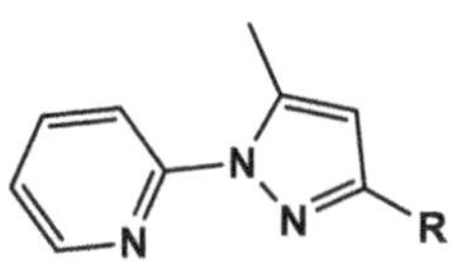 P1: R=-CH_2OH P2: R: -$COOC_2H_5$	1 M HCl	Mild steel	10^{-3} M/P1 (94%) > P2 (87%)	103
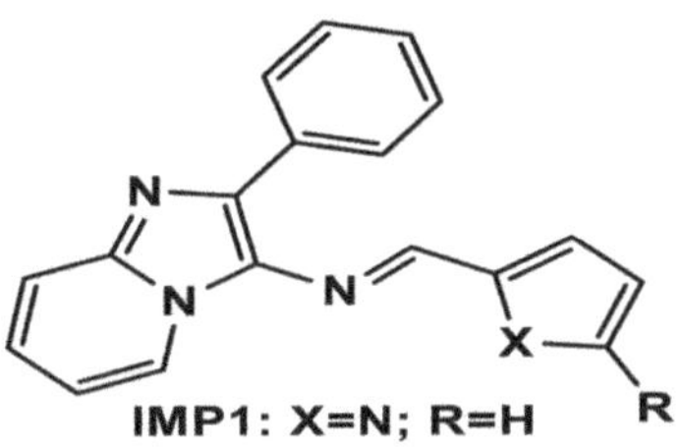 IMP1: X=N; R=H IMP2: X=S; R=H IMP3: X=S; R=NO_2	1 M HCl	Mild steel	10^{-3} M (IMP1)/97%	104

(*continued*)

Table 1.2 (*continued*)

Chem. Str. and abbreviation	Electrolytes	Metal	Conc./%IE	Ref.
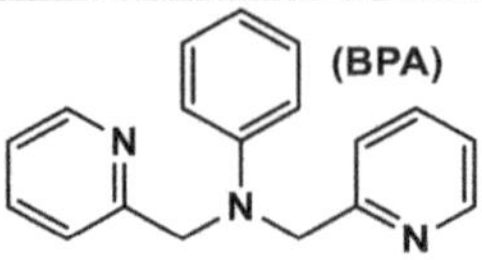 (BPA)	1 M HCl	Mild steel	3.0 mM/94.3%	105
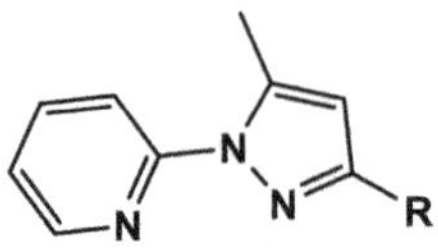 P1: R=-CH_2OH P2: R: -$COOC_2H_5$	1 M HCl	Mild steel	10^{-3} M/P1 (94%) > P2 (87%)	103
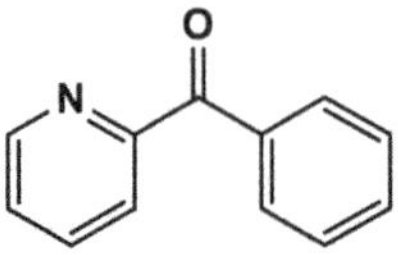 **2BP**	1 M HCl	Mild steel	—	108

1.4 Summary

The present chapter describes the basics of corrosion and corrosion prevention. Literature study indicates that various strategies have been implemented to mitigate corrosion as it causes huge safety and economic losses. Out of various methods of corrosion mitigation, the use of organic compounds, especially heterocyclic compounds, is one of the most common, effective and economical methods. These compounds become effective by adsorbing on the metallic surface, and literature analysis suggests that adsorption of heterocyclic corrosion inhibitors on the metallic surface mostly follows the Langmuir adsorption isotherm. Electrochemical analyses indicate the heterocyclic compound-based interface-type corrosion inhibitors become effective by adsorbing at the interface of the metal surface and electrolyte. The heterocyclic compounds also behave as mixed-type inhibitors *i.e.* become effective by retarding both anodic and cathodic Tafel reactions without changing the corrosion potential significantly. Adsorption behaviour of corrosion mitigation is extensively reported by SEM, AFM, EDX and XPS *etc.* methods. DFT based computational study indicates that heterocyclic corrosion inhibitors adsorb on metallic surfaces using their electron rich centres, and they interact with the surface using the donor–acceptor mechanism.

Chem. Str. and abbreviation	Electrolytes	Metal	%IE/Conc.	Ref.
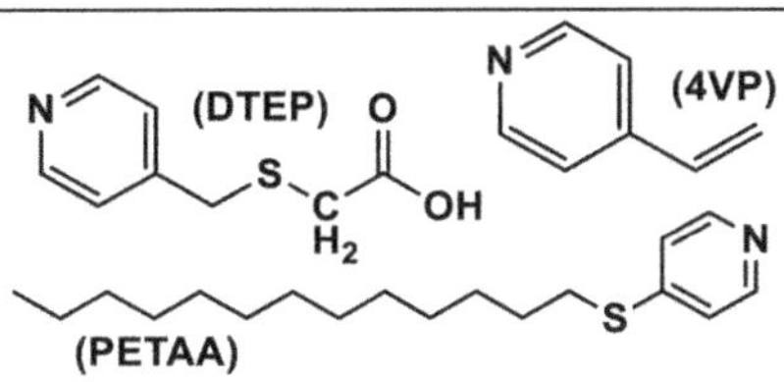	1 M HCl	Mild steel	10^{-3} M (DTEP)/93.3%	106
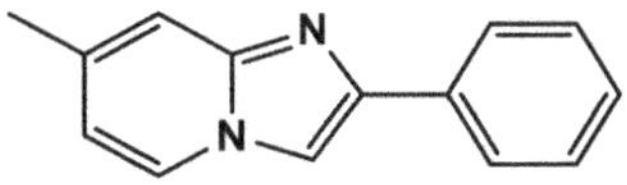 MPP	1.0 M HCl	Carbon steel	1×10^{-3} M/93.7%	107
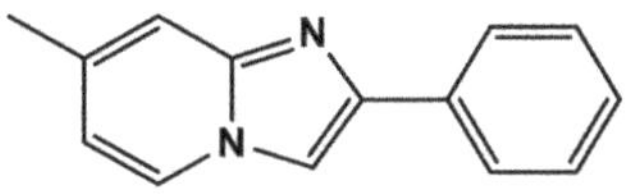 PPI	CS/2.0 M H_3PO_4	Carbon steel	10^{-3} M/91.1%	109

References

1. G. A. Jacobson, *Corrosion—a Natural but Controllable Process*, 2003.
2. G. Koch, Cost of corrosion, *Trends in Oil and Gas Corrosion Research and Technologies*, 2017, pp. 3–30.
3. C. Verma and M. Quraishi, Recent progresses in Schiff bases as aqueous phase corrosion inhibitors: Design and applications, *Coord. Chem. Rev.*, 2021, **446**, 214105.
4. D. K. Verma, Density functional theory (DFT) as a powerful tool for designing corrosion inhibitors in aqueous phase, ed. A. Ali, *Advanced Engineering Testing*, IntechOpen, London, 2018, p. 87.
5. B. E. Brycki, I. H. Kowalczyk, A. Szulc, O. Kaczerewska and M. Pakiet, Organic corrosion inhibitors, *Corrosion Inhibitors, Principles and Recent Applications*, 2018, vol. 3, p. 33.
6. M. Goyal, S. Kumar, I. Bahadur, C. Verma and E. E. Ebenso, Organic corrosion inhibitors for industrial cleaning of ferrous and non-ferrous metals in acidic solutions: A review, *J. Mol. Liq.*, 2018, **256**, 565–573.
7. B. Sanyal, Organic compounds as corrosion inhibitors in different environments—a review, *Prog. Org. Coat.*, 1981, **9**, 165–236.
8. C. Verma, L. O. Olasunkanmi, E. E. Ebenso, M. A. Quraishi and I. B. Obot, Adsorption behavior of glucosamine-based, pyrimidine-fused

heterocycles as green corrosion inhibitors for mild steel: experimental and theoretical studies, *J. Phys. Chem. C*, 2016, **120**, 11598–11611.
9. D. K. Yadav and M. A. Quraishi, Application of some condensed uracils as corrosion inhibitors for mild steel: gravimetric, electrochemical, surface morphological, UV–visible, and theoretical investigations, *Ind. Eng. Chem. Res.*, 2012, **51**, 14966–14979.
10. D. K. Yadav and M. Quraishi, Electrochemical investigation of substituted pyranopyrazoles adsorption on mild steel in acid solution, *Ind. Eng. Chem. Res.*, 2012, **51**, 8194–8210.
11. M. Rbaa, M. Fardioui, C. Verma, A. S. Abousalem, M. Galai, E. Ebenso, T. Guedira, B. Lakhrissi, I. Warad and A. Zarrouk, 8-Hydroxyquinoline based chitosan derived carbohydrate polymer as biodegradable and sustainable acid corrosion inhibitor for mild steel: Experimental and computational analyses, *Int. J. Biol. Macromol.*, 2020, **155**, 645–655.
12. R. Haldhar, D. Prasad, A. Saxena and P. Singh, Valeriana wallichii root extract as a green & sustainable corrosion inhibitor for mild steel in acidic environments: experimental and theoretical study, *Mater. Chem. Front.*, 2018, **2**, 1225–1237.
13. O. Dagdag, Z. Safi, H. Erramli, O. Cherkaoui, N. Wazzan, L. Guo, C. Verma, E. Ebenso and A. El Harfi, Adsorption and anticorrosive behavior of aromatic epoxy monomers on carbon steel corrosion in acidic solution: computational studies and sustained experimental studies, *RSC Adv.*, 2019, **9**, 14782–14796.
14. A. Marella, O. P. Tanwar, R. Saha, M. R. Ali, S. Srivastava, M. Akhter, M. Shaquiquzzaman and M. M. Alam, Quinoline: A versatile heterocyclic, *Saudi Pharm. J.*, 2013, **21**, 1–12.
15. R. Musiol, M. Serda, S. Hensel-Bielowka and J. Polanski, Quinoline-based antifungals, *Curr. Med. Chem.*, 2010, **17**, 1960–1973.
16. J. Alvárez-Builla and J. Barluenga, Heterocyclic compounds: An introduction, *Mod. Heterocycl. Chem.*, 2011, **1**, 1–9.
17. J. Alvarez-Builla, J. J. Vaquero and J. Barluenga, *Modern Heterocyclic Chemistry*, 4 Volume Set, John Wiley & Sons, 2011.
18. M. Orhan Puskullu, B. Tekiner and S. Suzen, Recent studies of antioxidant quinoline derivatives, *Mini-Rev. Med. Chem.*, 2013, **13**, 365–372.
19. C. Verma, M. Quraishi and E. E. Ebenso, Quinoline and its derivatives as corrosion inhibitors: A review, *Surf. Interfaces*, 2020, **21**, 100634.
20. C. Verma, M. Quraishi, L. Olasunkanmi and E. E. Ebenso, L-Proline-promoted synthesis of 2-amino-4-arylquinoline-3-carbonitriles as sustainable corrosion inhibitors for mild steel in 1 M HCl: experimental and computational studies, *RSC Adv.*, 2015, **5**, 85417–85430.
21. Ş. Erdoğan, Z. S. Safi, S. Kaya, D. Ö. Işın, L. Guo and C. Kaya, A computational study on corrosion inhibition performances of novel quinoline derivatives against the corrosion of iron, *J. Mol. Struct.*, 2017, **1134**, 751–761.
22. P. Singh, V. Srivastava and M. Quraishi, Novel quinoline derivatives as green corrosion inhibitors for mild steel in acidic medium:

electrochemical, SEM, AFM, and XPS studies, *J. Mol. Liq.*, 2016, **216**, 164–173.
23. H. Lgaz, R. Salghi, K. S. Bhat, A. Chaouiki and S. Jodeh, Correlated experimental and theoretical study on inhibition behavior of novel quinoline derivatives for the corrosion of mild steel in hydrochloric acid solution, *J. Mol. Liq.*, 2017, **244**, 154–168.
24. I. Obot, N. Ankah, A. Sorour, Z. Gasem and K. Haruna, 8-Hydroxyquinoline as an alternative green and sustainable acidizing oilfield corrosion inhibitor, *Sustainable Mater. Technol.*, 2017, **14**, 1–10.
25. S. Shen, Y. Zuo and X. Zhao, The effects of 8-hydroxyquinoline on corrosion performance of a Mg-rich coating on AZ91D magnesium alloy, *Corros. Sci.*, 2013, **76**, 275–283.
26. Q. Zong, L. Wang, W. Sun and G. Liu, Active deposition of bis (8-hydroxyquinoline) magnesium coating for enhanced corrosion resistance of AZ91D alloy, *Corros. Sci.*, 2014, **89**, 127–136.
27. M. El Faydy, M. Galai, A. El Assyry, A. Tazouti, R. Touir, B. Lakhrissi, M. E. Touhami and A. Zarrouk, Experimental investigation on the corrosion inhibition of carbon steel by 5-(chloromethyl)-8-quinolinol hydrochloride in hydrochloric acid solution, *J. Mol. Liq.*, 2016, **219**, 396–404.
28. S.-m. Li, H.-r. Zhang and J.-h. Liu, Corrosion behavior of aluminum alloy 2024-T3 by 8-hydroxy-quinoline and its derivative in 3.5% chloride solution, *Trans. Nonferrous Met. Soc. China*, 2007, **17**, 318–325.
29. M. El Faydy, R. Touir, M. E. Touhami, A. Zarrouk, C. Jama, B. Lakhrissi, L. Olasunkanmi, E. Ebenso and F. Bentiss, Corrosion inhibition performance of newly synthesized 5-alkoxymethyl-8-hydroxyquinoline derivatives for carbon steel in 1 M HCl solution: experimental, DFT and Monte Carlo simulation studies, *Phys. Chem. Chem. Phys.*, 2018, **20**, 20167–20187.
30. B. Lakhrissi, I. Warad, C. Verma, E.-S. M. Sherif, E. E. Ebenso and A. Zarrouk, Experimental and computational investigations on the anti-corrosive and adsorption behavior of 7-N, N'-dialkyaminomethyl-8-Hydroxyquinolines on C40E steel surface in acidic medium, *J. Colloid Interface Sci.*, 2020, **576**, 330–344.
31. Z. Rouifi, M. Rbaa, F. Benhiba, T. Laabaissi, H. Oudda, B. Lakhrissi, A. Guenbour, I. Warad and A. Zarrouk, Preparation and anti-corrosion activity of novel 8-hydroxyquinoline derivative for carbon steel corrosion in HCl molar: Computational and experimental analyses, *J. Mol. Liq.*, 2020, **307**, 112923.
32. M. Rbaa, A. S. Abousalem, M. E. Touhami, I. Warad, F. Bentiss, B. Lakhrissi and A. Zarrouk, Novel Cu (II) and Zn (II) complexes of 8-hydroxyquinoline derivatives as effective corrosion inhibitors for mild steel in 1.0 M HCl solution: computer modeling supported experimental studies, *J. Mol. Liq.*, 2019, **290**, 111243.
33. M. El Faydy, B. Lakhrissi, A. Guenbour, S. Kaya, F. Bentiss, I. Warad and A. Zarrouk, In situ synthesis, electrochemical, surface morphological, UV–visible, DFT and Monte Carlo simulations of novel

5-substituted-8-hydroxyquinoline for corrosion protection of carbon steel in a hydrochloric acid solution, *J. Mol. Liq.*, 2019, **280**, 341–359.
34. M. El Faydy, M. Galai, M. E. Touhami, I. B. Obot, B. Lakhrissi and A. Zarrouk, Anticorrosion potential of some 5-amino-8-hydroxyquinolines derivatives on carbon steel in hydrochloric acid solution: gravimetric, electrochemical, surface morphological, UV–visible, DFT and Monte Carlo simulations, *J. Mol. Liq.*, 2017, **248**, 1014–1027.
35. A. Eldesoky and S. Nozha, The adsorption and corrosion inhibition of 8-hydroxy-7-quinolinecarboxaldehyde derivatives on C-steel surface in hydrochloric acid, *Chin. J. Chem. Eng.*, 2017, **25**, 1256–1265.
36. M. El Faydy, F. Benhiba, B. Lakhrissi, M. E. Touhami, I. Warad, F. Bentiss and A. Zarrouk, The inhibitive impact of both kinds of 5-isothiocyanatomethyl-8-hydroxyquinoline derivatives on the corrosion of carbon steel in acidic electrolyte, *J. Mol. Liq.*, 2019, **295**, 111629.
37. M. Abou-Dobara, N. Omar, M. Diab, A. El-Sonbati, S. M. Morgan, O. Salem and A. Eldesoky, Polymer complexes. LXXV. Characterization of quinoline polymer complexes as potential bio-active and anti-corrosion agents, *Mater. Sci. Eng., C*, 2019, **103**, 109727.
38. M. El Faydy, B. Lakhrissi, C. Jama, A. Zarrouk, L. O. Olasunkanmi, E. E. Ebenso and F. Bentiss, Electrochemical, surface and computational studies on the inhibition performance of some newly synthesized 8-hydroxyquinoline derivatives containing benzimidazole moiety against the corrosion of carbon steel in phosphoric acid environment, *J. Mater. Res. Technol.*, 2020, **9**, 727–748.
39. M. Rbaa, F. Benhiba, I. Obot, H. Oudda, I. Warad, B. Lakhrissi and A. Zarrouk, Two new 8-hydroxyquinoline derivatives as an efficient corrosion inhibitors for mild steel in hydrochloric acid: synthesis, electrochemical, surface morphological, UV–visible and theoretical studies, *J. Mol. Liq.*, 2019, **276**, 120–133.
40. M. Rbaa, M. Galai, F. Benhiba, I. Obot, H. Oudda, M. E. Touhami, B. Lakhrissi and A. Zarrouk, Synthesis and investigation of quinazoline derivatives based on 8-hydroxyquinoline as corrosion inhibitors for mild steel in acidic environment: experimental and theoretical studies, *Ionics*, 2019, **25**, 3473–3491.
41. M. Rbaa, H. Lgaz, Y. El Kacimi, B. Lakhrissi, F. Bentiss and A. Zarrouk, Synthesis, characterization and corrosion inhibition studies of novel 8-hydroxyquinoline derivatives on the acidic corrosion of mild steel: experimental and computational studies, *Mater. Discovery*, 2018, **12**, 43–54.
42. M. El Faydy, M. Rbaa, L. Lakhrissi, B. Lakhrissi, I. Warad, A. Zarrouk and I. B. Obot, Corrosion protection of carbon steel by two newly synthesized benzimidazol-2-ones substituted 8-hydroxyquinoline derivatives in 1 M HCl: Experimental and theoretical study, *Surf. Interfaces*, 2019, **14**, 222–237.
43. M. Rbaa, A. S. Abousalem, Z. Rouifi, R. Benkaddour, P. Dohare, M. Lakhrissi, I. Warad, B. Lakhrissi and A. Zarrouk, Synthesis, antibacterial

study and corrosion inhibition potential of newly synthesis oxathiolan and triazole derivatives of 8-hydroxyquinoline: Experimental and theoretical approach, *Surf. Interfaces*, 2020, **19**, 100468.
44. M. Rbaa and B. Lakhrissi, Novel oxazole and imidazole based on 8-hydroxyquinoline as a corrosion inhibition of mild steel in HCl Solution: insights from Experimental and Computational Studies, *Surf. Interfaces*, 2019, **15**, 43–59.
45. D. Douche, H. Elmsellem, L. Guo, B. Hafez, B. Tüzün, A. El Louzi, K. Bougrin, K. Karrouchi and B. Himmi, Anti-corrosion performance of 8-hydroxyquinoline derivatives for mild steel in acidic medium: Gravimetric, electrochemical, DFT and molecular dynamics simulation investigations, *J. Mol. Liq.*, 2020, **308**, 113042.
46. R. S. Erami, M. Amirnasr, S. Meghdadi, M. Talebian, H. Farrokhpour and K. Raeissi, Carboxamide derivatives as new corrosion inhibitors for mild steel protection in hydrochloric acid solution, *Corros. Sci.*, 2019, **151**, 190–197.
47. C. Verma and M. Quraishi, 2-Amino-4-(2, 4-dihydroxyphenyl) quinoline-3-carbonitrile as sustainable corrosion inhibitor for SAE 1006 steel in 1 M HCl: Electrochemical and surface investigation, *J. Assoc. Arab Univ. Basic Appl. Sci.*, 2017, **23**, 29–36.
48. S. Chen, S. Chen and W. Li, Corrosion Inhibition Effect of a New Quinoline Derivative on Q235 Steel in H_2SO_4 Solution, *Int. J. Electrochem. Sci.*, 2019, **14**, 11419–11428.
49. D. Wang, D. Yang, D. Zhang, K. Li, L. Gao and T. Lin, Electrochemical and DFT studies of quinoline derivatives on corrosion inhibition of AA5052 aluminium alloy in NaCl solution, *Appl. Surf. Sci.*, 2015, **357**, 2176–2183.
50. C. Verma, L. Olasunkanmi, I. Obot, E. E. Ebenso and M. Quraishi, 5-Arylpyrimido-[4, 5-b] quinoline-diones as new and sustainable corrosion inhibitors for mild steel in 1 M HCl: a combined experimental and theoretical approach, *RSC Adv.*, 2016, **6**, 15639–15654.
51. M. R. E. S. Aly, H. Shokry, T. Sharshar and M. A. Amin, A newly synthesized sulphated 8-hydroxyquinoline derivative to effectively control aluminum corrosion in perchloric acid: Electrochemical and positron annihilation studies, *J. Mol. Liq.*, 2016, **214**, 319–334.
52. K. Ansari, S. Ramkumar, D. Nalini and M. Quraishi, Studies on adsorption and corrosion inhibitive properties of quinoline derivatives on N80 steel in 15% hydrochloric acid, *Cogent Chem.*, 2016, **2**, 1145032.
53. E. E. Ebenso, I. B. Obot and L. Murulana, Quinoline and its derivatives as effective corrosion inhibitors for mild steel in acidic medium, *Int. J. Electrochem. Sci.*, 2010, **5**, 1574–1586.
54. D. Zhang, L. Li, L. Cao, N. Yang and C. Huang, Studies of corrosion inhibitors for zinc–manganese batteries: quinoline quaternary ammonium phenolates, *Corros. Sci.*, 2001, **43**, 1627–1636.
55. J. Zhao, H. Duan and R. Jiang, Synergistic corrosion inhibition effect of quinoline quaternary ammonium salt and Gemini surfactant in H_2S and CO_2 saturated brine solution, *Corros. Sci.*, 2015, **91**, 108–119.

56. B. M. Mistry, S. K. Sahoo and S. Jauhari, Experimental and theoretical investigation of 2-mercaptoquinoline-3-carbaldehyde and its Schiff base as an inhibitor of mild steel in 1 M HCl, *J. Electroanal. Chem.*, 2013, **704**, 118–129.
57. L. Jiang, Y. Qiang, Z. Lei, J. Wang, Z. Qin and B. Xiang, Excellent corrosion inhibition performance of novel quinoline derivatives on mild steel in HCl media: experimental and computational investigations, *J. Mol. Liq.*, 2018, **255**, 53–63.
58. V. R. Saliyan and A. V. Adhikari, Quinolin-5-ylmethylene-3-{[8-(trifluoromethyl) quinolin-4-yl] thio} propanohydrazide as an effective inhibitor of mild steel corrosion in HCl solution, *Corros. Sci.*, 2008, **50**, 55–61.
59. V. V. Gite, P. D. Tatiya, R. J. Marathe, P. P. Mahulikar and D. G. Hundiwale, Microencapsulation of quinoline as a corrosion inhibitor in polyurea microcapsules for application in anticorrosive PU coatings, *Prog. Org. Coat.*, 2015, **83**, 11–18.
60. R. Marathe, P. Tatiya, A. Chaudhari, J. Lee, P. Mahulikar, D. Sohn and V. Gite, Neem acetylated polyester polyol—Renewable source based smart PU coatings containing quinoline (corrosion inhibitor) encapsulated polyurea microcapsules for enhance anticorrosive property, *Ind. Crops Prod.*, 2015, **77**, 239–250.
61. P. Mourya, P. Singh, A. Tewari, R. Rastogi and M. Singh, Relationship between structure and inhibition behaviour of quinolinium salts for mild steel corrosion: experimental and theoretical approach, *Corros. Sci.*, 2015, **95**, 71–87.
62. S. K. Saha, P. Ghosh, A. Hens, N. C. Murmu and P. Banerjee, Density functional theory and molecular dynamics simulation study on corrosion inhibition performance of mild steel by mercapto-quinoline Schiff base corrosion inhibitor, *Phys. E*, 2015, **66**, 332–341.
63. W. Zhang, R. Ma, H. Liu, Y. Liu, S. Li and L. Niu, Electrochemical and surface analysis studies of 2-(quinolin-2-yl) quinazolin-4 (3H)-one as corrosion inhibitor for Q235 steel in hydrochloric acid, *J. Mol. Liq.*, 2016, **222**, 671–679.
64. Z. M. Alamshany and A. A. Ganash, Synthesis, characterization, and anti-corrosion properties of an 8-hydroxyquinoline derivative, *Heliyon*, 2019, **5**, e02895.
65. E. F. Scriven and R. Murugan, Pyridine and pyridine derivatives, *Kirk-Othmer Encyclopedia of Chemical Technology*, 2000.
66. P. Patil, S. Sethy, T. Sameena and K. Shailaja, Pyridine and its biological activity: a review, *Asian J. Res. Chem.*, 2013, **6**, 888–899.
67. E. Abele, R. Abele and E. Lukevics, Pyridine oximes: Synthesis, reactions, and biological activity, *Chem. Heterocycl. Compd.*, 2003, **39**, 825–865.
68. C. Verma, K. Y. Rhee, M. Quraishi and E. E. Ebenso, Pyridine based N-heterocyclic compounds as aqueous phase corrosion inhibitors: A review, *J. Taiwan Inst. Chem. Eng.*, 2020, **117**, 265–277.
69. K. Ansari, M. Quraishi and A. Singh, Corrosion inhibition of mild steel in hydrochloric acid by some pyridine derivatives: an experimental and quantum chemical study, *J. Ind. Eng. Chem.*, 2015, **25**, 89–98.

70. D. Patil and A. Sharma, Inhibition of corrosion of aluminium in potassium hydroxide solution by pyridine derivatives, *Int. Scholarly Res. Not.*, 2014, **2014**, 154285.
71. B. D. Mert, A. O. Yüce, G. Kardaş and B. Yazıcı, Inhibition effect of 2-amino-4-methylpyridine on mild steel corrosion: experimental and theoretical investigation, *Corros. Sci.*, 2014, **85**, 287–295.
72. R. Karthik, G. Vimaladevi, S.-M. Chen, A. Elangovan, B. Jeyaprabha and P. Prakash, Corrosion inhibition and adsorption behavior of 4-amino acetophenone pyridine 2-aldehyde in 1 M hydrochloric acid, *Int. J. Electrochem. Sci.*, 2015, **10**, 4666–4681.
73. R. Yıldız, A. Döner, T. Doğan and İ. Dehri, Experimental studies of 2-pyridinecarbonitrile as corrosion inhibitor for mild steel in hydrochloric acid solution, *Corros. Sci.*, 2014, **82**, 125–132.
74. A. Al-Amiery and L. Shaker, Corrosion inhibition of mild steel using novel pyridine derivative in 1 M hydrochloric acid, *Koroze Ochr. Mater.*, 2020, **64**, 59–64.
75. R. Padash, E. Jamalizadeh and A. H. Jafari, Adsorption and corrosion inhibition behavior of aluminium by 2, 6-di methyl pyridine in distilled water, *Anti-Corros. Methods Mater.*, 2017, **64**, 550–554.
76. K. Ansari and M. Quraishi, Experimental and computational studies of naphthyridine derivatives as corrosion inhibitor for N80 steel in 15% hydrochloric acid, *Phys. E*, 2015, **69**, 322–331.
77. W. Zhang, H.-J. Li, Y. Wang, Y. Liu, Q.-Z. Gu and Y.-C. Wu, Gravimetric, electrochemical and surface studies on the anticorrosive properties of 1-(2-pyridyl)-2-thiourea and 2-(imidazol-2-yl)-pyridine for mild steel in hydrochloric acid, *New J. Chem.*, 2018, **42**, 12649–12665.
78. A. Al-Amiery, T. A. Salman, K. F. Alazawi, L. M. Shaker, A. A. H. Kadhum and M. S. Takriff, Quantum chemical elucidation on corrosion inhibition efficiency of Schiff base: DFT investigations supported by weight loss and SEM techniques, *Int. J. Low-Carbon Technol.*, 2020, **15**, 202–209.
79. K. Ansari, M. Quraishi and A. Singh, Pyridine derivatives as corrosion inhibitors for N80 steel in 15% HCl: Electrochemical, surface and quantum chemical studies, *Measurement*, 2015, **76**, 136–147.
80. H. Ju, X. Li, N. Cao, F. Wang, Y. Liu and Y. Li, Schiff-base derivatives as corrosion inhibitors for carbon steel materials in acid media: quantum chemical calculations, *Corros. Eng., Sci. Technol.*, 2018, **53**, 36–43.
81. H. Ashassi-Sorkhabi, B. Shaabani and D. Seifzadeh, Corrosion inhibition of mild steel by some Schiff base compounds in hydrochloric acid, *Appl. Surf. Sci.*, 2005, **239**, 154–164.
82. M. A. Quraishi, 2-Amino-3, 5-dicarbonitrile-6-thio-pyridines: new and effective corrosion inhibitors for mild steel in 1 M HCl, *Ind. Eng. Chem. Res.*, 2014, **53**, 2851–2859.
83. P. Dohare, M. Quraishi and I. Obot, A combined electrochemical and theoretical study of pyridine-based Schiff bases as novel corrosion inhibitors for mild steel in hydrochloric acid medium, *J. Chem. Sci.*, 2018, **130**, 8.

84. K. J. Thomas, Electro Analytical and Gravimetrical Investigations on Corrosion Inhibition Properties of Pyridine-Carbaldehyde Derivatives on Carbon Steel, *Chem. Sci. Rev. Lett.*, 2017, **6**, 2300–2308.
85. Y. Qiang, L. Guo, S. Zhang, W. Li, S. Yu and J. Tan, Synergistic effect of tartaric acid with 2, 6-diaminopyridine on the corrosion inhibition of mild steel in 0.5 M HCl, *Sci. Rep.*, 2016, **6**, 33305.
86. M. Murmu, S. K. Saha, N. C. Murmu and P. Banerjee, Effect of stereochemical conformation into the corrosion inhibitive behaviour of double azomethine based Schiff bases on mild steel surface in 1 mol L^{-1} HCl medium: An experimental, density functional theory and molecular dynamics simulation study, *Corros. Sci.*, 2019, **146**, 134–151.
87. C. Verma, L. O. Olasunkanmi, T. W. Quadri, E.-S. M. Sherif and E. E. Ebenso, Gravimetric, electrochemical, surface morphology, DFT, and Monte Carlo simulation studies on three N-substituted 2-aminopyridine derivatives as corrosion inhibitors of mild steel in acidic medium, *J. Phys. Chem. C*, 2018, **122**, 11870–11882.
88. B. Xu, Y. Ji, X. Zhang, X. Jin, W. Yang and Y. Chen, Experimental and theoretical studies on the corrosion inhibition performance of 4-amino-N, N-di-(2-pyridylmethyl)-aniline on mild steel in hydrochloric acid, *RSC Adv.*, 2015, **5**, 56049–56059.
89. M. Yadav and S. Kumar, Experimental, thermodynamic and quantum chemical studies on adsorption and corrosion inhibition performance of synthesized pyridine derivatives on N80 steel in HCl solution, *Surf. Interface Anal.*, 2014, **46**, 254–268.
90. Y. Ji, B. Xu, W. Gong, X. Zhang, X. Jin, W. Ning, Y. Meng, W. Yang and Y. Chen, Corrosion inhibition of a new Schiff base derivative with two pyridine rings on Q235 mild steel in 1.0 M HCl, *J. Taiwan Inst. Chem. Eng.*, 2016, **66**, 301–312.
91. R. Chaudhary, T. Namboodhiri, I. Singh and A. Kumar, Effect of pyridine and its derivatives on corrosion of 0040, 2826MB, and 2605–8–2 metallic glasses in sulphuric acid solution at 25° C, *Br. Corros. J.*, 1989, **24**, 273–278.
92. A. Dutta, S. K. Saha, P. Banerjee, A. K. Patra and D. Sukul, Evaluating corrosion inhibition property of some Schiff bases for mild steel in 1 M HCl: competitive effect of the heteroatom and stereochemical conformation of the molecule, *RSC Adv.*, 2016, **6**, 74833–74844.
93. M. Abdallah, M. Alfakeer, H. Altass, A. M. Alharbi, I. Althagafi, N. Hasan and E. Mabrouk, The polarographic and corrosion inhibition performance of some Schiff base compounds derived from 2-amino-3-hydroxypyridine in aqueous media, *Egypt. J. Pet.*, 2019, **28**, 393–399.
94. A. Kosari, M. H. Moayed, A. Davoodi, R. Parvizi, M. Momeni, H. Eshghi and H. Moradi, Electrochemical and quantum chemical assessment of two organic compounds from pyridine derivatives as corrosion inhibitors for mild steel in HCl solution under stagnant condition and hydrodynamic flow, *Corros. Sci.*, 2014, **78**, 138–150.

95. C. Varghese, K. Thomas, V. Raphael and K. Shaju, Corrosion inhibition capacity of two heterocyclic oximes on copper in nitric acid: electrochemical, quantum chemical and surface morphological investigations, *Curr. Chem. Lett.*, 2019, **8**, 1–12.
96. M. Abdallah, S. Ahmed, H. Altass, I. Zaafarany, M. Salem, A. Aly and E. Hussein, Competent inhibitor for the corrosion of zinc in hydrochloric acid based on 2, 6-bis-[1-(2-phenylhydrazono) ethyl] pyridine, *Chem. Eng. Commun.*, 2019, **206**, 137–148.
97. Y. Meng, W. Ning, B. Xu, W. Yang, K. Zhang, Y. Chen, L. Li, X. Liu, J. Zheng and Y. Zhang, Inhibition of mild steel corrosion in hydrochloric acid using two novel pyridine Schiff base derivatives: a comparative study of experimental and theoretical results, *RSC Adv.*, 2017, **7**, 43014–43029.
98. S. Gupta, A. Dandia, P. Singh and M. Quraishi, Green synthesis of pyrazolo [3, 4-b] pyridine derivatives by ultrasonic technique and their application as corrosion inhibitor for mild steel in acid medium, *J. Mater. Environ. Sci.*, 2015, **6**, 168–177.
99. O. El Khattabi, B. Zerga, M. Sfaira, M. Taleb, M. Ebn Touhami, B. Hammouti, L. Herrag and M. Mcharfi, On the adsorption properties of an imidazole-pyridine derivative as corrosion inhibitor of mild steel in 1 M HCL, *Pharma Chem.*, 2012, **4**, 1759–1768.
100. K. Bouayad, Y. K. Rodi, H. Elmsellem, I. Abdel-Rahman, E. El Ghadraoui, S. Chakroune and E. Essassi, Density-Functional Theory and Experimental Evaluation of Inhibition Mechanism of Novel Imidazo [4, 5-b] pyridine Derivatives,*J. Mater. Environ. Sci.*, 2018, **9**(92), 474–786.
101. P. Singh, M. Quraishi, S. Gupta and A. Dandia, Investigation of the corrosion inhibition effect of 3-methyl-6-oxo-4-(thiophen-2-yl)-4, 5, 6, 7-tetrahydro-2H-pyrazolo [3, 4-b] pyridine-5-carbonitrile (TPP) on mild steel in hydrochloric acid,*J. Taibah Univ. Sci.*, 2016, **10**, 139–147.
102. A. Khadiri, A. Ousslim, K. Bekkouche, A. Aouniti, I. Warad, A. Elidrissi, B. Hammouti, F. Bentiss, M. Bouachrine and A. Zarrouk, 4-(2-(2-(2-(2-(Pyridine-4-yl) ethylthio) ethoxy) ethylthio) ethyl) pyridine as new corrosion inhibitor for mild steel in 1.0 M HCl solution: experimental and theoretical studies,*J. Bio- Tribo-Corros.*, 2018, **4**, 64.
103. K. Tebbji, H. Oudda, B. Hammouti, M. Benkaddour, M. El Kodadi and A. Ramdani, Inhibition effect of two organic compounds pyridine–pyrazole type in acidic corrosion of steel, *Colloids Surf., A*, 2005, **259**, 143–149.
104. A. El Aatiaoui, M. Koudad, T. Chelfi, S. Erkan, M. Azzouzi, A. Aouniti, K. Savaş, M. Kaddouri, N. Benchat and A. Oussaid, Experimental and theoretical study of new Schiff bases based on imidazo (1, 2-a) pyridine as corrosion inhibitor of mild steel in 1M HCl,*J. Mol. Struct.*, 2021, **1226**, 129372.
105. B. Xu, Y. Ji, X. Zhang, X. Jin, W. Yang and Y. Chen, Experimental and theoretical evaluation of N, N-Bis (2-pyridylmethyl) aniline as a novel corrosion inhibitor for mild steel in hydrochloric acid, *J. Taiwan Inst. Chem. Eng.*, 2016, **59**, 526–535.

106. O. Krim, A. Elidrissi, B. Hammouti, A. Ouslim and M. Benkaddour, Synthesis, characterization, and comparative study of pyridine derivatives as corrosion inhibitors of mild steel in HCl medium, *Chem. Eng. Commun.*, 2009, **196**, 1536–1546.
107. A. Ghazoui, R. Saddik, N. Benchat, M. Guenbour, B. Hammouti, S. Al-Deyab and A. Zarrouk, Comparative study of pyridine and pyrimidine derivatives as corrosion inhibitors of C38 steel in molar HCl, *Int. J. Electrochem. Sci.*, 2012, **7**, 7080–7097.
108. A. James and N. Oforka, 2-Benzoyl pyridine: An inhibitor for mild steel corrosion in hydrochloric acid solution, *Adv. Appl. Sci. Res.*, 2014, **5**, 1–6.
109. D. B. Hmamou, R. Salghi, A. Zarrouk, H. Zarrok, B. Hammouti, S. Al-Deyab, A. El Assyry, N. Benchat and M. Bouachrine, Electrochemical and gravimetric evaluation of 7-methyl-2-phenylimidazo [1, 2-α] pyridine of carbon steel corrosion in phosphoric acid solution, *Int. J. Electrochem. Sci.*, 2013, **8**, 11526–11545.

CHAPTER 2

Green Corrosion Inhibition: Concepts of Green Chemistry

2.1 Introduction

2.1.1 Corrosion Inhibition: Traditional to Green Corrosion Inhibitors

The use of organic compounds, especially heterocyclic compounds, is one of the most effective, economical and popular methods of corrosion mitigation.[1,2] However, currently the use of traditional organic compounds as corrosion inhibitors is limited because of increasing ecological awareness and increasing demands for green chemistry.[3–5] Inorganic compounds such as chromates, borates, silicates, zinc salts, molybdates and nitriles were utilized as the oldest corrosion inhibitors before 1960. These inorganic compounds form a highly stable and effective (1st E) protective oxide or passive layer; therefore, they are also known as passivators. In spite of the high effectiveness of inorganic corrosion inhibitors, they were replaced by relatively more economical alternatives (2nd E), including gluconates, phosphonates, polyphosphates, phosphonic acid, surface active chelates, polyacrylates and carboxylates. These compounds were successfully utilized as effectual corrosion inhibitors during 1960–1980. Generally, these compounds precipitate at the metal surface and form a protective layer. Therefore, these alternatives are known as precipitators. Nevertheless, after 1980, ecological considerations (3rd E) came into play, and economical alternatives were replaced by eco-friendly alternatives of natural and biological origin, such as bio-surfactants, vitamins, tannins, bio-polymers and natural compounds. These eco-friendly alternatives of natural origin were effectively and successfully utilized as

Green Chemistry Series No. 73
Green Corrosion Inhibition: Fundamentals, Design, Synthesis and Applications
By Chandrabhan Verma, Mumtaz A. Quraishi and Dheeraj Singh Chauhan

Published by the Royal Society of Chemistry, www.rsc.org

corrosion inhibitors during 1980–1995. However, recently the concept of Eco-friendliness (4th E) is also considered during the designing and synthesis of effective corrosion inhibitors (Figure 2.1).

Due to increasing demands for green chemistry and sustainable developments, corrosion scientists and engineers are trying to develop new eco-friendly alternatives in order to replace traditionally existing toxic corrosion inhibitors.[3,6] Environmental friendliness or toxicity of a chemical compound can be assessed by its bioaccumulation, biodegradability and effect on the environment. The Oslo and Paris Commission (OSPAR) and Registration, Evaluation, Authorisation and Restriction of Chemicals (REACH) set some references to assess these parameters.[7–9] Noticeably, REACH is a EUR (European Union Regulation) that came into force on 7 June 2007. The parameters set by these commissions are used to describe the biodegradability, toxicity and bioaccumulation limits and their effect on the environment and human health. Toxicity of a chemical compound can be determined by its LC_{50} (lethal concentration) and EC_{50} (effective concentration) value. Obviously, LC_{50} represents the concentration of the chemical compounds that kills 50% of the total living population,[10–12] whereas EC_{50} represents the concentration of the chemical compound that adversely affects the growth of 50% of the total living population. Obviously, a chemical compound with a lower value of EC_{50}/LC_{50} is associated with high toxicity. Chemical compounds having EC_{50}/LC_{50} values of more than 10 mg kg^{-1} would be considered nontoxic or sustainable. Microorganisms possess the ability to degrade chemical species by a process known as decomposition. Therefore, these

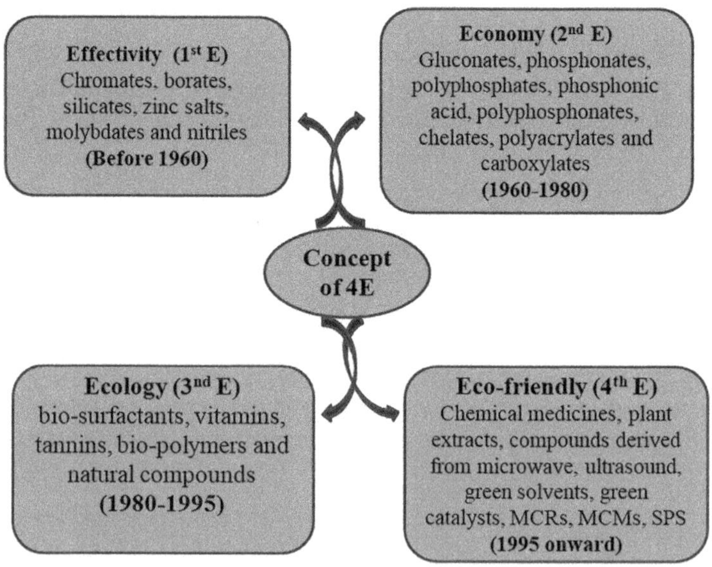

Figure 2.1 Development of corrosion inhibitors and concept of 4E.

microorganisms are known as decomposers. A chemical species (corrosion inhibitor) would be regarded as eco-friendly when it decomposes by greater than 60% after exposure to the environment for 28 days.[13–15] The non-eco-friendly chemical compounds are those that take more than 28 days, usually months and years, for decomposition. Bioaccumulation is the ability of any chemical species to accumulate in the body of a living organism.[16–18] In general, bioaccumulation is analysed in terms of the partition coefficient ($\log K_{OW}$ or D^{OW}). Bioaccumulation is mainly measured in a mixture of water and octanol. An eco-friendly chemical is one that has a $\log K_{OW}$ or D^{OW} value of 3 or less.[19]

2.2 Different Sources of Green Corrosion Inhibitors

2.2.1 Synthetic Green Corrosion Inhibitors

2.2.1.1 Multicomponent Reactions

Multicomponent reactions (MCRs) are chemical transformation reactions in which one or more reactants combine together to give a product containing significant elements of all the reactants (Figure 2.2a).[20] Various organic compounds, especially heterocyclic compounds, useful for numerous industrial and biological applications, are synthesized using MCRs (Figure 2.2b).[21] The MCRs possess numerous advantages over the traditional multistep reactions (MSRs).[22,23] Some of the major advantages of MCRs are as follows:

(i) Ease of performance and purification
(ii) Simplicity and speedy access to large libraries of organic molecules
(iii) Low waste production and discharge
(iv) Operational simplicity and high atom economy

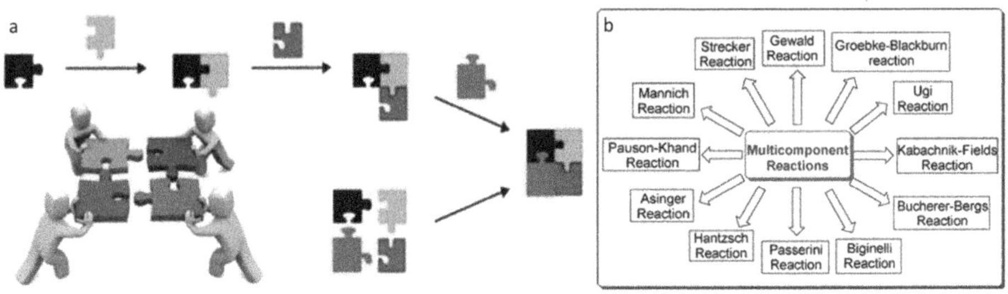

Figure 2.2 (a) Diagrammatic illustration of multistep (upper) and multicomponent reaction (lower).[20] Reproduced from ref. 20 with permission from the Royal Society of Chemistry. (b) Various named reactions catalysed by MCRs.[21] Reproduced from ref. 21 with permission from the Royal Society of Chemistry.

(v) Fewer purification and work-up steps
(vi) Minimal use of toxic and expensive catalysts
(vii) Highly specific, chemo- and regio-selectivity
(viii) Use of conventional and eco-friendly solvents such as water, amines and alcohols for purification
(ix) Simple to highly complex molecules can be synthesized using MCRs
(x) Generally, use of simple, readily and economically/commercially obtainable starting materials
(xi) Facile automation
(xii) Minimal adverse effects on the environment
(xiii) Cost effective and time saving
(xiv) Can be catalysed using nonconventional heating such as microwave and ultrasounds.

Literature investigation suggests that numerous organic compounds, particularly heterocyclic compounds derived from multicomponent reactions (MCRs), are successfully used as green corrosion inhibitors for different metals and alloys in various electrolytes.[24,25] Generally, these compounds become effective by adsorbing on the metal surface and blocking the active sites present on the metal surface. They inhibit anodic and cathodic Tafel reactions and act as mixed-type corrosion inhibitors. Their adsorption mostly follows the Langmuir adsorption isotherm model. Through surface investigations such as SEM, EDX, AFM, XPS and XRD *etc.* it can be assessed that these compounds build a protective film through their adsorption. This film isolates the metal from the aggressive environment and protects it from corrosion.

2.2.1.2 Nonconventional Heating

In most chemical transformations, activation of reactant molecules requires heating. One of the oldest methods of heating is hot plate heating, which is associated with several shortcomings, including slow and non-uniform activation.[26,27] Hot plate heating is mostly called conventional heating. This causes activation of the molecules near the wall of the vessel. However, because of the increasing demands for fast activation and uniform heating, microwave (MW) and ultrasound (US) irradiations have emerged as the most powerful nonconventional heating sources for traditional hot plate heating reactions.[28,29] MW and US heating are associated with numerous advantages that make them ideal and eco-friendly heating sources. Literature survey suggests that compounds derived from MW and US irradiations are commonly used for different biological and industrial applications. Noticeably, MW and US heating are used to catalyse MCRs where three or more reactants combine together in a single step.[30–32] Obviously, MW and US cause instantaneous heating of the reaction vessel and therefore activation of the reactant molecules (Figure 2.3).

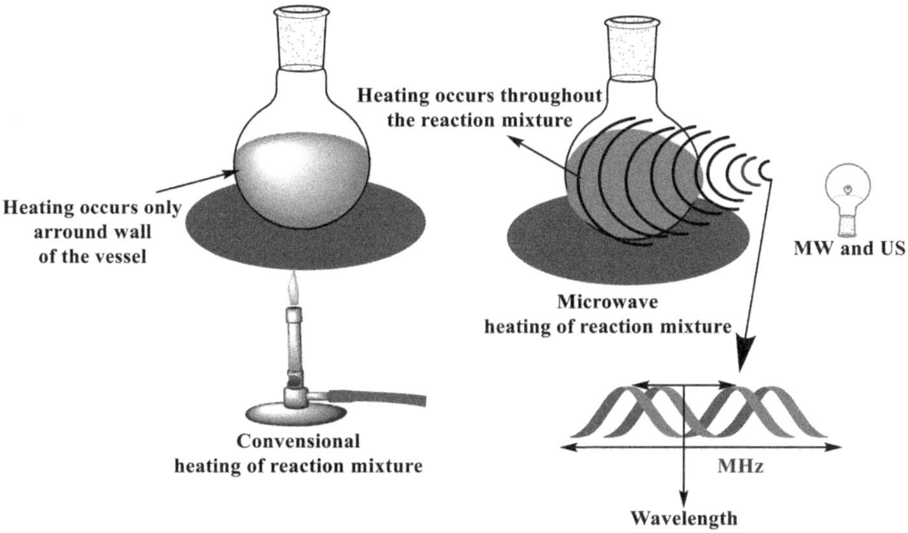

Figure 2.3 Activation mechanism of reactant molecules using conventional and microwave heating.

Some of the major advantages of MW and US catalysed reactions are as follows:

(i) Instantaneous heating and activation of reactants
(ii) Rapid and uniform heating throughout the reaction vessel
(iii) Increase in reaction rate (up to 1000 fold)
(iv) High chemo- and regio-selectivity
(v) High atom economy (high synthetic efficiency)
(vi) Minimal use, production and discharge of waste product(s)
(vii) Facile automation and selectivity
(viii) Development of highly pure compounds
(ix) Very high reproducibility
(x) Avoids unwanted heating of the reaction vessel
(xi) Avoids the unnecessary heating of the atmosphere
(xii) Ease and minimal number of purification and work-up steps
(xiii) Cost effective and time saving

A number of chemical transformations have been carried out using MW and US heatings (Figure 2.4). Organic compounds derived from MW and US irradiations are extensively used as corrosion inhibitors for different metals and alloys.[33,34] These compounds mainly act as mixed- and interface-type corrosion inhibitors. Their adsorption mainly follows the Langmuir adsorption isotherm model. Computational studies indicate that corrosion inhibitors derived from MW and US irradiations interact with the metal surface using donor–acceptor interactions. These compounds

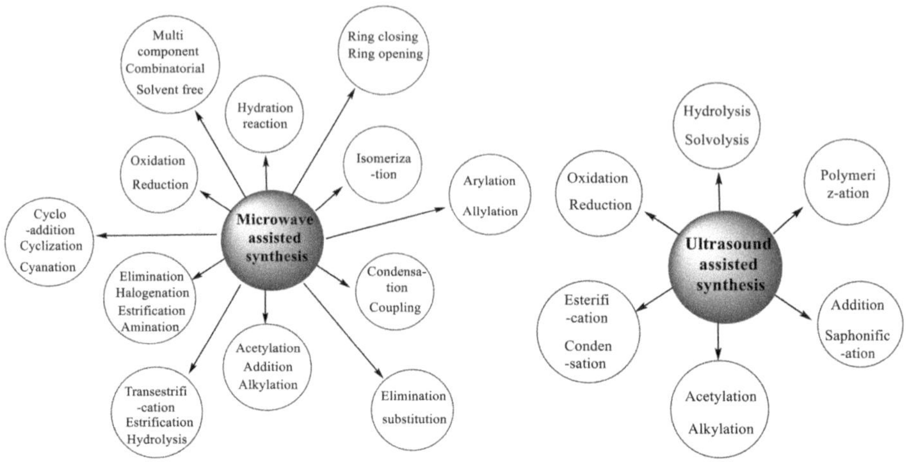

Figure 2.4 Some common reactions catalysed by microwave (left side) and ultrasound (right side) heating.

acquire flat or nearly flat orientations and provide maximum surface coverage and protection.

2.2.1.3 *Use of Eco-friendly Solvents*

Solvents play a crucial role in chemical transformations.[35,36] They solubilize reactant molecules and provide a medium for chemical changes. These mediums acquire various salient features that make them useful for chemical transformations. Nevertheless, most of the traditional solvents are toxic and non-eco-friendly. Their use and discharge are generally hazardous for living organisms and to the environment. A large amount of solvents is also used for works-up and purification processes such as for solvent extracts and column chromatography.[35,37] Their discharge to the surroundings causes huge safety problems. In view of the high toxicity of traditional solvents, some new environmentally friendly alternatives have been developed and used as solvents for green transformations. Some of the common green solvents are water,[38,39] ionic liquids,[40,41] supercritical CO_2[42,43] and polyethylene glycol (PEG).[44,45]

Water is known as a universal and primary solvent as it has the ability to dissolve relatively more chemicals compared to other solvents. Water has a molecular formula and molecular weight of H_2O and 18.01528(33) g mol^{-1}, respectively.[46,47] Water gains the highest density of 1000 kg m^{-3} at 4 °C.[48] Water is a highly polar molecule with a dipole moment of 1.85 D and therefore it is associated with a very high magnitude of dielectric constant.[49,50] Water serves as an excellent solvent only for polar reactant molecules. Non-polar or highly complex molecules containing aromatic rings show limited or no solubility in the water. Some of the organic compounds require cosolvent for their solubility in the water. Water has been established as one of the greenest solvents because of its following properties:[51,52]

(i) High liquidus range (0–100 °C)
(ii) Excellent polarity and dielectric constant
(iii) Solubilizes a wide range of organic and inorganic compounds
(iv) High heat capacity and thermal stability
(v) High heat of vaporization
(vi) Significantly high adhesive and cohesive forces
(vii) Non-flammable and nontoxic
(viii) Bio-renewable and eco-friendly
(ix) Water in liquid form is denser than in its solid form
(x) Commercially available and cost effective
(xi) High redox stability and unique redox potential
(xii) CO_2 dissolving capability
(xiii) Ability to form hydrogen bonding with the polar compounds.

Nevertheless, the use of water as a solvent is associated with some disadvantages, such as many reagents are water sensitive and degrade in water, water is limited for polar molecules, and the removal of excessive water needs distillation. Water is also a poor solvent for organic compounds and can change the acidity and/or basicity of reactant molecules. Despite its disadvantages, water is the best solvent for green transformation. Numerous chemical transformations for the compounds of industrial and biological applications have been carried out in water. Some of the major applications of ionic liquids (ILs) are illustrated in Figure 2.5. Organic compounds derived from water are widely used as anticorrosive materials for various systems. These compounds become effective by adsorbing on the metallic surface. Mainly their adsorption follows the Langmuir adsorption isotherm.

ILs are green alternatives to be used as solvents for numerous chemical transformations.[53,54] Obviously, ionic liquids are organic salts that are liquid below 100 °C.[53–56] A few ionic liquids are liquid at or below room temperature.[57,58] Ionic liquids are associated with various benefits that make them suitable solvents for the chemical transformations of compounds having industrial and biological uses. Some common properties of ionic liquids are as follows:[59,60]

(i) Negligible or very low vapour pressure
(ii) Easy and economical synthesis
(iii) Versatile roles *i.e.* behave as reagents, solvents and catalysts
(iv) Designer chemicals *i.e.* properties such as viscosity *etc.* can be suitably modified
(v) Very high liquidus range
(vi) High polarity and dielectric constant *i.e.* excellent dissolving tendency
(vii) Recyclability and reusability
(viii) High thermal and chemical stability

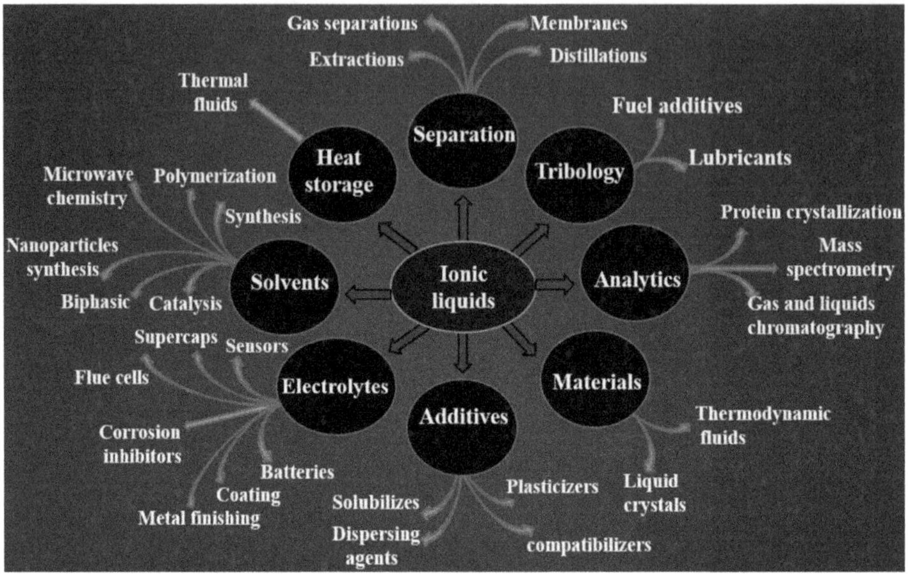

Figure 2.5 Some major industrial applications of ionic liquids.[67] Reproduced from ref. 67 with permission from Springer Nature, Copyright 2019.

(ix) Versatile solvents *i.e.* ILs tend to dissolve organic as well as inorganic chemicals
(x) Besides behaving as solvents, catalysts and reagents, ILs act as strong corrosion inhibitors.

However, ILs also acquire some disadvantages, such as being non-biodegradable, expensive and toxic. Literature investigation suggests that ILs serve as excellent green solvents for the synthesis of a variety of chemicals. Some of these compounds act as corrosion inhibitors for different metal/electrolyte systems.[61,62] Similar to ILs, PEG[63] and supercritical CO_2[64–66] have been established as green solvents and various chemical transformations have been carried out in these solvents.

2.2.2 Natural Green Corrosion Inhibitors

2.2.2.1 Plant Extracts as Green Corrosion Inhibitors

Plants possess the ability to convert radiant energy (sunlight) into glucose by the reaction of CO_2 and H_2O. They are the primary source of natural chemicals on earth. Literature study suggests that organic and aqueous extracts derived from plants are widely used as corrosion inhibitors for numerous metal/electrolyte systems.[6,68] Obviously, each plant extract contains numerous phytochemicals, such as flavonoids, saponins, alkaloids, steroids, anthocyanins, flavones, terpenoids, tannins, aromatic acids and tricarboxylic acids *etc.* (Figure 2.6).[6] These phytochemicals contain numerous electron rich

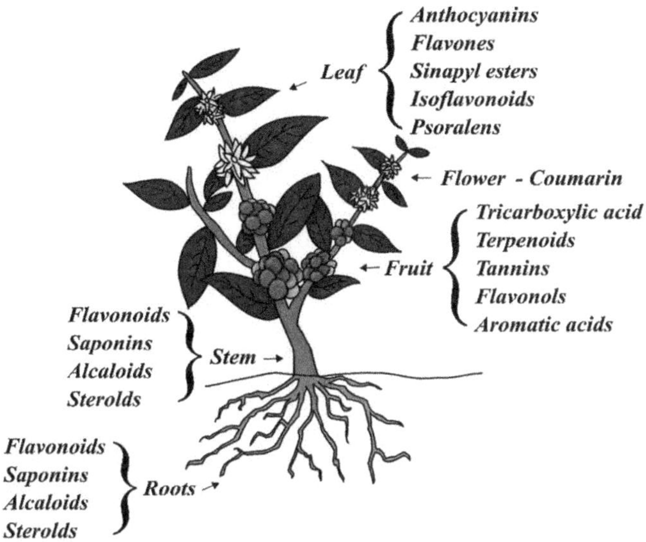

Figure 2.6 Diagrammatic illustration of phytochemicals present in the various portions of a plant extract to be used as corrosion inhibitors.[68] Reproduced from ref. 68, https://doi.org/10.3390/pr8080942, under the terms of the CC BY 4.0 licence https://creativecommons.org/licenses/by/4.0/.

polar functional groups having π- and non-bonding electrons through which they easily get adsorbed on the metal surface and form corrosion protective film. The corrosion inhibition effectiveness of phytochemicals depends upon their electronic nature. The greenness of the plant extracts is based on the fact that they are derived from natural and commercial resources. Extracts derived from different portions of the plant, including root, leaf, stem, fruit, flower and bark *etc.* are widely used as effective anticorrosive materials. Noticeably, leaf extracts exhibit relatively higher corrosion inhibition potential compared to other extracts of the same plant.

The use of plant extracts as corrosion inhibitors in place of traditional organic corrosion inhibitors is associated with numerous advantages. Some of the common advantages are as follows:

(i) Unlike traditional organic corrosion inhibitors, plant extracts are derived from natural resources *i.e.* from plants.
(ii) Plant extracts are biodegradable, nontoxic, non-bioaccumulative and non-biomagnificable.
(iii) Plant extracts are easy to process and apply.
(iv) Plant extracts generally show high corrosion protection at relatively low concentrations.
(v) A single plant extract can be used for different metal/electrolyte systems.
(vi) The use of plant extracts as corrosion inhibitors is a green and cost effective approach.

Nevertheless, not all plant extracts can be regarded as green as the extraction of some plant extracts requires highly toxic organic solvents. More so, some of the plant extracts are not very effective and require reasonably high concentrations for a good corrosion inhibition potential. Some plant extracts are very specific and useful only for one metal/electrolyte system. The extract may be prepared in aqueous or organic phases, and are called aqueous and organic extracts, respectively. Obviously, organic extracts contain relatively more phytochemicals compared to aqueous extracts, however, organic extracts experience solubility problems in aqueous electrolytes. Preparation of both aqueous and organic extracts involves common steps (Figure 2.7). Preparation of a plant extract involves the following steps:

Step 1: Collection, washing and cleaning of the plant materials.
Step 2: Drying and crushing of the plant materials.
Step 3: Stirring of the powdered plant materials in a particular solvent or a mixture of solvents (biphasic separation).
Step 4: Separation and concentration of the respective plant extract.
Step 5: Storage of the extract and subsequent use as a corrosion inhibitor.

Literature investigation reveals that plant extracts behave as reasonably good corrosion inhibitors, and their anticorrosive effects are attributed to the presence of phytochemicals.[6] Through polarization investigation, it is

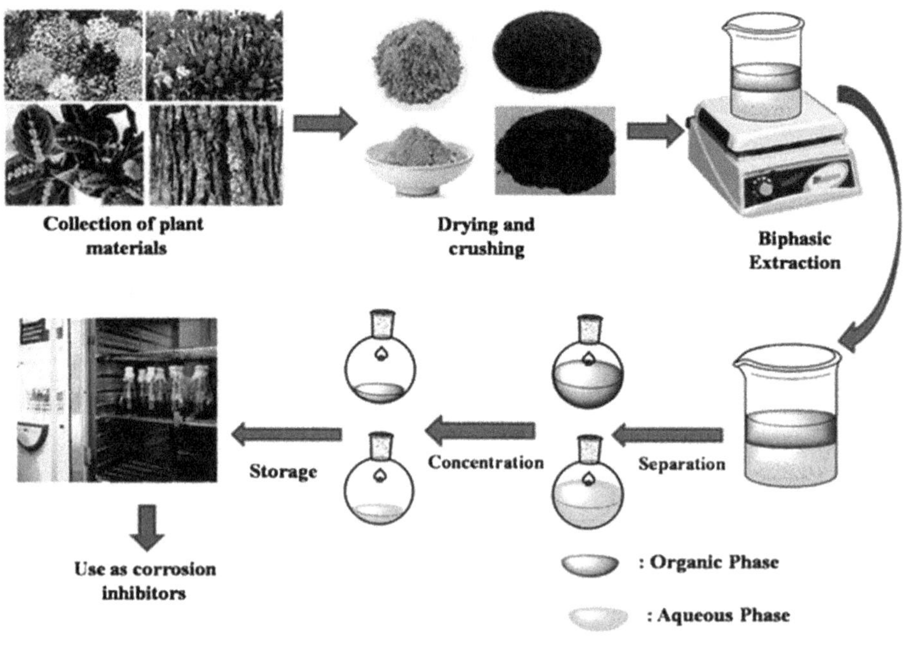

Figure 2.7 Illustration for the preparation of a plant extract.

derived that most plant extracts act as mixed-type corrosion inhibitors and they become effective by blocking the surface active sites.[68] SEM, EDX, AFM and XPS analyses have also been used to demonstrate the corrosion inhibition and adsorption behaviour of phytochemicals present in the extracts. Computational modellings, especially DFT, MDS and MCS, are also used to describe the interactions between phytochemicals and metal surfaces.

2.2.2.2 *Chemical Medicines (Drugs)*

Chemical medicines, commonly known as drugs, are another biological active class of organic compounds that are widely used as corrosion inhibitors for numerous metal/electrolyte systems.[69,70] It is extensively reported that chemical medicines are green alternatives to be used as corrosion inhibitors; however, not all drugs are green. The chemical medicines derived from natural resources such as plants (ayurvedic medicines) of course can be regarded as green, but allopathic medicines are synthesized in the laboratory using toxic and expensive starting materials, solvents and catalysts and cannot be regarded as green. Obviously, these medicines are synthesized using multistep reactions and are associated with numerous works-up and purification steps. Therefore, the syntheses of these medicines are associated with the use and discharge of huge amounts of environmentally malignant solvents that adversely affect human health and other living microorganisms in the surroundings. Nevertheless, in general, all drugs, independent of their nature of being natural or synthetic, are biodegradable and non-bioaccumulative. In this sense, allopathic medicines used as corrosion inhibitors can also be regarded as green alternatives. Various classifications of drugs are proposed based on their various properties (Figure 2.8).

Recently, the use of expired drugs as corrosion inhibitors is gaining particular attention, as most fresh drugs are quite expensive.[71–73] Generally, both expired and fresh drugs are quite complex molecules and contain numerous electron rich sites such as multiple bonds and polar functional

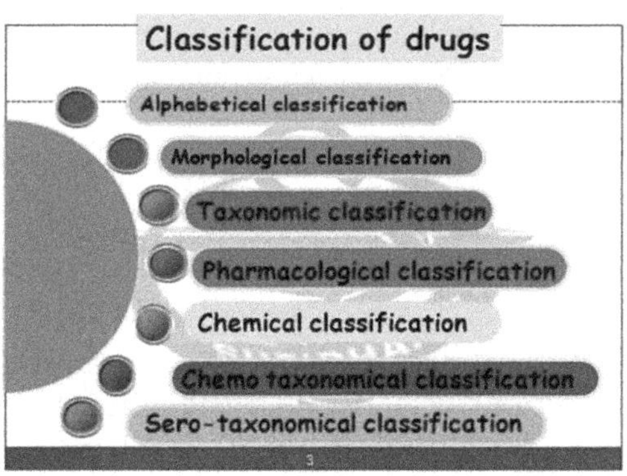

Figure 2.8 Various criteria for the classification of chemical medicines (drugs).

groups through which they easily get adsorbed and provide excellent surface protection. Most drugs behave as mixed-type corrosion inhibitors as they do not shift corrosion potential significantly and block the surface active sites responsible for corrosion. Adsorption behaviour of drug molecules is greatly supported by surface analyses, especially through SEM, EDX and AFM studies. Interactions of drug molecules with the metal surfaces are also reinforced by computational studies, especially using DFT, MDS and MCS methods.

2.2.2.3 *Natural Resources and Their Derivatives*

Nowadays, the use of natural resources such as carbohydrates,[74,75] amino acids (AAs)[76,77] and their derivatives as corrosion inhibitors is gaining particular attention. The use of natural resources has the following advantages:[74–77]

- (i) Natural resources are biodegradable, biotolerable and non-bioaccumulative.
- (ii) Natural resources exhibit excellent anticorrosive activity.
- (iii) Natural resources contain numerous adsorption sites/electron rich centres.
- (iv) Natural resources are commercially and economically available.
- (v) Natural resources are easy to process, isolate, purify and use.
- (vi) Natural resources are useful for different metal/electrolyte systems *i.e.* have versatile applications.
- (vii) Natural resources are environmentally benign and safe for living organisms, including human beings.
- (viii) Natural resources have minimized adverse effects on the environment.
- (ix) Natural resources have the possibility of chemical modification for improvement in corrosion inhibition potential.
- (x) The solubility and anticorrosive acidities of natural resources can be enhanced by proper functionalization.
- (xi) Natural resources are nontoxic and non-inflammatory.

Nevertheless, the implementation of natural resources is linked to some disadvantages, such as their limited solubility in a polar electrolyte and expensive isolation, and their isolation sometimes utilizes toxic and expensive solvents *etc.* More so, in general, natural resources are quite sensitive to heat and easily undergo hydrolysis, degradation and decomposition, particularly at elevated temperatures. Carbohydrates, AAs and their derivatives are extensively used as corrosion inhibitors. Mostly, they behave as mixed- and interface-type corrosion inhibitors.[74,76,78] Their adsorption on the metallic surface follows the Langmuir adsorption isotherm model.

2.3 Conclusion

Because of the growing interest and demand for green chemistry and sustainable developments, the use of traditional organic corrosion inhibitors is limited due to their association with various shortcomings, including toxicity. In view of these issues, numerous green alternatives of synthetic and natural origin have been developed and implemented as effective corrosion inhibitors. Among the natural corrosion inhibitors, plant extracts, chemical medicines (ayurvedic medicines), carbohydrates, amino acids (AAs) and their derivatives are commonly used. Because of their natural origin, these materials are nontoxic, eco-friendly, non-bioaccumulative and biodegradable in nature. These compounds contain various electron rich sites through which they easily get adsorbed on the metallic surface. In the synthetic green corrosion inhibitors, compounds derived from multicomponent reactions (MCRs), mechanochemical mixing, microwave (MW) and ultrasound (US) have gained particular attention. These synthetic techniques are associated with various advantages such as shorter reaction time, high atom economy, high selectivity, ease of operation, uniform, instantaneous activation and many more. The green corrosion inhibitors of natural and synthetic origins mostly behave as mixed-type corrosion inhibitors and their adsorption follows the Langmuir adsorption isotherm. Their adsorption mode of corrosion prevention is greatly reinforced by surface and computational analyses.

References

1. M. A. Quraishi, D. S. Chauhan, and V. S. Saji, *Heterocyclic Organic Corrosion Inhibitors: Principles and Applications*, Elsevier, 2020.
2. C. Verma, M. H. Abdellattif, A. Alfantazi and M. Quraishi, N-Heterocycle Compounds as Aqueous Phase Corrosion Inhibitors: A Robust, Effective and Economic Substitute, *J. Mol. Liq.*, 2021, **340**, 117211.
3. V. S. Sastri, *Green Corrosion Inhibitors: Theory and Practice*, John Wiley & Sons, 2012.
4. J. Haque, V. Srivastava, D. S. Chauhan, H. Lgaz and M. A. Quraishi, Microwave-induced synthesis of chitosan Schiff bases and their application as novel and green corrosion inhibitors: experimental and theoretical approach, *ACS Omega*, 2018, **3**, 5654–5668.
5. S. H. Zaferani, M. Sharifi, D. Zaarei and M. R. Shishesaz, Application of eco-friendly products as corrosion inhibitors for metals in acid pickling processes–A review, *J. Environ. Chem. Eng.*, 2013, **1**, 652–657.
6. C. Verma, E. E. Ebenso, I. Bahadur and M. Quraishi, An overview on plant extracts as environmental sustainable and green corrosion inhibitors for metals and alloys in aggressive corrosive media, *J. Mol. Liq.*, 2018, **266**, 577–590.
7. C. T. Moermond, M. P. Janssen, J. A. de Knecht, M. H. Montforts, W. J. Peijnenburg, P. G. Zweers and D. T. Sijm, PBT assessment using the

revised annex XIII of REACH: a comparison with other regulatory frameworks, *Integr. Environ. Assess. Manage.*, 2012, **8**, 359–371.
8. M. Pavan and A. P. Worth, Review of estimation models for biodegradation, *QSAR Comb. Sci.*, 2008, **27**, 32–40.
9. R. Lagoni, *Regional Protection of the Marine Environment in the Northeast Atlantic Under the OSPAR Convention of 1992, The Stockholm Declaration and Law of the Marine Environment*, Brill Nijhoff, 2003, pp. 183–203.
10. C. E. Boyd, LC50 calculations help predict toxicity, *Global Aquacult. Advocate*, 2005, **8**, 84–87.
11. V. S. Leblond and A. Hontela, Effects ofin VitroExposures to Cadmium, Mercury, Zinc, and 1-(2-Chlorophenyl)-1-(4-chlorophenyl)-2, 2-dichloroethane on Steroidogenesis by Dispersed Interrenal Cells of Rainbow Trout (Oncorhynchus mykiss), *Toxicol. Appl. Pharmacol.*, 1999, **157**, 16–22.
12. P. Vivekanandhan, R. Venkatesan, G. Ramkumar, S. Karthi, S. Senthil-Nathan and M. S. Shivakumar, Comparative analysis of major mosquito vectors response to seed-derived essential oil and seed pod-derived extract from Acacia nilotica, *Int. J. Environ. Res. Public Health*, 2018, **15**, 388.
13. C. Verma, E. E. Ebenso, M. Quraishi and C. M. Hussain, Recent developments in sustainable corrosion inhibitors: design, performance and industrial scale applications, *Mater. Adv.*, 2021, **2**, 3806–3850.
14. M. D. Summers, S. L. Blunk, and B. M. Jenkins, *How Straw Decomposes: Implications for Straw Bale Construction*, Ecological Building Network, 2003.
15. R. Wedmann, I. Ivanovic-Burmazovic and M. R. Filipovic, Nitrosopersulfide ($SSNO^-$) decomposes in the presence of sulfide, cyanide or glutathione to give HSNO/SNO^-: consequences for the assumed role in cell signalling, *Interface Focus*, 2017, **7**, 20160139.
16. R. L. Williamson Jr, D. T. Burton, J. H. Clarke and L. E. Fleming, Gathering danger: the urgent need to regulate toxic substances that can bioaccumulate, *Ecol. Law Q.*, 1993, **20**, 605.
17. S. A. van der Heijden and M. T. Jonker, Evaluation of liposome – water partitioning for predicting bioaccumulation potential of hydrophobic organic chemicals, *Environ. Sci. Technol.*, 2009, **43**, 8854–8859.
18. J. M. Conder, R. A. Hoke, W. d. Wolf, M. H. Russell and R. C. Buck, Are PFCAs bioaccumulative? A critical review and comparison with regulatory criteria and persistent lipophilic compounds, *Environ. Sci. Technol.*, 2008, **42**, 995–1003.
19. W. M. Meylan, P. H. Howard, R. S. Boethling, D. Aronson, H. Printup and S. Gouchie, Improved method for estimating bioconcentration/bioaccumulation factor from octanol/water partition coefficient, *Environ. Toxicol. Chem.*, 1999, **18**, 664–672.
20. H. G. Alvim, E. N. da Silva Junior and B. A. Neto, What do we know about multicomponent reactions? Mechanisms and trends for the Biginelli, Hantzsch, Mannich, Passerini and Ugi MCRs, *RSC Adv.*, 2014, **4**, 54282–54299.

21. M. S. Singh and S. Chowdhury, Recent developments in solvent-free multicomponent reactions: a perfect synergy for eco-compatible organic synthesis, *RSC Adv.*, 2012, **2**, 4547–4592.
22. W. S. Bremner and M. G. Organ, Multicomponent reactions to form heterocycles by microwave-assisted continuous flow organic synthesis, *J. Comb. Chem.*, 2007, **9**, 14–16.
23. D. J. Ramón and M. Yus, Asymmetric multicomponent reactions (AMCRs): the new frontier, *Angew. Chem., Int. Ed.*, 2005, **44**, 1602–1634.
24. C. Verma, J. Haque, M. Quraishi and E. E. Ebenso, Aqueous phase environmental friendly organic corrosion inhibitors derived from one step multicomponent reactions: a review, *J. Mol. Liq.*, 2019, **275**, 18–40.
25. C. Verma, and M. Quraishi, Green corrosion inhibitors derived through one-step multicomponent reactions: recent developments, *Environmentally Sustainable Corrosion Inhibitors*, 2022, pp. 289–302.
26. A. Barge, S. Tagliapietra, L. Tei, P. Cintas and G. Cravotto, Pd-catalyzed reactions promoted by ultrasound and/or microwave irradiation, *Curr. Org. Chem.*, 2008, **12**, 1588–1612.
27. P. Cintas, A. Barge, S. Tagliapietra, L. Boffa and G. Cravotto, Alkyne–azide click reaction catalyzed by metallic copper under ultrasound, *Nat. Protoc.*, 2010, **5**, 607–616.
28. R. N. Baig and R. S. Varma, Alternative energy input: mechanochemical, microwave and ultrasound-assisted organic synthesis, *Chem. Soc. Rev.*, 2012, **41**, 1559–1584.
29. G. Cravotto and P. Cintas, The combined use of microwaves and ultrasound: improved tools in process chemistry and organic synthesis, *Chem. – Eur. J.*, 2007, **13**, 1902–1909.
30. F. Penteado, B. Monti, L. Sancineto, G. Perin, R. G. Jacob, C. Santi and E. J. Lenardão, Ultrasound-Assisted Multicomponent Reactions, Organometallic and Organochalcogen Chemistry, *Asian J. Org. Chem.*, 2018, **7**, 2368–2385.
31. A. Dandia, R. Singh and S. Bhaskaran, Multicomponent reactions and ultrasound: a synergistic approach for the synthesis of bioactive heterocycles, *Curr. Green Chem.*, 2014, **1**, 17–39.
32. R. S. Varma, "Greener" chemical syntheses using mechanochemical mixing or microwave and ultrasound irradiation, *Green Chem. Lett. Rev.*, 2007, **1**, 37–45.
33. C. Verma, M. Quraishi and E. E. Ebenso, Microwave and ultrasound irradiations for the synthesis of environmentally sustainable corrosion inhibitors: An overview, *Sustainable Chem. Pharm.*, 2018, **10**, 134–147.
34. A. Dandia, S. Gupta, P. Singh and M. Quraishi, Ultrasound-assisted synthesis of pyrazolo [3, 4-b] pyridines as potential corrosion inhibitors for mild steel in 1.0 M HCl, *ACS Sustainable Chem. Eng.*, 2013, **1**, 1303–1310.
35. H. C. Hailes, Reaction solvent selection: the potential of water as a solvent for organic transformations, *Org. Process Res. Dev.*, 2007, **11**, 114–120.
36. M. Liu, J. Zhao, Z. Luo, Z. Sun, N. Pan, H. Ding and X. Wang, Unveiling solvent-related effect on phase transformations in CsBr–PbBr2 system: coordination and ratio of precursors, *Chem. Mater.*, 2018, **30**, 5846–5852.

37. V. Polshettiwar and R. S. Varma, Microwave-assisted organic synthesis and transformations using benign reaction media, *Acc. Chem. Res.*, 2008, **41**, 629–639.
38. Y. Marcus, *Supercritical Water: A Green Solvent: Properties and Uses*, John Wiley & Sons, 2012.
39. X. Cao, G. Zhang, L. Jiang, Y. Cai, Y. Gao, W. Yang, X. He, Q. Zeng, G. Xing and Y. Jia, Water, a green solvent for fabrication of high-quality CsPbBr3 films for efficient solar cells, *ACS Appl. Mater. Interfaces*, 2019, **12**, 5925–5931.
40. T. Welton, Ionic liquids in green chemistry, *Green Chem.*, 2011, **13**, 225.
41. R. D. Rogers, K. R. Seddon, Division of Industrial Engineering Chemistry, Staff American Chemical Society, and Meeting Staff American Chemical Society, *Ionic Liquids as Green Solvents*, ACS Publications, 2003.
42. D. Lozowski, Supercritical CO_2: A green solvent, *Chem. Eng.*, 2010, **117**, 15.
43. S. P. Nalawade, F. Picchioni and L. Janssen, Supercritical carbon dioxide as a green solvent for processing polymer melts: Processing aspects and applications, *Prog. Polym. Sci.*, 2006, **31**, 19–43.
44. E. Kianpour and S. Azizian, Polyethylene glycol as a green solvent for effective extractive desulfurization of liquid fuel at ambient conditions, *Fuel*, 2014, **137**, 36–40.
45. M. Vafaeezadeh and M. M. Hashemi, Polyethylene glycol (PEG) as a green solvent for carbon–carbon bond formation reactions, *J. Mol. Liq.*, 2015, **207**, 73–79.
46. L. Torres-Ronda and X. S. i del Alcázar, The properties of water and their applications for training, *J. Hum. Kinet.*, 2014, **44**, 237.
47. P. Ball, Water—an enduring mystery, *Nature*, 2008, **452**, 291–292.
48. M. Tanaka, G. Girard, R. Davis, A. Peuto and N. Bignell, Recommended table for the density of water between 0 °C and 40 °C based on recent experimental reports, *Metrologia*, 2001, **38**, 301.
49. J. Gregory, D. Clary, K. Liu, M. Brown and R. Saykally, The water dipole moment in water clusters, *Science*, 1997, **275**, 814–817.
50. S. A. Clough, Y. Beers, G. P. Klein and L. S. Rothman, Dipole moment of water from Stark measurements of H_2O, HDO, and D_2O, *J. Chem. Phys.*, 1973, **59**, 2254–2259.
51. R. Cooke and I. Kuntz, The properties of water in biological systems, *Annu. Rev. Biophys. Bioeng.*, 1974, **3**, 95–126.
52. D. Eisenberg, and W. Kauzmann, *The Structure and Properties of Water*, Oxford University Press on Demand, 2005.
53. R. L. Vekariya, A review of ionic liquids: Applications towards catalytic organic transformations, *J. Mol. Liq.*, 2017, **227**, 44–60.
54. R. D. Rogers and K. R. Seddon, Ionic liquids--solvents of the future?, *Science*, 2003, **302**, 792–793.
55. S. A. Forsyth, J. M. Pringle and D. R. MacFarlane, Ionic liquids—an overview, *Aust. J. Chem.*, 2004, **57**, 113–119.
56. M. Galiński, A. Lewandowski and I. Stępniak, Ionic liquids as electrolytes, *Electrochim. Acta*, 2006, **51**, 5567–5580.

57. K. N. Marsh, J. A. Boxall and R. Lichtenthaler, Room temperature ionic liquids and their mixtures—a review, *Fluid Phase Equilib.*, 2004, **219**, 93–98.
58. T. Welton, Room-temperature ionic liquids. Solvents for synthesis and catalysis, *Chem. Rev.*, 1999, **99**, 2071–2084.
59. S. Zhang, N. Sun, X. He, X. Lu and X. Zhang, Physical properties of ionic liquids: database and evaluation, *J. Phys. Chem. Ref. Data*, 2006, **35**, 1475–1517.
60. K. Mihkel, Physical and chemical properties of ionic liquids based on the dialkylimidazolium cation, *Est Proc. Acad. Sci. Chem*, 2000, **49**, 145–155.
61. M. Deyab, Understanding the anti-corrosion mechanism and performance of ionic liquids in desalination, petroleum, pickling, de-scaling, and acid cleaning applications, *J. Mol. Liq.*, 2020, **309**, 113107.
62. C. Verma, S. H. Alrefaee, M. Quraishi, E. E. Ebenso and C. M. Hussain, Recent developments in sustainable corrosion inhibition using ionic liquids: A review, *J. Mol. Liq.*, 2021, **321**, 114484.
63. M. Ludwiczak, J. Szyling, A. Garbicz, T. Sokolnicki, J. Pyziak and J. d. Walkowiak, Application of green solvents: PEG and $scCO_2$ in the mono-or biphasic catalytic systems for the repetitive batch coupling of vinylsilanes with vinyl boronates toward 1-boryl-1-silylethenes, *Inorg. Chem.*, 2020, **59**, 17555–17564.
64. K. Shanab, C. Neudorfer, E. Schirmer and H. Spreitzer, Green solvents in organic synthesis: an overview, *Curr. Org. Chem.*, 2013, **17**, 1179–1187.
65. L. A. Blanchard, D. Hancu, E. J. Beckman and J. F. Brennecke, Green processing using ionic liquids and CO_2, *Nature*, 1999, **399**, 28–29.
66. W. Leitner, Supercritical carbon dioxide as a green reaction medium for catalysis, *Acc. Chem. Res.*, 2002, **35**, 746–756.
67. C. Verma and E. E. Ebenso, Ionic liquid-mediated functionalization of graphene-based materials for versatile applications: a review, *Graphene Technol.*, 2019, **4**, 1–15.
68. A. Miralrio and A. Espinoza Vázquez, Plant extracts as green corrosion inhibitors for different metal surfaces and corrosive media: A review, *Processes*, 2020, **8**, 942.
69. G. Gece, Drugs: A review of promising novel corrosion inhibitors, *Corros. Sci.*, 2011, **53**, 3873–3898.
70. M. Abdallah, Rhodanine azosulpha drugs as corrosion inhibitors for corrosion of 304 stainless steel in hydrochloric acid solution, *Corros. Sci.*, 2002, **44**, 717–728.
71. N. Vaszilcsin, V. Ordodi and A. Borza, Corrosion inhibitors from expired drugs, *Int. J. Pharm.*, 2012, **431**, 241–244.
72. N. K. Gupta, C. Gopal, V. Srivastava and M. Quraishi, Application of expired drugs in corrosion inhibition of mild steel, *Int. J. Pharm. Chem. Anal.*, 2017, **4**, 8–12.
73. H. I. Al-Shafey, R. A. Hameed, F. Ali, A. e.-A. S. Aboul-Magd and M. Salah, Effect of expired drugs as corrosion inhibitors for carbon steel in 1M HCL solution, *Int. J. Pharm. Sci. Rev. Res.*, 2014, **27**, 146–152.

74. S. A. Umoren and U. M. Eduok, Application of carbohydrate polymers as corrosion inhibitors for metal substrates in different media: a review, *Carbohydr. Polym.*, 2016, **140**, 314–341.
75. M. Yadav, T. Sarkar and I. Obot, Carbohydrate compounds as green corrosion inhibitors: electrochemical, XPS, DFT and molecular dynamics simulation studies, *RSC Adv.*, 2016, **6**, 110053–110069.
76. B. El Ibrahimi, A. Jmiai, L. Bazzi and S. El Issami, Amino acids and their derivatives as corrosion inhibitors for metals and alloys, *Arabian J. Chem.*, 2020, **13**, 740–771.
77. L. Hamadi, S. Mansouri, K. Oulmi and A. Kareche, The use of amino acids as corrosion inhibitors for metals: a review, *Egypt. J. Pet.*, 2018, **27**, 1157–1165.
78. M. Shahini, B. Ramezanzadeh and H. E. Mohammadloo, Recent advances in biopolymers/carbohydrate polymers as effective corrosion inhibitive macro-molecules: A review study from experimental and theoretical views, *J. Mol. Liq.*, 2020, 115110.

CHAPTER 3

Designing of Green Corrosion Inhibitors: Computational and Experimental Approaches

3.1 Introduction

An organic corrosion inhibitor molecule, when added to a corrosive medium, adsorbs on a target metallic substrate. The adsorption of a corrosion inhibitor on a metal surface is controlled by a number of parameters such as inhibitor concentration, solution concentration, and pH of the solution, temperature, flow rate, presence of corrosion product, *etc.* Therefore, to gather a thorough understanding of the adsorption and inhibition behaviour, certain measurements are required to be undertaken. These tests are designed in accordance with the standards purported by the American Society for Testing and Materials (ASTM), and the National Association of Corrosion Engineers (NACE). The ASTM standards present a primary reference guide for the detailed procedures.[1] The overall goal of these measurements is to collect information on the thermodynamic and kinetic parameters governing the inhibitor adsorption and inhibition performance to obtain a detailed understanding of the inhibitor adsorption mechanism.

This chapter provides a brief overview of some of the modern experimental and computational techniques that are being used to correlate the inhibitor structural aspects with the physicochemical nature of inhibitor adsorption. The major parameters elucidated with the chemical,[2–4] electrochemical,[5–9] surface analytical,[2,9,10] and computational techniques[11,12] are described

Green Chemistry Series No. 73
Green Corrosion Inhibition: Fundamentals, Design, Synthesis and Applications
By Chandrabhan Verma, Mumtaz A. Quraishi and Dheeraj Singh Chauhan

Published by the Royal Society of Chemistry, www.rsc.org

herein. The techniques and the derived parameters are expected to guide the readers in understanding the nature of the inhibitor function described in the subsequent chapters.

3.2 Methods of Evaluation of Corrosion Inhibitors

3.2.1 Chemical and Electrochemical Techniques

As mentioned above, the organic corrosion inhibitors generally function by adsorption over a target metallic surface. To understand the adsorption and the corrosion inhibition of an inhibitor molecule, it is necessary to perform some specific experimental measurements. These include the quantitative estimation of the corrosion inhibition efficiency and the evaluation of several thermodynamic/kinetic parameters, which elucidate the adsorption and inhibition mechanism of an organic molecule on a given metallic surface. These tests can be divided into different categories based on the operating principle involved, such as (i) immersion tests (gravimetric measurements), (ii) electrochemical analyses, and (iii) surface characterization. These tests are carried out following specific standards as prescribed using the ASTM procedures.[1] The immersion and the electrochemical tests can be performed at different inhibitor concentrations, for different periods, and at varying temperatures, and the obtained data can be used to plot adsorption isotherms and provides the desired adsorption parameters. While the immersion tests provide the information on the overall corrosion rate, the electrochemical tests can provide the instantaneous corrosion rates, the electrochemical mechanism of corrosion, and the inhibitor adsorption.

The immersion (or the so-called weight loss) tests are the simplest and yet provide the most reliable information. The corrosion rates, optimum inhibitor concentration, coverage of the metal surface by inhibitor, and thermodynamic and kinetic parameters of adsorption are obtained from this test. Typically, a metal sample is pre-cleaned following standard methods and weighed before and after immersing into a corrosive medium, without and with the corrosion inhibitor. After a certain period of time, the sample is taken out and weighed again to check the difference in the weights, which provides an estimation of the corrosion inhibition efficiency and other parameters.[19–22] Corrosion rate (C_R) is the speed at which the metallic sample degrades when immersed in a specific corrosive solution. It can be given by:[23]

$$C_R = \frac{87.6w}{AtD} \tag{3.1}$$

where, C_R = corrosion rate (mm y^{-1}), w = average weight loss (mg), A = sample surface area (cm^2), t = exposure time (h), D = density of metal specimen (g cm^{-3}). The surface coverage (θ) and the corrosion inhibition efficiency (η%) can be determined from the values of the C_R:[2,5,8,24]

$$\theta = \frac{C_R - C_{R(i)}}{C_R} \tag{3.2}$$

$$\eta\% = \frac{C_{\mathrm{R}} - C_{\mathrm{R(i)}}}{C_{\mathrm{R}}} \times 100 \tag{3.3}$$

where C_{R} and $C_{\mathrm{R(i)}}$ = corrosion rates for blank and inhibited metal samples, respectively.

The most common electrochemical techniques are potentiostatic electrochemical impedance spectroscopy (EIS),[25,26] electrochemical frequency modulation (EFM),[27,28] linear polarization resistance (LPR), potentiodynamic polarization tests (PDP),[29] and cyclic voltammetry (CV). Electrochemical methods, such as EIS, EFM, and LPR, can be categorized as non-destructive techniques, whereas PDP and CV are destructive techniques.[5–9] These techniques elucidate the electrochemical parameters that provide a piece of quantitative information on the adsorption of inhibitors and the protective film formation at the metal/solution interface. The EIS can provide information of the electrochemical mechanism of the corrosion and inhibition process, *i.e.*, whether the process is capacitive controlled (Figure 3.1a), diffusion controlled (Figure 3.1b), or has an inductive behaviour (Figure 3.1c). The PDP can provide information on the anodic/cathodic/mixed type of inhibition behaviour (Figure 3.1e). In addition, the PDP technique can also shed light on the desorption of the inhibitor film and the mechanism of inhibitor adsorption and hydrogen evolution. The EIS, EFM, and the LPR are considered as the non-destructive techniques, which do not cause any damage to the metal surface. The PDP and the CV cause the polarization of the surface, and after either of these techniques, no further experiments can be done on the surface to obtain any reliable information. In addition, certain advanced techniques such as the dynamic EIS (DEIS), EFM trends, scanning electrochemical microscopy (SECM), electrochemical quartz crystal microbalance or nanobalance (EQCM/EQCN), localized EIS (LEIS) are also performed as per the requirement. The DEIS and the EFM trends are some of the recently used techniques in the area of corrosion inhibition. Table 3.1 lists the major electrochemical techniques with the significance of each of them. Equations for corrosion inhibition efficiency calculation *via* various electrochemical methods:

From EIS:

$$\eta_{\mathrm{EIS}}\% = \frac{R_{\mathrm{ct}}^{\mathrm{inh}} - R_{\mathrm{ct}}}{R_{\mathrm{ct}}^{\mathrm{inh}}} \times 100 \tag{3.4}$$

From LPR:

$$\eta_{\mathrm{LPR}}\% = \frac{R_{\mathrm{P}}^{\mathrm{inh}} - R_{\mathrm{P}}}{R_{\mathrm{P}}^{\mathrm{inh}}} \times 100 \tag{3.5}$$

where the symbols R_{ct}, R_{P} and $R_{\mathrm{ct}}^{\mathrm{inh}}$, $R_{\mathrm{p}}^{\mathrm{inh}}$ denote the charge transfer resistance in the presence and the absence of the inhibitor, respectively.

From EFM:

$$\eta_{\mathrm{EFM}}\% = \frac{i_{\mathrm{corr}} - i_{\mathrm{corr}}^{\mathrm{inh}}}{i_{\mathrm{corr}}} \times 100 \tag{3.6}$$

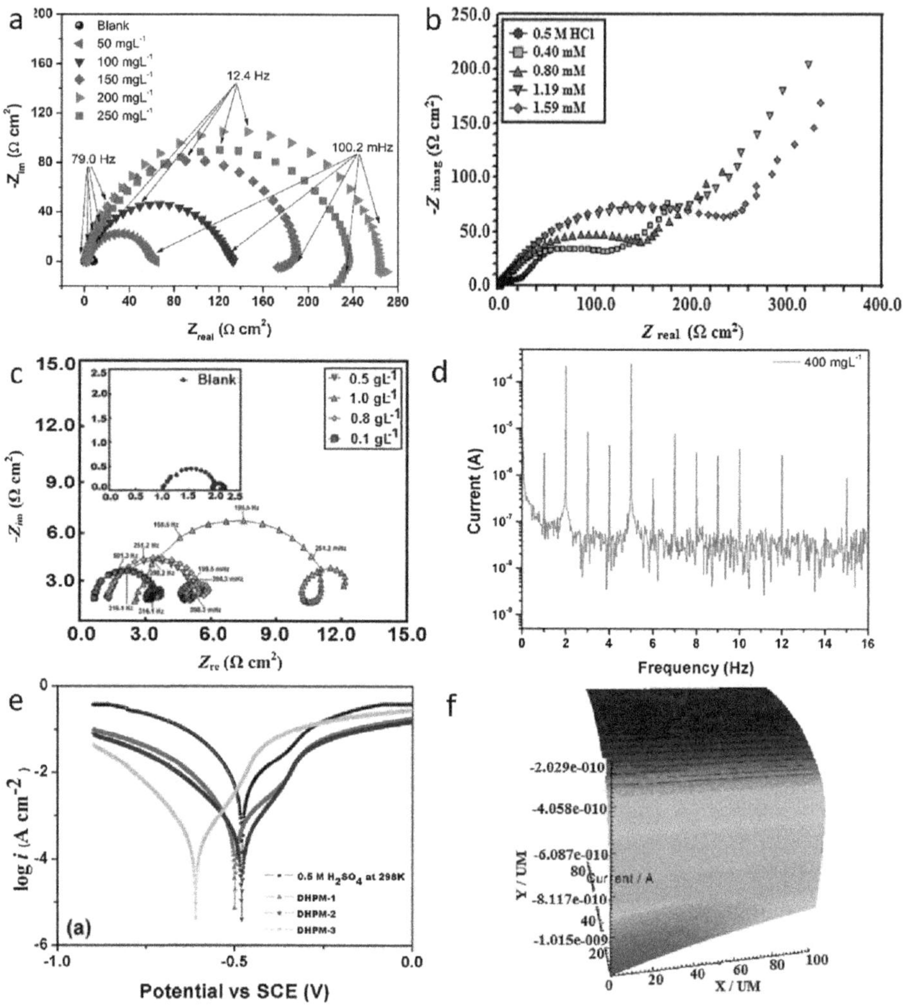

Figure 3.1 Electrochemical characterization of corrosion inhibitor adsorption by EIS (a) capacitive control[13] (b) diffusion control.[14] Adapted from ref. 13 and reproduced from ref. 14 with permission from the Royal Society of Chemistry. (c) Appearance of inductive loops.[15] Reproduced from ref. 15, https://doi.org/10.1007/s40090-016-0103-y, under the terms of the CC BY 4.0 licence https://creativecommons.org/licenses/by/4.0/. (d) EFM plot.[16] Reproduced from ref. 16 with permission from Springer Nature, Copyright 2020. (e) PDP curves;[17] (f) SECM micrograph.[18] Reproduced from ref. 17 and 18 with permission from Elsevier, Copyright 2019 and 2015.

From PDP:

$$\eta_{\mathrm{PDP}}\% = \frac{i_{\mathrm{corr}} - i_{\mathrm{corr}}^{\mathrm{inh}}}{i_{\mathrm{corr}}} \times 100 \tag{3.7}$$

Table 3.1 Electroanalytical techniques for corrosion inhibitor evaluation.

Technique/major parameters derived	Symbol	Unit	Significance	Ref.
Electrochemical impedance spectroscopy (EIS)				
Solution resistance/charge transfer resistance/film resistance/inductive resistance	$R_s/R_{ct}/R_f/R_L$	$\Omega\ cm^2$	• Non-destructive technique • Used to probe the inhibitor film formation at the metal surface • Indicates whether that electrochemical mechanism is capacitive or diffusion controlled • Provides information on the electrochemical equivalent circuit formed by inhibitor adsorption at the electrochemical interface • Goodness of fit parameter validates the fitting of EIS data	30–33
Double layer capacitance/film capacitance	C_{dl}/C_f	$\mu F\ cm^{-2}$		
Phase shift	n	—		
Relaxation time constant	τ	s		
Phase angle	α	°		
Inductance	L	$H\ cm^2$		
Warburg impedance	W	$\Omega^{-1}\ cm^{-2}\ s^{1/2}$		
Goodness of fit	χ^2	—		
Inhibition efficiency	η	%		
Dynamic electrochemical impedance spectroscopy (DEIS)				
Same as in EIS	—	—	• Provides EIS parameters with a long time range	34 and 35
Electrochemical frequency modulation (EFM)				
Anodic/cathodic Tafel slopes	β_a/β_c	$V\ dec^{-1}$	• Non-destructive and fast technique • Provides the mechanism of corrosion and inhibition • Provides the corrosion current densities and corrosion rate • Causality factors can verify the validity of the electrochemical data	27 and 36
Corrosion current density	i_{corr}	$A\ cm^{-2}$		
Corrosion rate	C_R	mmpy		
Inhibition efficiency	η	%		
Electrochemical frequency modulation trend (EFM trend)				
Same as in EFM	—	—	• Monitors changes in corrosion rate with time	27 and 36
Linear polarization resistance (LPR)				
Anodic/cathodic Tafel slopes	β_a/β_c	$V\ dec^{-1}$	• Non-destructive and fast technique • Provides the mechanism of corrosion and inhibition • Provides the corrosion rate	37–39
Corrosion rate	C_R	mmpy		

(*continued*)

Table 3.1 (*continued*)

Technique/major parameters derived	Symbol	Unit	Significance	Ref.
Inhibition efficiency	η	%		
Potentiodynamic polarization (PDP)				
Anodic/cathodic Tafel slopes	β_a/β_c	V dec^{-1}	• Destructive technique • Provides information on adsorption and inhibition mechanism • Corrosion current densities and corrosion rate can be obtained	37–39
Corrosion potential	E_{corr}	V		
Corrosion current density	i_{corr}	A cm^{-2}		
Corrosion rate	C_R	mmpy		
Inhibition efficiency	η	%		
Scanning electrochemical microscopy (SECM)				
Current density	I	A cm^{-2}	• Provides surface imaging of the sample in terms of current profile • Can probe in details of pitting corrosion • Takes a long time • Difficult to perform SECM in concentrated acids	40 and 41
Surfaces images				
Pit depth				

where the symbols i_{corr} and i_{corr}^{inh} denote the charge transfer resistance in the presence and the absence of the inhibitor, respectively.

Classification of inhibitors based on electrochemical nature:

- Anodic inhibitors that retard the anodic reactions by the formation of passive films (*e.g.* phosphate, molybdate, benzoate *etc.*).
- Cathodic inhibitors that retard cathodic reactions by forming thick films (*e.g.* Ca, Mg, Ce salts *etc.*).
- Mixed inhibitors that inhibit both anodic and cathodic corrosion processes.

3.2.2 Surface Analysis of Corrosion Inhibitor Adsorption

The third class of the corrosion tests includes the direct examination of the metal surface with or without an adsorbed film of corrosion inhibitor to check the variation in surface hydrophobicity/wettability (water contact angle), roughness (SEM, AFM), SEM/AFM and includes the imaging of the metal surface at the microscopic scales. Contact angle measurements provide information on the improvement in the hydrophobic nature indicating the inhibitor adsorption. In the example (Figure 3.2), three imidazolidine derivatives were analysed for sweet corrosion of J55 steel samples in CO_2 saturated 3.5% NaCl.[42] It can be observed that the incorporation of electron donating group ($-OCH_3$) and the electron-withdrawing groups ($-NO_2$) led to a difference in the surface hydrophobicity of the steel samples as revealed by the WCA measurements. The corresponding SEM images support the improvement in the surface homogeneity upon introduction of the corrosion inhibitor AT-1. Quantitative information on the images derived *via* AFM indicates a lowering in the average surface roughness from 178 nm to 11 nm in the presence of the adsorbed AT-1 inhibitor film.

On the other hand, XPS and ToF-SIMS are highly sensitive surface analytical techniques that can indicate the presence of any metal–inhibitor complexes formed on the surface; shift in the inhibitor peaks compared to that of the pure inhibitor can indicate the chemical binding of inhibitor, *etc.* The ToF-SIMS is a newly explored and highly sensitive instrumental technique in the area of corrosion inhibitor adsorption. This method can afford information on the atoms, functional groups, and rings of the inhibitor molecule that are involved with the metal surface. Furthermore, this can also elucidate information on the number of metal surface atoms taking part in the complex formation with a single inhibitor molecule. Therefore, the surface characterization can also contribute to the elucidation of the inhibitor adsorption and metal–inhibitor interaction. Table 3.2 lists the major techniques of surface analysis of corrosion inhibitor film with the significance of each of them.

In Figure 3.3(a), we have shown a schematic of the construction and working principles of a ToF-SIMS spectrometer.[43] Figure 3.3(b–e) show the structure of a corrosion inhibitor AMTT analysed for Cu surface in 3.5% NaCl and the results of the surface analysis of the same *via* ToF-SIMS.[10] A comparison

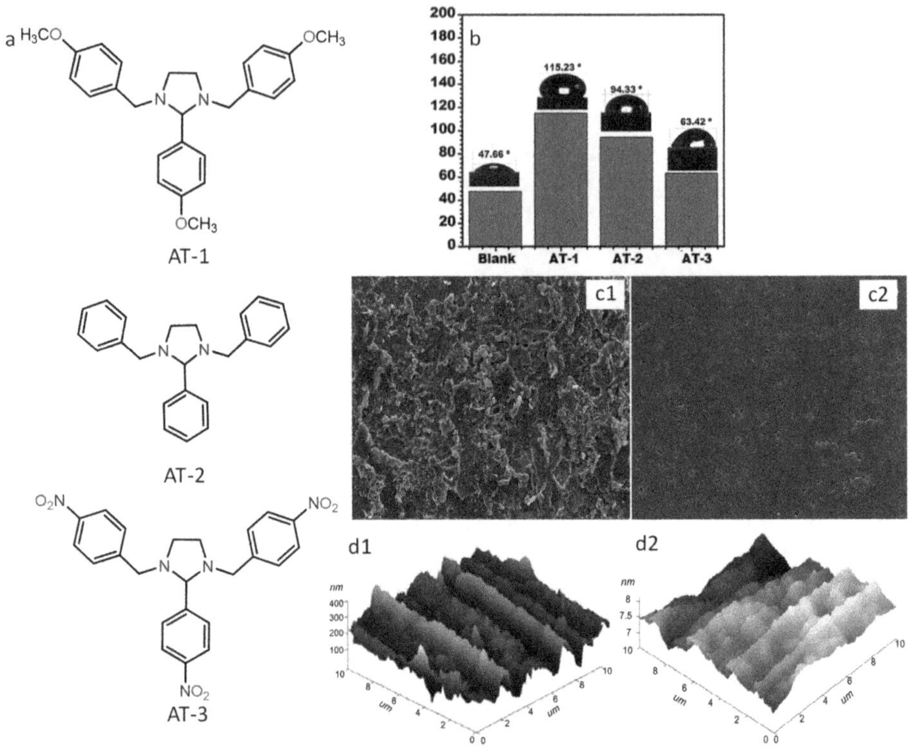

Figure 3.2 Techniques for surface analysis of corrosion inhibitor films: (a) Structures of imidazolidine derivatives having different electron donating (AT-1) to electron-withdrawing (AT-2) substituents; (b) corresponding WCA of J55 steel samples in CO_2 saturated 3.5% NaCl in the absence and presence of inhibitors; (c1 and c2) SEM of blank and AT-1 inhibited J55 steel samples; (d1 and d2) corresponding AFM images.[42] Reproduced from ref. 42 with permission from Elsevier, Copyright 2019.

of the global negative mass ion spectra in Figure 3.3(c) shows the presence of different organic adsorbates on Cu substrate in various mass regions. The presence of adsorbed inhibitor is further highlighted in Figure 3.3(d) wherein the AMTT-adsorbed Cu sample shows characteristic intense negative ions: $C_3H_5N_4S^-$ (m/z = 129.0260), $C_3H_4N_4S^-$ (m/z = 128.0213) and $C_3H_3N_4S^-$ (m/z = 127.0120), as expected. Ionic species pertaining to inhibitors, such as CN^-, S^- (having 90 times more intensity compared to blank samples), and CN_2^- are also identified. Evidence of metal–inhibitor binding is given by Figure 3.3(e) by the presence of fragment ions identifiable as $C_3H_4N_4SCu^-$ (m/z = 190.9529), corresponding to the inhibitor bonded to Cu minus 1 H, $C_3H_2N_4SCu^-$ (m/z = 188.9270) and $C_3HN_4SCu^-$ (m/z = 187.9253). These results provide evidence for the bonding of the corrosion inhibitor to the Cu substrate *via* its –SH and/or –NH_2 groups (or triazole ring). The results of XPS analysis performed for the mild steel samples post immersion in 1 M HCl

Table 3.2 Major surface analytical techniques.

Technique/major parameters derived	Symbol	Unit	Significance	Ref.
Water contact angle (WCA)				
Water contact angle	—	°	• Surface hydrophobicity of metal in the absence/presence of adsorbed inhibitor	45 and 46
Scanning electron microscopy (SEM)				
Surface micrographs	—	—	• Surface morphology of metal samples in the absence/presence of adsorbed inhibitor film • Microstructure of adsorbed corrosion inhibitor film	4 and 47
Energy dispersive X-ray analysis (EDX)				
Energy *vs.* intensity plots	—	—	• Surface chemical composition • Identifies suppression of corrosion product peaks in the presence of inhibitor	5, 9, 48 and 49
Atomic force microscopy (AFM)				
3-D surface topography	—	—	• Quantitative surface roughness of metal sample in the absence/presence of adsorbed inhibitor film	9, 24, 47 and 50
X-ray diffraction (XRD)				
Typical diffractogram is a characteristic 2θ values *vs.* peak intensity according to Bragg's law	Intensity 2θ	a.u. °	• Identify which corrosion products have formed • Suppression of corrosion product peaks in the presence of corrosion inhibitor indicates adsorption	51
X-ray photoelectron spectroscopy (XPS)				
Survey scan spectra in the form of binding energy *vs.* peak intensity curves	Intensity 2θ	a.u. °	• Identifies which corrosion products have formed • Indicates metal–inhibitor interaction • Takes a long time	44 and 52
Time of flight secondary-ion mass spectrometry (ToF-SIMS)				
Provides mass (u) *vs.* intensity (counts) curves	Mass	a.u.	• Highly sensitive technique • Indicates inhibitor complexes formed with the metal surface atoms • Can provide surface chemical composition images (although time taking)	10

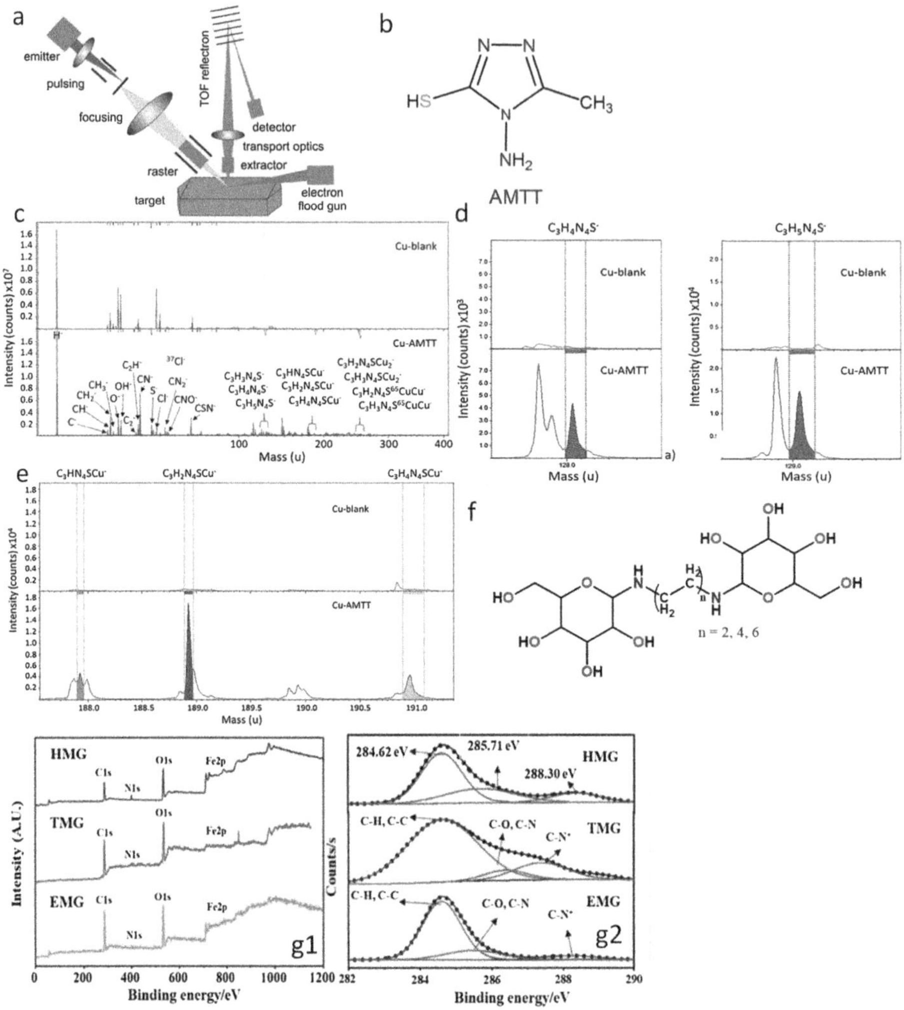

Figure 3.3 (a) Schematic of the major components of a ToF-SIMS spectrometer.[43] Reproduced from ref. 43, https://doi.org/10.1039/C3CP54337D, under the terms of the CC BY 3.0 licence https://creativecommons.org/licenses/by/3.0/. (b) Structure of aminomercaptotriazole (AMTT) adsorption on Cu substrate being analysed by ToF-SIMS;[10] (c) comparison of global negative ion mass spectra on blank and AMTT-adsorbed Cu surface;[10] (d) different types of AMTT complexes;[10] and (e) different types of AMTT-Cu complexes formed on Cu surface;[10] (f) structures of three glucose derivatives differing in the length of the alkyldiamine chain in between;[44] (g1) XPS survey scan spectra for the adsorption of the glucose derivatives on mild steel in 1 M HCl; and (g2) corresponding C 1s spectra of these inhibitors.[44] Reproduced from ref. 10 and 44 with permission from Elsevier, Copyright 2019 and 2020.

containing three glucose derivatives joined by alkyldiamine chains having increasing length (EMG, TMG, HMG) [structures shown in Figure 3.3(f)] are shown in Figure 3.3(g1 and g2).[44] The XPS study supports the formation of the metal–inhibitor linkage.

3.2.3 Computational Methods of Analysis

The conventionally performed techniques of exhaustive experimental measurements often result in an insufficient amount of information on the inhibitor adsorption mechanism.[53–55] On the other hand, the advancements in computational research and simulations have provided a mighty tool to provide this information.[11,12] While the experimental methods provide considerable information on the adsorption and inhibition performance, these methods require the use of sophisticated instrumentation and consume a lot of time. Still, accurate information on the correlation of the structural aspects of a given corrosion inhibitor with the adsorption behaviour is often lacking. Computational techniques have emerged to provide support for the experimental measurements carried out for inhibitor adsorption in corrosion inhibition studies.[11,12] These techniques involve the estimation of the quantum chemical parameters, which correlate the structural aspects of an organic molecule with the inhibition performance.[54,56,57] Some of the commonly used software for computational studies are listed in Table 3.3.

An organic corrosion inhibitor, when introduced to a corrosive environment, can undergo a structural change such as protonation/deprotonation, bending/twisting *etc.* which can considerably influence the adsorption behaviour. The pK_a analysis can check if there is a protonation in the inhibitor structure depending upon the experimental pH range. The semi-empirical and density functional theory (DFT) based computations analyse the structure of an inhibitor and correlate with the adsorption behaviour in terms of the computationally derived energy parameters such as the frontier orbital energies, molecular orbital energy gap, total energy, *etc.* In such computations, an important aspect is the computation time required, which strongly depends upon the complexity of the inhibitor structure, the chemical composition of the solvent *etc.* if that is also being considered.

However still, the orientation of the inhibitor on a given metal surface is an important aspect that is required to be understood. In this case, molecular simulation methods such as the Monte Carlo (MC) and the Molecular dynamics (MD) simulations can shed light on the inhibitor adsorption on a given metallic substrate. The different energy parameters calculated considering the solvent/electrolyte environment involved can also provide knowledge on the adsorption of the neutral/protonated inhibitor forms on a given metallic surface. These techniques can estimate the distance between the atoms of the inhibitor from a selected crystal plane of the metal or metal oxide, can indicate the type of bonding (electrostatic, coordinate, H-bond formation) *etc.* Furthermore, these methods can also provide information on the diffusivity of the corrosive electrolyte ions through the adsorbed film

Table 3.3 Major computational analysis techniques.

Technique/major parameters derived	Symbol	Unit	Significance	Ref.
			pK_a analysis	
Microspecies distribution at different pH	—	—	• Shows different protonated/deprotonated inhibitor structures at varying pH values	6 and 10
Density functional theory (DFT) based quantum chemical parameters				
Molecular orbital energy gap	ΔE	eV	• Inhibition efficiency can be predicted based on quantum chemical parameters • Lower ΔE indicates higher reactivity • High E_{HOMO} indicates electron donating tendency; High E_{LUMO} indicates electron accepting tendency • Comparison of parameters in neutral/protonated form • Comparison of parameters on different substituents/functional groups attached	55 and 58
Energy of the highest occupied molecular orbital	E_{HOMO}	eV		
Energy of the lowest unoccupied molecular orbital	E_{LUMO}	eV		
Electronegativity	χ	eV		
Global hardness	η	eV		
Global softness	σ	—		
Electrophilicity index	ω	—		
Nucleophilicity index	ε	—		
Fraction of electrons transferred	ΔN	—		
Monte Carlo (MC) simulations				
Total energy/adsorption energy	E_{total}/E_{ads}	kcal mol^{-1}	• Energy parameters derived for adsorption of a corrosion inhibitor on a preferred metal surface plane • Indicate strength of adsorption	53
Molecular dynamics (MD) simulations				
Binding energy/adsorption energy/deformation energy/solvation energy	$E_{bind.}/E_{ads.}/E_{deform.}/E_{solv.}$	kcal mol^{-1}	• Molecular level insight on solvent and ion transport • Comparison of adsorption of neutral/protonated forms and in gaseous/aqueous medium	53
Mean square displacement (MSD); diffusion coefficient	MSD	Å	• Shows effectiveness of inhibitor film in preventing the migration of corrosive species	41
Radial distribution function (RDF) $g(r)$ *vs.* distance plots	X	Å	• Distance between inhibitor and metal surface atoms • Indicates the likelihood of physical/chemical adsorption	41 and 52

of the inhibitor. Figure 3.4 presents the results obtained using some of the computational methods of corrosion inhibitor evaluation.

In Figure 3.4a, the protonated structure of inhibitor hexamethylene-1,6-bis(*N*-D-glucopyranosylamine) is depicted. It can be seen that in the experimental pH range (pH ~ 4) this is the most abundant protonated form. The complete pK_a analysis is given in ref. 6. It is obvious that at the experimental pH of 4, the inhibitor is likely to undergo protonation at the N atoms. From the results of the DFT derived electrostatic potential maps of the neutral and the protonated forms of the inhibitor shown in Figure 3.4b, it is clear that in the neutral condition, the inhibitor shows a high negative charge at the heteroatoms (indicating electron donating sites), whereas in the protonated condition, there is a blue colour appearing in the ESP indicating that now the corrosion inhibitor can accept the electrons at these sites. Figure 3.4c shows the DFT derived HOMO/LUMO distributions and the molecular orbital energy gap of the corrosion inhibitor (Chitosan Schiff base). It can be observed that after protonation, there is a lowering in the energy gap, indicating that in the protonated condition, the inhibitor shows a greater tendency to adsorb at the metal surface. The charge density binding analysis shown in Figure 3.4d provides support for the chemical adsorption of MET tautomers *via* Fe–C and Fe–N bonding interactions. In most cases, the Fe–C and Fe–N bond distances are around 2.0 Å, which supports the covalent nature of MET-Fe interaction. Figure 3.4e shows the results of the MD simulation study of the adsorption of the imidazolidine-based inhibitor on Fe (110) plane. This indicates that all the tested molecules undergo nearly flat adsorption on the metal surface allowing maximum coverage of the metal surface area. This allows strong metal–inhibitor interactions to occur *via* electron donation/acceptance, back-donation and through the π electrons of the heterocyclic and phenyl rings. In this way, it can be concluded that the inhibitor molecules cover the entire exposed area of the metal surface and therefore prevent the corrosion of metal from aggressive solution. Figure 3.5 shows the DFT results obtained for two series of corrosion inhibitors based on dihydropyrimidinones and naphthyridines. The corresponding data of molecular structures, inhibition efficiencies, and energy gap are shown in Table 3.4.

3.3 Conclusions and Prospects

This chapter presents an outline of the major experimental and computational techniques generally carried out to assess the inhibitor adsorption and protection performance. The experimental and computational techniques discussed herein facilitate the development of a mechanistic model for metal–inhibitor interaction *via* electron donation/acceptance *etc.* The thermodynamic and kinetic parameters allow the understanding of the physical, chemical or mixed physical–chemical type of adsorption *via* the choice of suitable adsorption isotherms. The variation in the corrosion rate with a rise in inhibitor concentration, solution temperature *etc.* allows

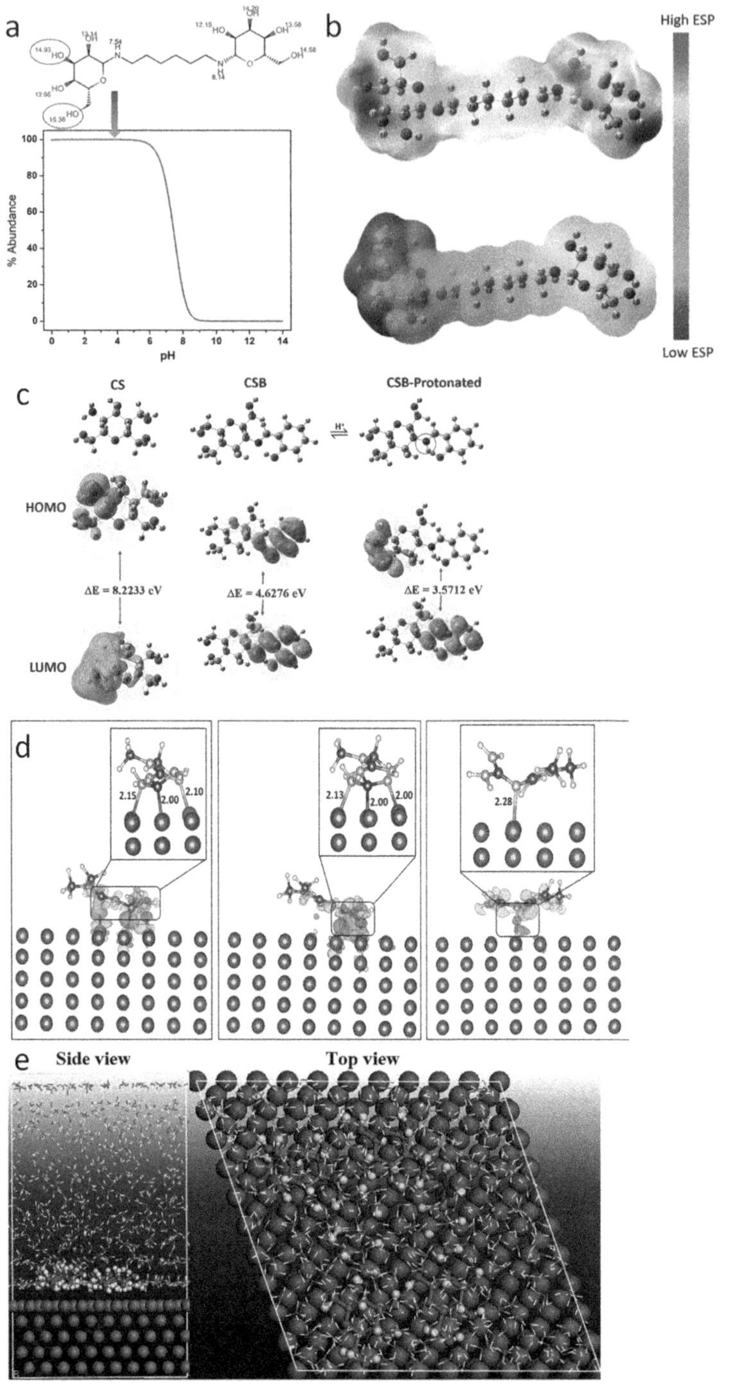

Figure 3.4 (a) pK_a analysis of hexamethylene-1,6-bis(*N*-D-glucopyranosylamine), and (b) electrostatic potential maps of same in neutral and protonated condition;[6] adapted from ref. 6 with permission from the Royal Society

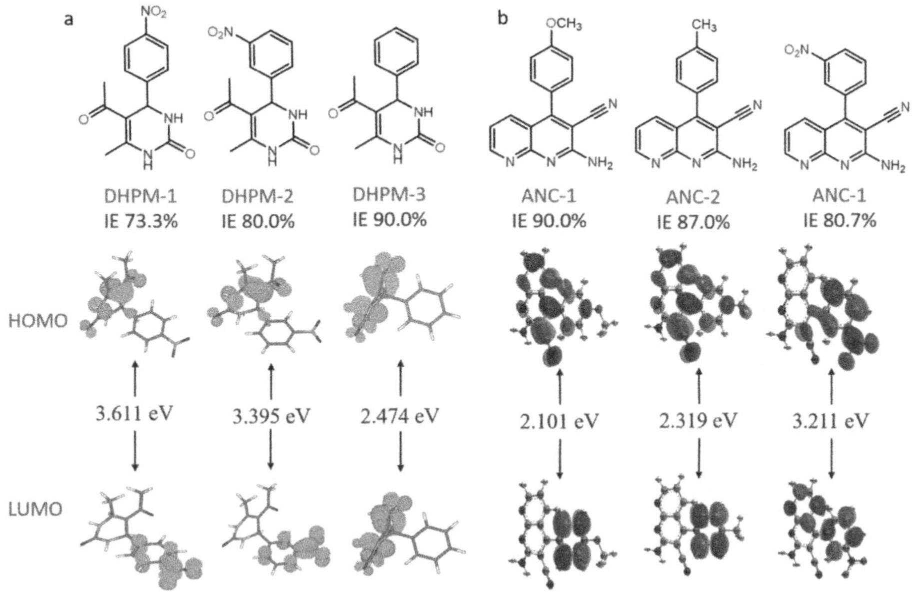

Figure 3.5 Optimized molecular structures and HOMO/LUMO distributions of (a) DHPM[61] and (b) ANC[62] derivatives showing molecular orbital energy gap. Reproduced from ref. 61 and 62 with permission from Elsevier, Copyright 2010 and 2015.

the understanding of adsorption behaviour. Further, gravimetric (weight loss) and electrochemical (EFM, LPR, EIS, PDP *etc.*) examinations afford the understanding of charge transfer or diffusion control, anodic/cathodic predominance *etc.* in the inhibition behaviour. Computational techniques provide a theoretical correlation of the molecular structure to the experimentally determined inhibition behaviour.

Further studies should focus on the detailed simulation involving the effect of variation in acid/inhibitor concentrations, and solution temperatures *etc.* on the inhibition behaviour. The detailed experimental measures should include high throughput screening of inhibitors having the same molecular framework but differing in the presence of functional groups, π-bonds, phenyl rings *etc.* Simultaneous computational techniques should be carried out to rapidly screen *via* reactivity indices.

---

of Chemistry. (c) DFT derived HOMO/LUMO structures and molecular orbital energy gap of chitosan and chitosan-salicylaldehyde Schiff base (in neutral and protonated conditions).[59] Reproduced from ref. 59 with permission from Elsevier, Copyright 2020. (d) DFT simulations of expired metformin drug adsorption in sweet environment.[60] Reproduced from ref. 60 with permission from American Chemical Society, Copyright 2021. (e) Molecular dynamics simulations in aqueous medium of Chitosan-thiocarbohydrazide.[2] Reproduced from ref. 2 with permission from Elsevier, Copyright 2018.

Table 3.4 Correlation of quantum chemical parameters and experimentally determined corrosion inhibition efficiency for the DHPM[61] and ANC[62] derivatives. Reproduced from ref. 61 and 62 with permission from Elsevier, Copyright 2010 and 2015.

Inhibitors	E_{HOMO} (eV)	E_{LUMO} (eV)	$\Delta E_{LUMO\text{-}HOMO}$ (eV)	IE (expt)
DHPM-1	−6.273	−2.662	3.611	73.3
DHPM-2	−6.237	−2.841	3.395	80.0
DHPM-3	−6.224	−3.750	2.474	99.0
ANC-1 (−OCH_3)	−4.218	−2.117	2.101	90.0
ANC-2 (−CH_3)	−4.424	−2.105	2.319	87.0
ANC-3 (−H)	−5.295	−2.084	3.211	80.7

References

1. R. Baboian, *Corrosion Tests and Standards: Application and Interpretation*, ASTM international, 2005.
2. D. S. Chauhan, K. R. Ansari, A. A. Sorour, M. A. Quraishi, H. Lgaz and R. Salghi, Thiosemicarbazide and thiocarbohydrazide functionalized chitosan as ecofriendly corrosion inhibitors for carbon steel in hydrochloric acid solution, *Int. J. Biol. Macromol.*, 2018, **107**, 1747–1757.
3. F. El-Hajjaji, R. Salim, M. Messali, B. Hammouti, D. S. Chauhan, S. Almutairi and M. A. Quraishi, Electrochemical studies on new pyridazinium derivatives as corrosion inhibitors of carbon steel in acidic medium, *J. Bio- Tribo-Corros.*, 2019, **5**, 4.
4. N. Baig, D. S. Chauhan, T. A. Saleh and M. A. Quraishi, Diethylenetriamine functionalized graphene oxide as a novel corrosion inhibitor for mild steel in hydrochloric acid solutions, *New J. Chem.*, 2019, **43**, 2328–2337.
5. J. Haque, V. Srivastava, D. S. Chauhan, H. Lgaz and M. A. Quraishi, Microwave-induced synthesis of chitosan Schiff bases and their application as novel and green corrosion inhibitors: Experimental and theoretical approach, *ACS Omega*, 2018, **3**, 5654–5668.
6. I. B. Onyeachu, D. S. Chauhan, K. R. Ansari, I. Obot, M. A. Quraishi and A. H. Alamri, Hexamethylene-1, 6-bis (N–D-glucopyranosylamine) as a novel corrosion inhibitor for oil and gas industry: Electrochemical and computational analysis, *New J. Chem.*, 2019, **43**, 7282–7293.
7. P. Singh, D. S. Chauhan, K. Srivastava, V. Srivastava and M. A. Quraishi, Expired atorvastatin drug as corrosion inhibitor for mild steel in hydrochloric acid solution, *Int. J. Ind. Chem.*, 2017, **8**, 363–372.
8. D. K. Yadav, D. S. Chauhan, I. Ahamad and M. A. Quraishi, Electrochemical behavior of steel/acid interface: adsorption and inhibition effect of oligomeric aniline, *RSC Adv.*, 2013, **3**, 632–646.
9. P. Marcus, and F. B. Mansfeld, *Analytical Methods in Corrosion Science and Engineering*, CRC press, 2005.
10. D. S. Chauhan, M. A. Quraishi, C. Carrière, A. Seyeux, P. Marcus and A. Singh, Electrochemical, ToF-SIMS and computational studies of

4-amino-5-methyl-4H-1, 2, 4-triazole-3-thiol as a novel corrosion inhibitor for copper in 3.5% NaCl, *J. Mol. Liq.*, 2019, **289**, 111113.
11. M. Frisch, G. Trucks, H. B. Schlegel, G. Scuseria, M. Robb, J. Cheeseman, G. Scalmani, V. Barone, B. Mennucci, and G. Petersson, *Gaussian 09, Revision A. 02*, Inc., Wallingford, CT, 200, 2009.
12. P. Geerlings, F. De Proft and W. Langenaeker, Conceptual density functional theory, *Chem. Rev.*, 2003, **103**, 1793–1874.
13. D. S. Chauhan, M. A. Quraishi, A. A. Sorour, S. K. Saha and P. Banerjee, Triazole-modified chitosan: a biomacromolecule as a new environmentally benign corrosion inhibitor for carbon steel in a hydrochloric acid solution, *RSC Adv.*, 2019, **9**, 14990–15003.
14. Sudheer and M. A. Quraishi, The corrosion inhibition effect of aryl pyrazolo pyridines on copper in hydrochloric acid system: computational and electrochemical studies, *RSC Adv.*, 2015, **5**, 41923–41933.
15. N. Chaubey, V. K. Singh and M. A. Quraishi, Electrochemical approach of Kalmegh leaf extract on the corrosion behavior of aluminium alloy in alkaline solution, *Int. J. Ind. Chem.*, 2017, **8**, 75–82.
16. M. A. Quraishi, K. R. Ansari, D. S. Chauhan, S. A. Umoren and M. A. J. Mazumder, Vanillin modified chitosan as a new bio-inspired corrosion inhibitor for carbon steel in oil-well acidizing relevant to petroleum industry, *Cellulose*, 2020, **27**, 6425–6443.
17. P. Singh, D. S. Chauhan, S. S. Chauhan, G. Singh and M. A. Quraishi, Bio-inspired synergistic formulation from dihydropyrimdinones and iodide ions for corrosion inhibition of carbon steel in sulphuric acid, *J. Mol. Liq.*, 2019, **298**, 112051.
18. A. Singh, Y. Lin, I. Obot, E. E. Ebenso, K. Ansari and M. A. Quraishi, Corrosion mitigation of J55 steel in 3.5% NaCl solution by a macrocyclic inhibitor, *Appl. Surf. Sci.*, 2015, **356**, 341–347.
19. S. Wade and Y. Lizama, Clarke's solution cleaning used for corrosion product removal: effects on carbon steel substrate, *Corros. Prev.*, 2015, 050.
20. *A.S.f. Testing, Materials, ASTM G1-03: Standard Practice for Preparing, Cleaning, and Evaluating Corrosion Test Specimens*, ASTM, 2004.
21. A. Kina and J. Ponciano, Inhibition of carbon steel CO_2 corrosion in high salinity solutions, *Int. J. Electrochem. Sci.*, 2013, **8**, 12600–12612.
22. M. Finšgar and J. Jackson, Application of corrosion inhibitors for steels in acidic media for the oil and gas industry: a review, *Corros. Sci.*, 2014, **86**, 17–41.
23. P. Singh, E. E. Ebenso, L. O. Olasunkanmi, I. Obot and M. A. Quraishi, Electrochemical, theoretical, and surface morphological studies of corrosion inhibition effect of green naphthyridine derivatives on mild steel in hydrochloric acid, *J. Phys. Chem. C*, 2016, **120**, 3408–3419.
24. V. Srivastava, D. S. Chauhan, P. G. Joshi, V. Maruthapandian, A. A. Sorour and M. A. Quraishi, PEG-functionalized chitosan: A biological macromolecule as a novel corrosion inhibitor, *ChemistrySelect*, 2018, **3**, 1990–1998.

25. S. Dhillon and R. Kant, Theory for electrochemical impedance spectroscopy of heterogeneous electrode with distributed capacitance and charge transfer resistance, *J. Chem. Sci.*, 2017, **129**, 1277–1292.
26. J.-B. Jorcin, M. E. Orazem, N. Pébère and B. J. E. A. Tribollet, CPE analysis by local electrochemical impedance spectroscopy, *Electrochim. Acta*, 2006, **51**, 1473–1479.
27. I. Obot and I. B. Onyeachu, Electrochemical frequency modulation (EFM) technique: Theory and recent practical applications in corrosion research, *J. Mol. Liq.*, 2018, **249**, 83–96.
28. K. Khaled, Application of electrochemical frequency modulation for monitoring corrosion and corrosion inhibition of iron by some indole derivatives in molar hydrochloric acid, *Mater. Chem. Phys.*, 2008, **112**, 290–300.
29. R. Solmaz, Investigation of corrosion inhibition mechanism and stability of Vitamin B1 on mild steel in 0.5 M HCl solution, *Corros. Sci.*, 2014, **81**, 75–84.
30. G. Walter, A review of impedance plot methods used for corrosion performance analysis of painted metals, *Corros. Sci.*, 1986, **26**, 681–703.
31. F. Mansfeld, Recording and analysis of AC impedance data for corrosion studies, *Corrosion*, 1981, **37**, 301–307.
32. S. Satpati, S. K. Saha, A. Suhasaria, P. Banerjee and D. Sukul, Adsorption and anti-corrosion characteristics of vanillin Schiff bases on mild steel in 1 M HCl: experimental and theoretical study, *RSC Adv.*, 2020, **10**, 9258–9273.
33. M. Özcan, I. Dehri and M. Erbil, Organic sulphur-containing compounds as corrosion inhibitors for mild steel in acidic media: correlation between inhibition efficiency and chemical structure, *Appl. Surf. Sci.*, 2004, **236**, 155–164.
34. M. M. Solomon, H. Gerengi and S. A. Umoren, Carboxymethyl cellulose/silver nanoparticles composite: synthesis, characterization and application as a benign corrosion inhibitor for St37 steel in 15% H_2SO_4 medium, *ACS Appl. Mater. Interfaces*, 2017, **9**, 6376–6389.
35. H. Gerengi, M. Mielniczek, G. Gece and M. M. Solomon, Experimental and quantum chemical evaluation of 8-hydroxyquinoline as a corrosion inhibitor for copper in 0.1 M HCl, *Ind. Eng. Chem. Res.*, 2016, **55**, 9614–9624.
36. E. Kuş and F. Mansfeld, An evaluation of the electrochemical frequency modulation (EFM) technique, *Corros. Sci.*, 2006, **48**, 965–979.
37. G. S. Frankel, and M. Rohwerder, Electrochemical techniques for corrosion, *Encyclopedia of Electrochemistry: Online*, 2007.
38. D. K. Yadav, M. A. Quraishi and B. Maiti, Inhibition effect of some benzylidenes on mild steel in 1 M HCl: an experimental and theoretical correlation, *Corros. Sci.*, 2012, **55**, 254–266.
39. E. McCafferty, Validation of corrosion rates measured by the Tafel extrapolation method, *Corros. Sci.*, 2005, **47**, 3202–3215.
40. A. Singh, K. Ansari, A. Kumar, W. Liu, C. Songsong and Y. Lin, Electrochemical, surface and quantum chemical studies of novel imidazole

derivatives as corrosion inhibitors for J55 steel in sweet corrosive environment, *J. Alloys Compd.*, 2017, **712**, 121–133.

41. A. Singh, K. R. Ansari, Y. Lin, M. A. Quraishi, H. Lgaz and I.-M. Chung, Corrosion inhibition performance of imidazolidine derivatives for J55 pipeline steel in acidic oilfield formation water: Electrochemical, surface and theoretical studies, *J. Taiwan Inst. Chem. Eng.*, 2019, **95**, 341–356.
42. A. Singh, K. Ansari, Y. Lin, M. A. Quraishi, H. Lgaz and I.-M. Chung, Corrosion inhibition performance of imidazolidine derivatives for J55 pipeline steel in acidic oilfield formation water: Electrochemical, surface and theoretical studies, *J. Taiwan Inst. Chem. Eng.*, 2019, **95**, 341–356.
43. J. P. Hofmann, M. Rohnke and B. M. Weckhuysen, Recent advances in secondary ion mass spectrometry of solid acid catalysts: large zeolite crystals under bombardment, *Phys. Chem. Chem. Phys.*, 2014, **16**, 5465–5474.
44. J. Haque, V. Srivastava, D. S. Chauhan, M. A. Quraishi, A. M. Kumar and H. Lgaz, Electrochemical and surface studies on chemically modified glucose derivatives as environmentally benign corrosion inhibitors, *Sustainable Chem. Pharm.*, 2020, **16**, 100260.
45. M. Murmu, S. K. Saha, N. C. Murmu and P. Banerjee, Effect of stereochemical conformation into the corrosion inhibitive behaviour of double azomethine based Schiff bases on mild steel surface in 1 mol L^{-1} HCl medium: An experimental, density functional theory and molecular dynamics simulation study, *Corros. Sci.*, 2019, **146**, 134–151.
46. A. Singh, Y. Lin, K. Ansari, M. A. Quraishi, E. E. Ebenso, S. Chen and W. Liu, Electrochemical and surface studies of some Porphines as corrosion inhibitor for J55 steel in sweet corrosion environment, *Appl. Surf. Sci.*, 2015, **359**, 331–339.
47. P. Singh, D. S. Chauhan, S. S. Chauhan, G. Singh and M. A. Quraishi, Chemically modified expired Dapsone drug as environmentally benign corrosion inhibitor for mild steel in sulphuric acid useful for industrial pickling process, *J. Mol. Liq.*, 2019, **286**, 110903.
48. K. R. Ansari, M. A. Quraishi and A. Singh, Schiff's base of pyridyl substituted triazoles as new and effective corrosion inhibitors for mild steel in hydrochloric acid solution, *Corros. Sci.*, 2014, **79**, 5–15.
49. J. I. Goldstein, D. E. Newbury, J. R. Michael, N. W. Ritchie, J. H. J. Scott, and D. C. Joy, *Scanning Electron Microscopy and X-Ray Microanalysis*, Springer, 2017.
50. P. Dohare, D. S. Chauhan, A. A. Sorour and M. A. Quraishi, DFT and experimental studies on the inhibition potentials of expired Tramadol drug on mild steel corrosion in hydrochloric acid, *Mater. Discovery*, 2017, **9**, 30–41.
51. R. Karthik, P. Muthukrishnan, S.-M. Chen, B. Jeyaprabha and P. Prakash, Anti-corrosion inhibition of mild steel in 1 M hydrochloric acid solution by using tiliacoraaccuminata leaves extract, *Int. J. Electrochem. Sci.*, 2015, **10**, 3707–3725.
52. J. Haque, V. Srivastava, M. A. Quraishi, D. S. Chauhan, H. Lgaz and I.-M. Chung, Polar group substituted imidazolium zwitterion as eco-friendly corrosion inhibitors for mild steel in acid solution, *Corros. Sci.*, 2020, **172**, 108665.

53. I. Obot, K. Haruna and T. Saleh, Atomistic simulation: A unique and powerful computational tool for corrosion inhibition research, *Arabian J. Sci. Eng.*, 2019, **44**, 1–32.
54. C. Verma, H. Lgaz, D. Verma, E. E. Ebenso, I. Bahadur and M. A. Quraishi, Molecular dynamics and Monte Carlo simulations as powerful tools for study of interfacial adsorption behavior of corrosion inhibitors in aqueous phase: a review, *J. Mol. Liq.*, 2018, **260**, 99–120.
55. I. Obot, D. Macdonald and Z. Gasem, Density functional theory (DFT) as a powerful tool for designing new organic corrosion inhibitors. Part 1: an overview, *Corros. Sci.*, 2015, **99**, 1–30.
56. F. Neese, Software update: the ORCA program system, version 4.0, *Wiley Interdiscip. Rev.: Comput. Mol. Sci.*, 2018, **8**, e1327.
57. I. B. Obot, E. E. Ebenso and M. M. Kabanda, Metronidazole as environmentally safe corrosion inhibitor for mild steel in 0.5 M HCl: experimental and theoretical investigation, *J. Environ. Chem. Eng.*, 2013, **1**, 431–439.
58. G. Gece, The use of quantum chemical methods in corrosion inhibitor studies, *Corros. Sci.*, 2008, **50**, 2981–2992.
59. K. R. Ansari, D. S. Chauhan, M. A. Quraishi, M. A. J. Mazumder and A. Singh, Chitosan Schiff base: an environmentally benign biological macromolecule as a new corrosion inhibitor for oil & gas industries, *Int. J. Biol. Macromol.*, 2020, **144**, 305–315.
60. I. B. Onyeachu, S. Abdel-Azeim, D. S. Chauhan and M. A. Quraishi, Electrochemical and computational insights on the application of expired Metformin drug as a novel inhibitor for the Sweet corrosion of C1018 steel, *ACS Omega*, 2021, **6**, 65–76.
61. D. K. Yadav, B. Maiti and M. A. Quraishi, Electrochemical and quantum chemical studies of 3, 4-dihydropyrimidin-2 (1H)-ones as corrosion inhibitors for mild steel in hydrochloric acid solution, *Corros. Sci.*, 2010, **52**, 3586–3598.
62. K. R. Ansari and M. A. Quraishi, Experimental and computational studies of naphthyridine derivatives as corrosion inhibitor for N80 steel in 15% hydrochloric acid, *Phys. E*, 2015, **69**, 322–331.

CHAPTER 4

Ionic Liquids as Green Corrosion Inhibitors

4.1 Introduction

4.1.1 Ionic Liquids: Basics, Properties and Classification

Ionic liquids (ILs) are organic salts that are liquid at temperatures below 100 °C (212 °F).[1,2] Unlike ordinary liquids such as gasoline and water, which are mainly made of electrically neutral molecules, ionic liquids are predominantly composed of ions.[3] Ionic liquids are also known as fused salts, ionic melts, liquid electrolytes, ionic glasses, liquid salts, or ionic fluids. Some of the ionic liquids are liquid at or below room temperature (RT) *e.g.* 1-ethyl-3-methylimidazolium dicyanamide and 1-butyl-3,5-dimethylpyridinium bromide melt at −21 °C (−6 °F) and −24 °C (−11 °F), respectively.[4] Ionic liquids possess potential applications.[5–7] They are used as solvents, starting materials and catalysts in many chemical transformations. They can also be used as electrolytes. The low melting point ionic liquids are widely employed in electric battery applications. Because of their various salient features, ionic liquids are developed as one of the greenest and environmentally friendly substitutes for different industrial and biological applications.[8,9] Some of the common salient features of ionic liquids include low vapour pressure, low melting point, high solubility, extremely high thermal and chemical stability, low viscosity, excellent solvation property and high electrical conductivity.[10–13]

Green Chemistry Series No. 73
Green Corrosion Inhibition: Fundamentals, Design, Synthesis and Applications
By Chandrabhan Verma, Mumtaz A. Quraishi and Dheeraj Singh Chauhan

Published by the Royal Society of Chemistry, www.rsc.org

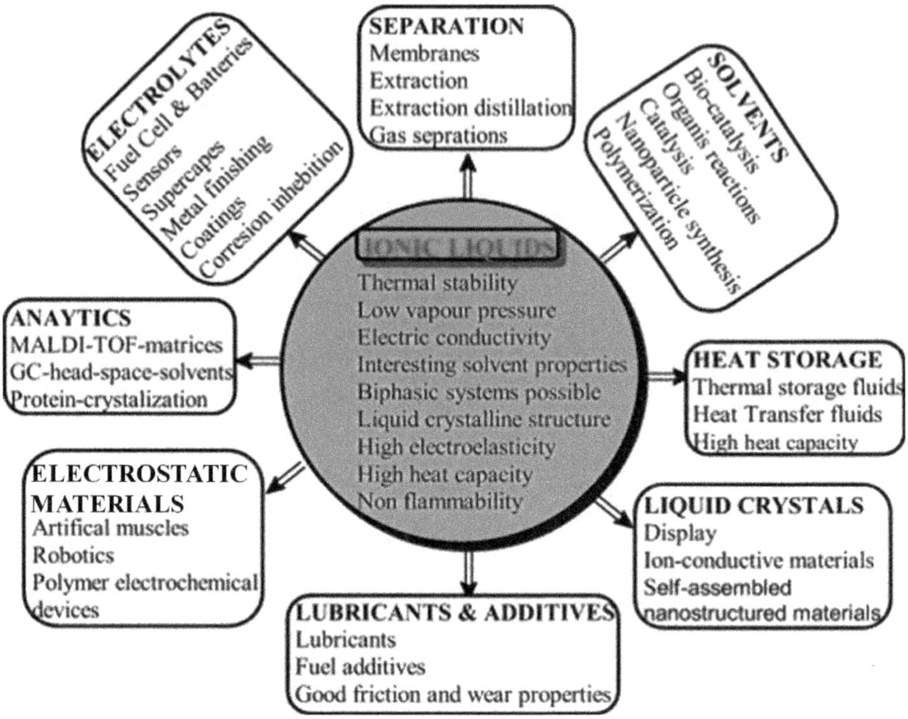

Figure 4.1 Properties and applications of ionic liquids. Reproduced from ref. 12 with permission from Elsevier, Copyright 2011.

One of the greatest advantages of ionic liquids is that their physiochemical properties can be suitably modified by changing the nature of the cations and/or anions. Because of this property, ionic liquids are called designer chemicals. Some of the major properties and applications of ionic liquids are illustrated in Figure 4.1. Ionic liquids can be classified based on various criteria *i.e.* acidic, basic, functionalized and supported ionic liquids (Figure 4.2). Ionic liquids are also widely used as corrosion inhibitors for various metal/electrolyte systems.

4.2 Mechanism of Corrosion Inhibition

Literature survey shows that ionic liquids become effective by adsorbing on the metallic surface.[14,15] Through their adsorption, ionic liquids block the active sites of the metal surface responsible for corrosive damage. Similar to traditional organic corrosion inhibitors, adsorption of ionic liquids may follow physisorption, chemisorption or physiochemisorption

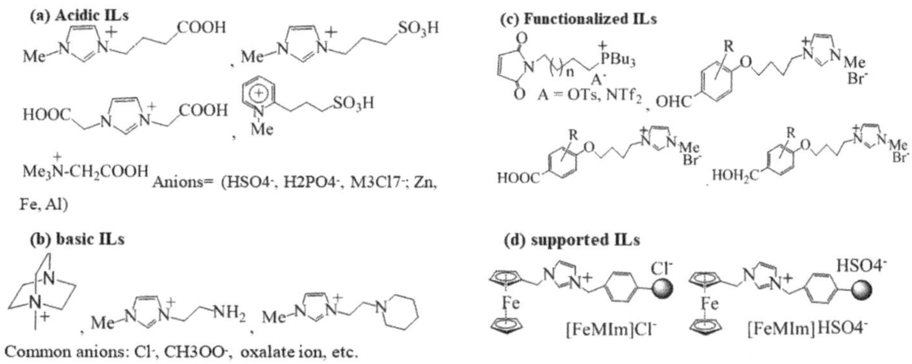

Figure 4.2 Classification of ionic liquids and examples of (a) acidic, (b) basic, (c) functionalized and (d) supported ionic liquids.

mechanisms. Ionic liquids adsorb on the metal surface through their electron rich centres called adsorption centres or sites. Noticeably, the interaction of ionic liquids with metallic surfaces involves a charge, *i.e.* electron, sharing mechanism.[16,17] In this process, electron dense site(s) of inhibitor molecules/ionic liquids transfer their electrons in the empty d-orbital of the surface metallic atoms through the process known as donation. However, metals are electron rich dense species, and so this type of electron transfer causes a state of interelectronic repulsion. This causes the metal surface to transfer its extra electrons in the empty p-orbital through the process called retro-donation. Donation and retro-donation strengthen each other through synergism. Obviously, donation and retro-donation involve chemical bonding between ILs and the metallic surface. It is important to mention that in acidic electrolytes, heteroatoms of the ILs easily undergo protonation and get converted into their cationic *i.e.* protonated forms. On the other hand, the metallic surface becomes negatively charged due to the accumulation of counter ions of the electrolyte and anions of ionic liquids. The opposite charged species attract each other through electrostatic force of attraction *i.e.* ionic bonding. After getting adsorbed, ionic liquids retard both anodic and cathodic Tafel reactions. The mechanism of anodic and cathodic reactions leading to the retardation of the corrosion of a divalent metal using ionic liquids in sulfuric (H_2SO_4) and hydrochloric (HCl) acids is proposed by Likhanova *et al.*:[18]

(a) Anodic reactions without ionic liquids (ILs)

$$M + nH_2O \leftrightarrow M(H_2O)_{n(ads)} \tag{4.1}$$

$$M(H_2O)_{n(ads)} + SO_4^{2-}/2Cl^- \leftrightarrow M[(H_2O)_nSO_4^{2-}/2Cl^-]_{ads} \tag{4.2}$$

$$M[(H_2O)_nSO_4^{2-}/2Cl^-]_{ads}^- \leftrightarrow M[(H_2O)_nSO_4/2Cl]_{ads} + 2e^- \tag{4.3}$$

$$M[(H_2O)_nSO_4/2Cl]_{ads} + 2e^- \leftrightarrow M^{2+} + OH^- + H^+ + SO_4^{2-}/2Cl^- \tag{4.4}$$

As the metallic surface becomes negatively charged due to the adsorption of counter ions (chloride and sulfate ions), ionic liquid cations preferably get adsorbed on anodic sites. The mechanism of ILs^+ adsorption is presented below:

(b) Anodic reactions with ionic liquids (ILs)

$$M + nH_2O \leftrightarrow M(H_2O)_{n(ads)} \tag{4.5}$$

$$M(H_2O)_{n(ads)} + SO_4^{2-}/2Cl^- \leftrightarrow M[(H_2O)_nSO_4^{2-}/2Cl^-]_{ads} \tag{4.6}$$

$$M[(H_2O)_nSO_4^{2-}/2Cl^-]_{ads} + ILs^+ \rightarrow M(H_2O)_nSO_4/2Cl - ILsC]_{ads}^- \tag{4.7}$$

$$\begin{aligned} &M[(H_2O)_nSO_4/2Cl - ILsC]_{ads}^- + 2IL^+ + SO_4^{2-}/2Cl^- \rightarrow \\ &M[(H_2O)_nSO_4/2Cl - ILsC]_{ads}^- + [2IL]^+SO_4^{2-}/2Cl^- \end{aligned} \tag{4.8}$$

These reactions can be simplified as follows:

$$M + X^- \leftrightarrow (MX)_{ads}^- \tag{4.9}$$

$$(MX)_{ads}^- + IL^+ \leftrightarrow (MX^-IL^+)_{ads} \tag{4.10}$$

(c) Cathodic reactions without ionic liquids (ILs)

$$M + H_3O^+ \leftrightarrow M(H_3O)_{ads}^+ \tag{4.11}$$

$$M(H_3O)_{ads}^+ \leftrightarrow M(H_3O)_{ads} + e^- \tag{4.12}$$

$$M(H_3O)_{ads} + e^- + H^+ \leftrightarrow M + H_2O + H_2 \tag{4.13}$$

(d) Cathodic reactions with ionic liquids (ILs)

$$M + IL^+ \leftrightarrow M(IL)^+_{ads} \tag{4.14}$$

$$M(IL)^+_{ads} + e^- \leftrightarrow M(IL)_{ads} \tag{4.15}$$

Retardation of anodic and cathodic Tafel reactions and adsorption of ionic liquids cations on anodic and cathodic sites of a metal surface is illustrated in Figure 4.3.

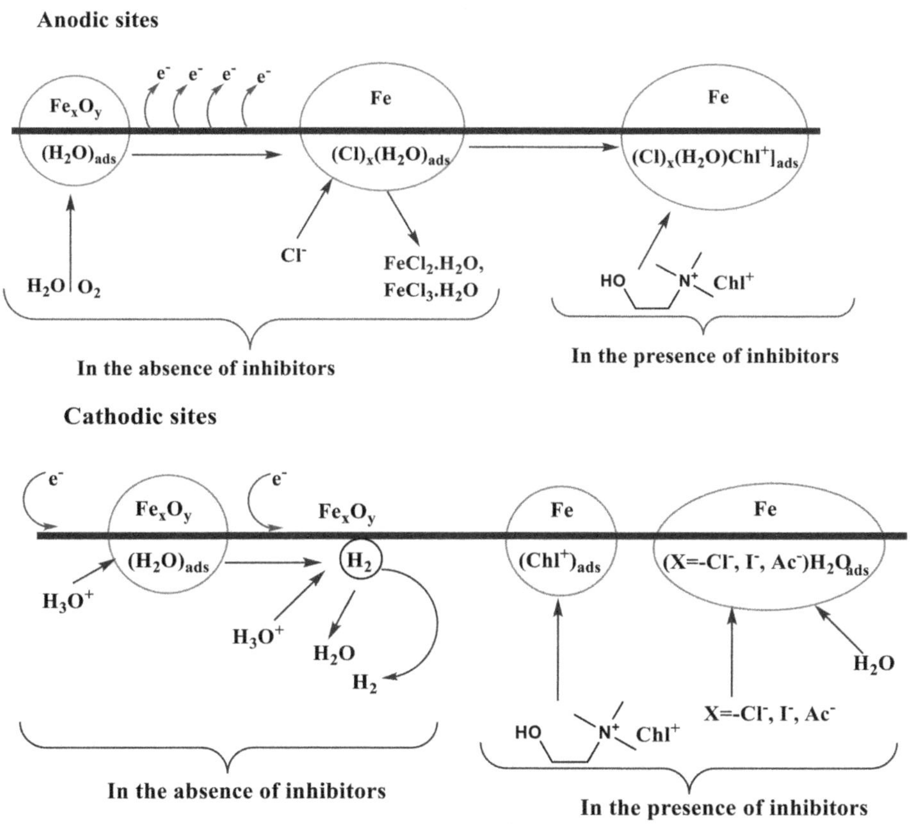

Figure 4.3 Diagrammatic illustration of anodic and cathodic reactions with and without an ionic liquid.[16] Reproduced from ref. 16 with permission from Elsevier, Copyright 2018.

4.3 Ionic Liquids as Corrosion Inhibitors: Literature Survey

4.3.1 Ionic Liquids as Corrosion Inhibitors in HCl

Different concentrations of HCl solution are widely used as electrolytes for various industrial and educational purposes.[19,20] Obviously, lower concentrations of HCl solution are used for educational research, while highly concentrated HCl solutions are extensively used for industrial cleaning processes such as acid picking, acid descaling and oil-well acidification.[20,21] Generally, these processes are adopted to clean a metallic surface by removing surface impurities such as surface metal oxides, rusts and scales. However, a huge amount of metallic structures is lost during these cleaning processes. Therefore, some external additives are added to the cleaning solutions to avoid undesirable metallic losses. The use of organic compounds is one of the most popular, economical and effective methods of corrosion mitigation.[22,23] Nevertheless, some organic compounds, principally those that contain two or more aromatic rings, experience solubility problems in polar electrolytes. Therefore, several attempts are made to enhance the solubility of organic corrosion inhibitors, including chemical derivatization and synergism. More so, these compounds are toxic in nature because of their volatile nature. Therefore, the use of ionic liquids is gaining particular attention due to their high solubility, eco-friendly nature and negligible vapour pressure. Literature investigation suggests that out of various classes of ionic liquids, imidazolium based ionic liquids are most frequently utilized as corrosion inhibitors. In such ionic liquids, the imidazole ring is directed toward the metal surface and its hydrophobic hydrocarbon chains are oriented toward the electrolyte solution (Figure 4.4.).

Chemical structures of some major ionic liquids used as corrosion inhibitors for different metals and alloys along with the nature of the electrolyte are

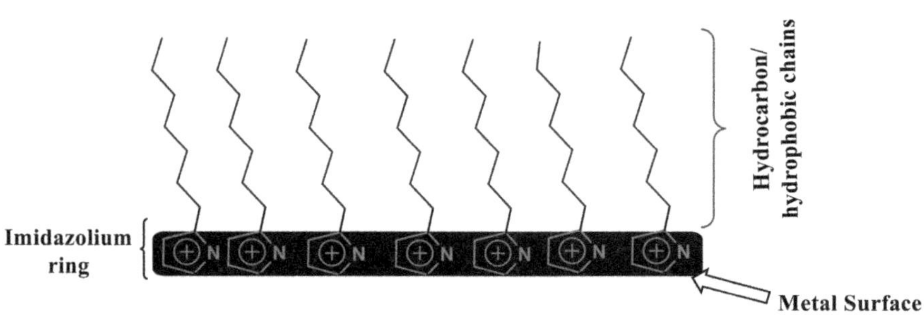

Figure 4.4 Illustration of adsorption of imidazolium based ionic liquids on the metallic surface.

illustrated in Table 4.1. Noticeably, in HCl based electrolytes, imidazolium based ionic liquids are most frequently used as metallic corrosion inhibitors. They become effectual by adsorbing and forming a protective film at the interface of the metal and the environment. Recently, numerous research and review reports have been published describing the corrosion inhibition potential of various classes of ionic liquids in HCl solutions. Kannan *et al.*[24] reported the corrosion inhibition potential of phenyltrimethyl ammonium tetrachloroaluminate $[PTMA]^+[AlCl_4]^-$ for carbon steel in 1 M HCl solution. Results showed that the tested ionic liquid exhibits an inhibition efficiency of 96% at 1.3 mM concentration. Adsorption of the ionic liquids on the metallic surface followed the Langmuir adsorption isotherm model. Authors proposed that $[PTMA]^+[AlCl_4]^-$ forms a protective film and avoids contact of corrosive species, *i.e.* HCl and H_2O molecules, with the metallic surface (Figure 4.5).

Electrochemical impedance spectroscopy (EIS) analysis suggests that $[PTMA]^+[AlCl_4]^-$ behaves as an interface-type corrosion inhibitor as it becomes effective by adsorbing at the interface. $[PTMA]^+[AlCl_4]^-$ retards both anodic and cathodic Tafel reactions and acts as a mixed-type corrosion inhibitor. However, some anodic predominance was also observed. Adsorption of $[PTMA]^+[AlCl_4]^-$ on the metallic surface was also supported by SEM and AFM analyses. These analyses suggest that presence of the ionic liquid causes a significant reduction in the roughness of the metal surface through the formation of corrosion protective film (Figure 4.6) DFT study was also conducted to describe the nature of interactions between $[PTMA]^+[AlCl_4]^-$ and metallic surface. Results demonstrate that $[PTMA]^+[AlCl_4]^-$ interacts with the metal surface using the charge sharing mechanism. The highest occupied molecular orbital (HOMO) and lowest unoccupied molecular orbital (LUMO) were mainly distributed over the aromatic ring, indicating that the aromatic ring mainly participates in charge sharing (Figure 4.6). Similar studies are extensively reported in the literature.

4.3.2 Ionic Liquids as Corrosion Inhibitors in H_2SO_4 and Other Electrolytes

Sulfuric acid (H_2SO_4) is another important electrolyte, and solutions of different concentrations are extensively used for cleaning different metal and alloy surfaces.[69,70] Obviously, similar to HCl solutions, highly concentrated H_2SO_4 solutions are used for acidic pickling, acid descaling, acid cleaning and oil-well acidification processes,[71,72] whereas lower concentrations are useful for academic research.[73] The use of organic compounds, especially those containing heteroatoms in the form of polar functional groups and multiple bonds, is one of the most effective methods of corrosion mitigation in H_2SO_4 solutions. Using their electron rich centres, called adsorption centres, these compounds adsorb on the metal surface

Table 4.1 Nature of metal and HCl solutions and chemical structure of ionic liquids used as corrosion inhibitors.

Chemical str. of ILs	System	Reference
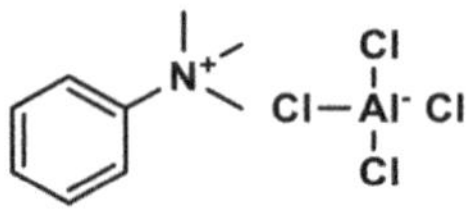 $[PTMA]^+[AlCl_4]^-$	Fe/1 M HCl	24
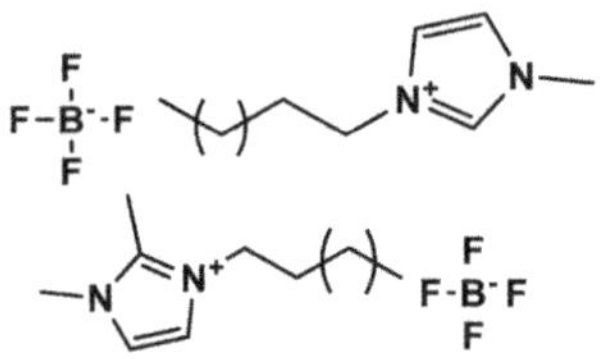 IL1: n=7 IL2: n=9 IL3: n=7 IL4: =9	Fe/1 M HCl	26
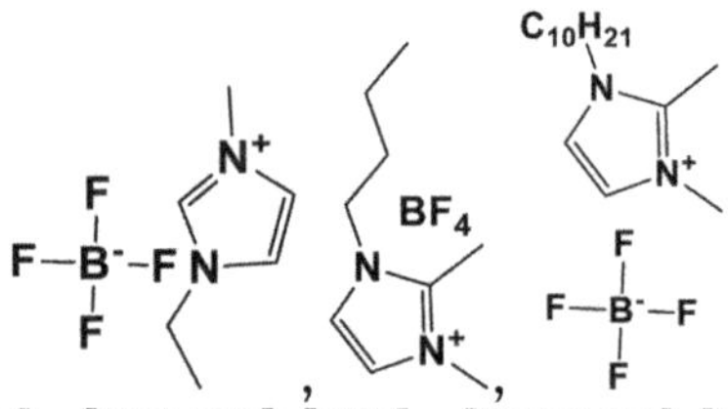 [EMIM]+[BF4]−, [BDMIM]+[BF4]−, [C10MIM]+[BF4]−	Fe/1 M HCl	28 and 29
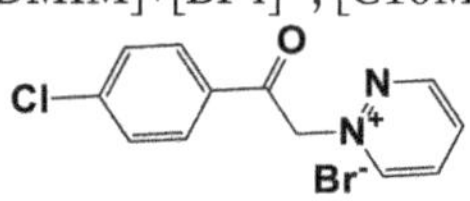 (CPEPB)	Fe/1 M HCl	31
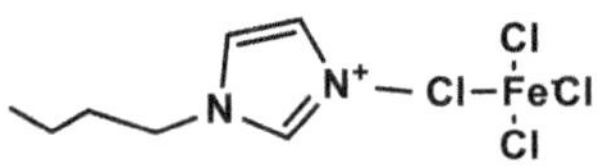 ([C4C1im][FeCl4])	Fe/1 M HCl	33
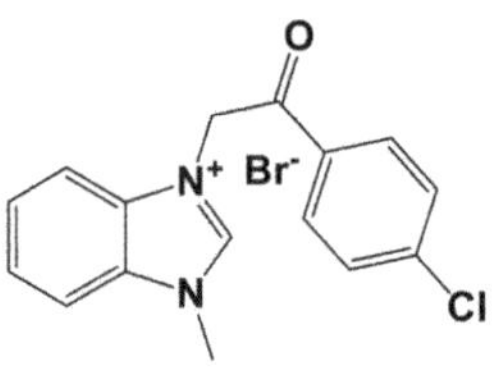 ([BMMB]+Br-)	Fe/1 M HCl	35
[C16M1Im] [Br] 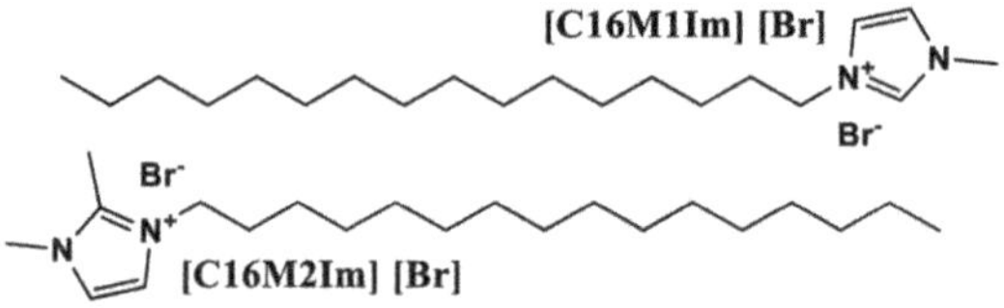 [C16M2Im] [Br]	Fe/1 M HCl	37

Chemical str. of ILs	System	Reference
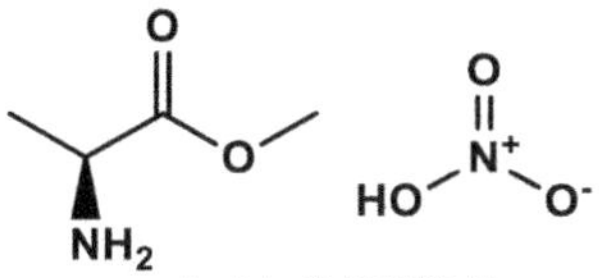 L-AlaC1NO3 IL	Mild steel/1 M HCl	25
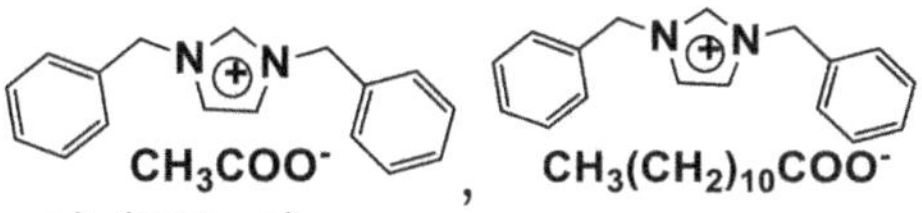 (DBImL), (DBImA)	Fe/1 M HCl & H_2SO_4	27
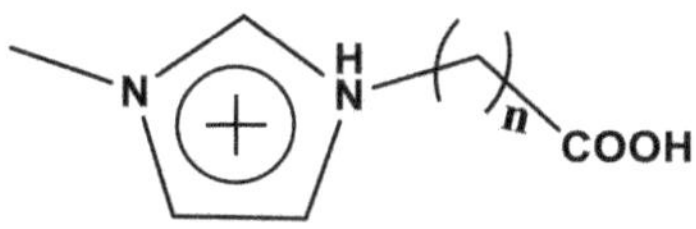 A: n=1; B= n=2;C=n=3	Fe/0.5 M HCl	30
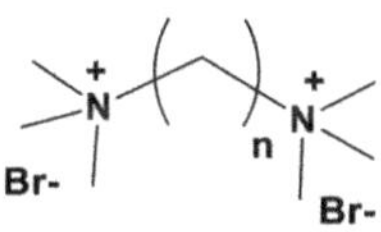 (G2IL): n = 2; (G3IL): n = 3; (G6IL): n = 6	Fe/1 M HCl	32
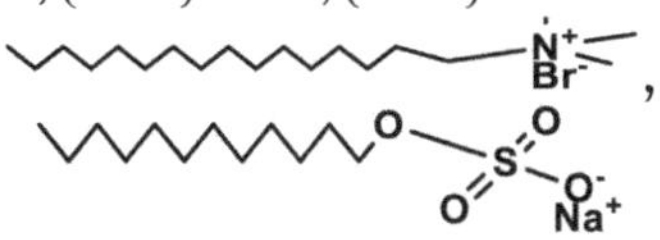 (CTAB), (SDS)	Fe/2 M HCl	34
MPIMI: n=1 BMIMI: n=2 HMIMI: n=3 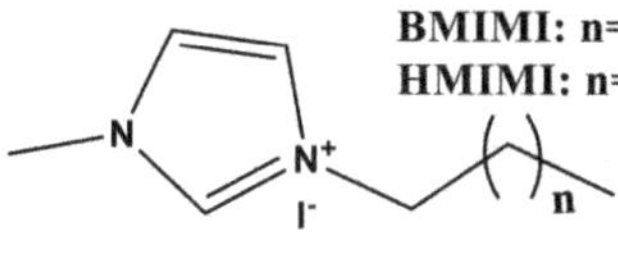	Fe/1 M HCl	36
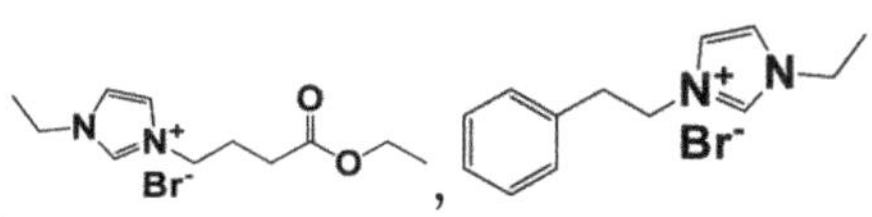 (I) (II) 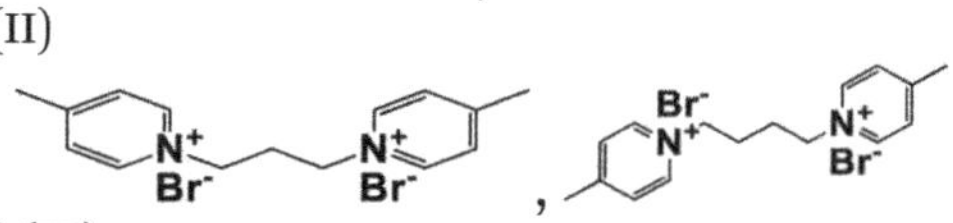 (III) (IV)	Fe/1 M HCl	38

(*continued*)

Table 4.1 (*continued*)

Chemical str. of ILs	System	Reference
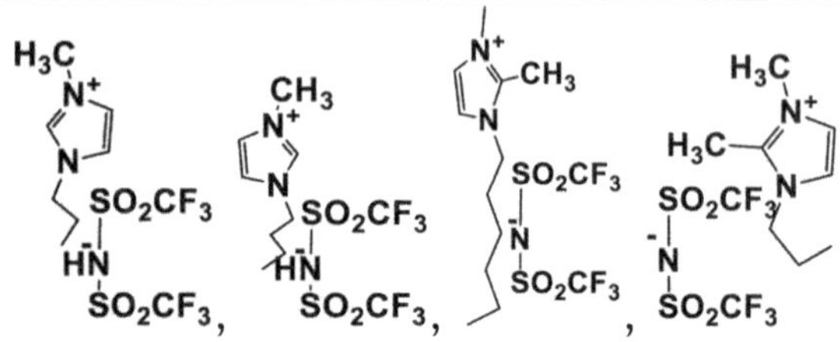	Fe/1 M HCl	39
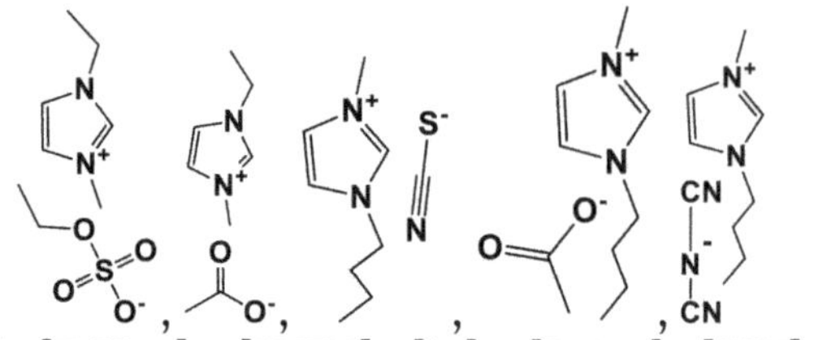 [EMIM] + [EtSO 4] −, [EMIM] + [Ac] −, [BMIM] + [SCN] −, [BMIM] + [Ac] −, [BMIM] + [DCA] −	Fe/1 M HCl	41
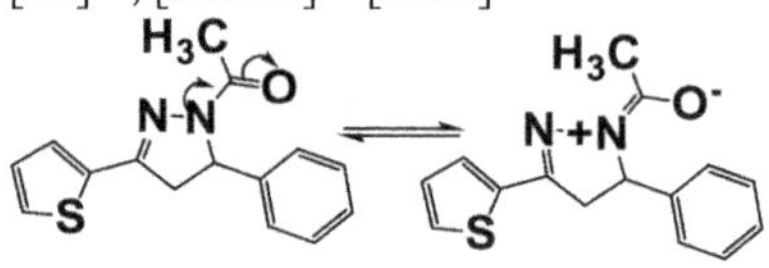	Fe/1 M HCl & H_2SO_4	43
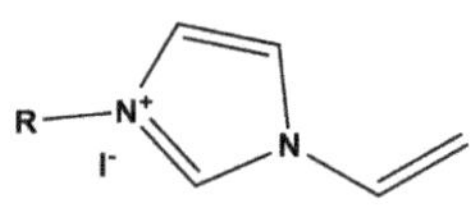 **[VMIM]I: R=-Me, [VPIM]I: R=-C_3H_7; [VBIM]I R=-C_4H_9**	X70 steel/0.5 M HCl	45
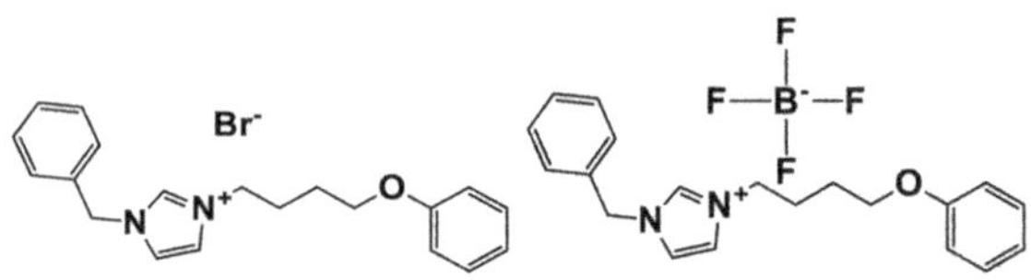 (MA1) (MA2)	Fe/1 M HCl	47
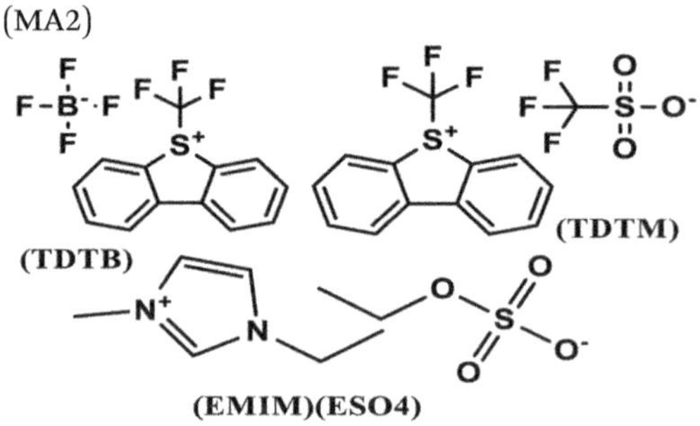	Fe/1 M HCl	49

Chemical str. of ILs	System	Reference
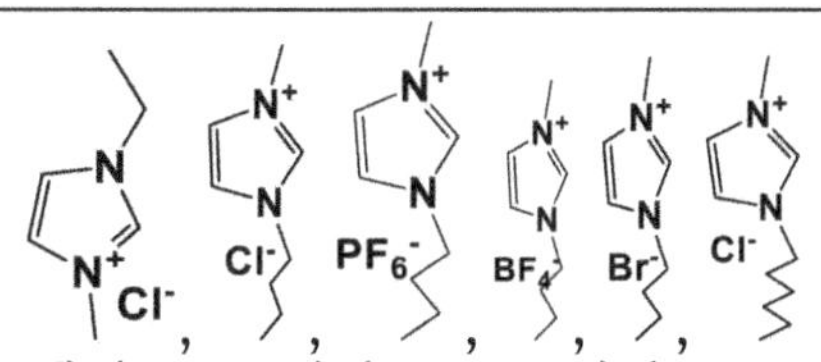 (EMIm Cl), (BMIm Cl), (BMIm PF6), (BMIm BF4), (BMIm Br), (HMIm Cl)	Fe/2 M HCl	40
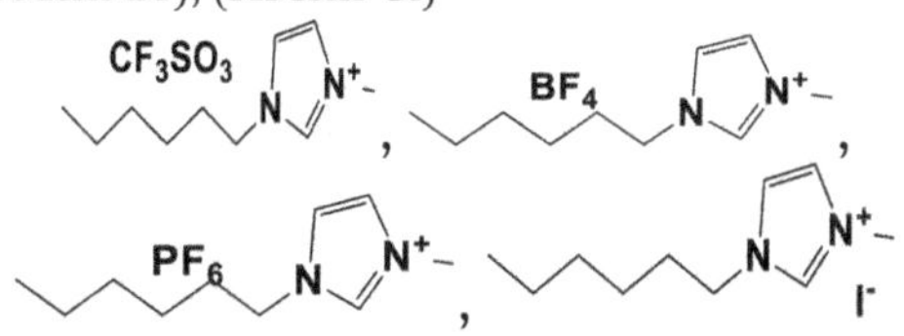 [HMIM][TfO] [HMIM][BF4], [HMIM][PF6] & [HMIM] [I]	Fe/1 M HCl	42
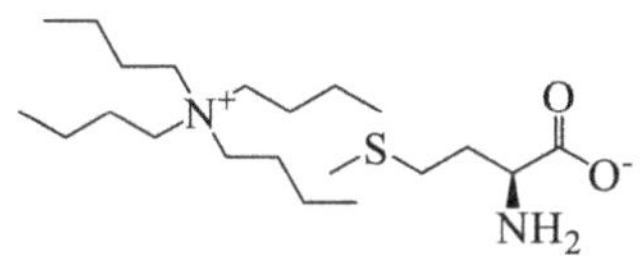	Fe/1 M HCl	44
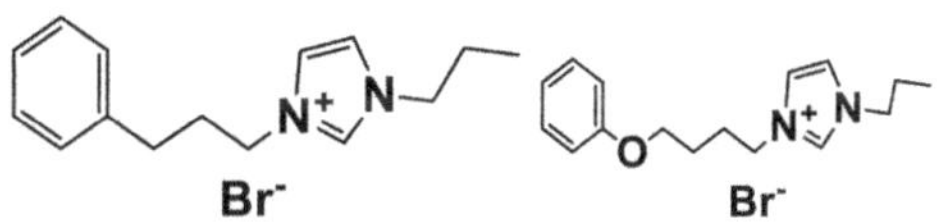 (PPIB1) (PPIB4)	Fe/1 M HCl	46
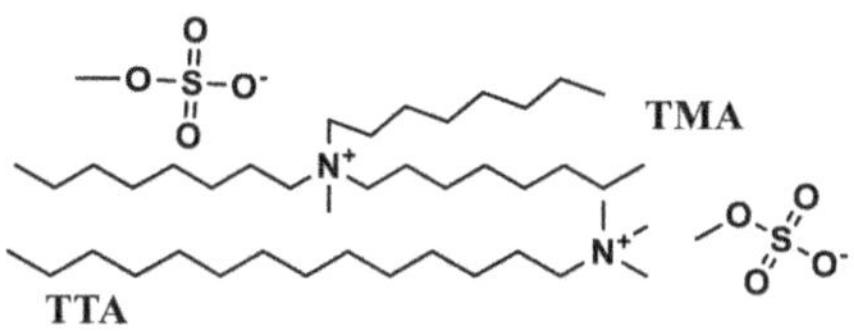	API-X52 Fe/1 M HCl	48
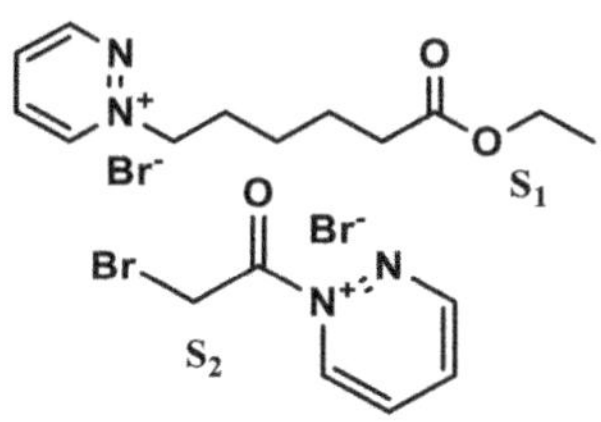	Fe/1 M HCl	50

(*continued*)

Table 4.1 (*continued*)

Chemical str. of ILs	System	Reference
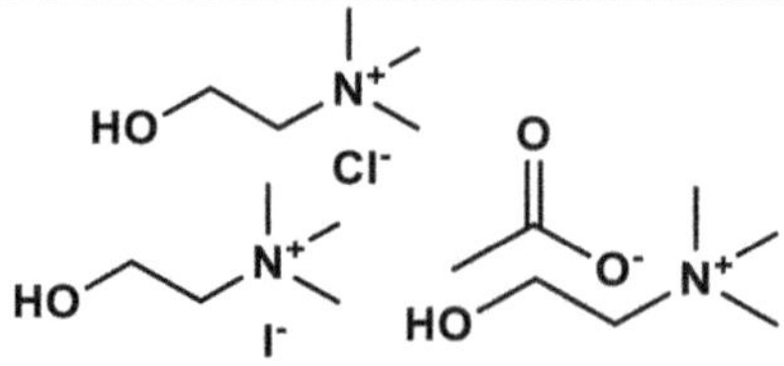 [Chl][Cl] [Chl][I] [Chl][Ac]	Fe/1 M HCl	16
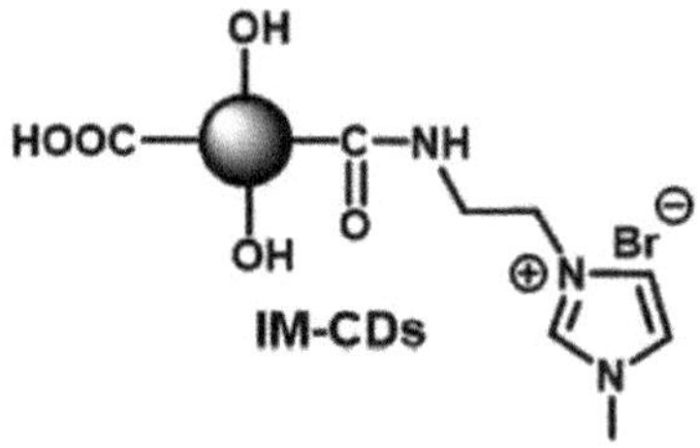	Q235 steel/1 M HCl	52
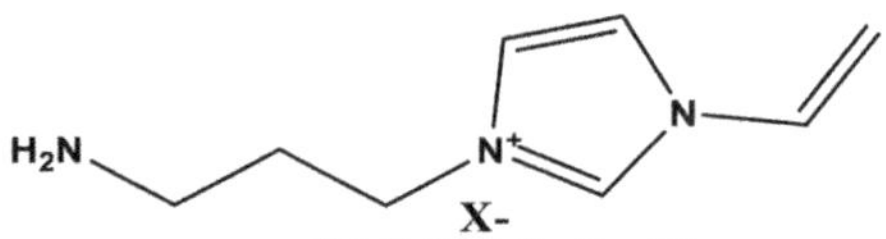 [VAIM][PF6]: X^-= PF_6 [VAIM][BF4]: X^-=BF_4	Fe/1 M HCl	54
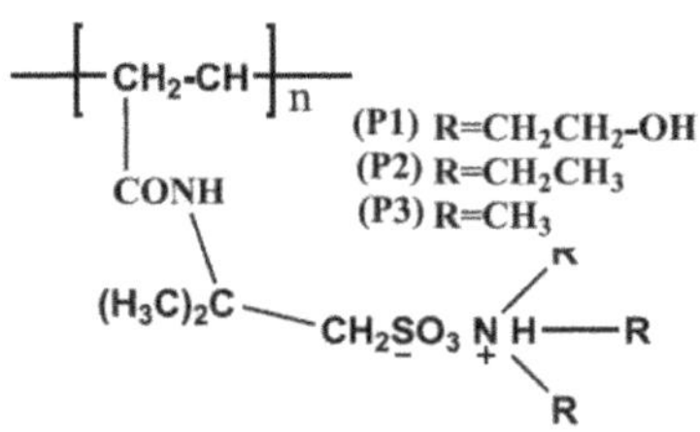	Fe/1 M HCl	56
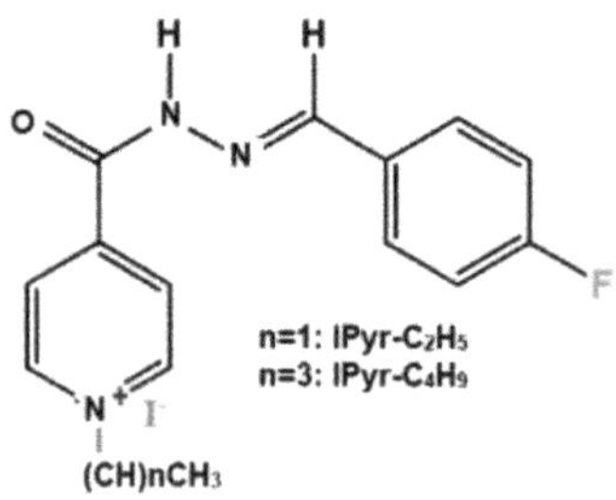	Fe/1 M HCl	58

Chemical str. of ILs	System	Reference
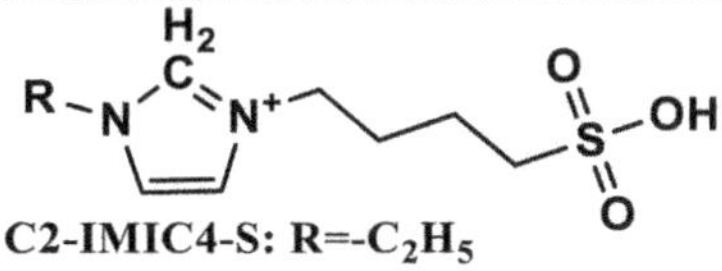 C2-IMIC4-S: R=-C_2H_5 C10-IMIC4-S: R=-$C_{10}H_{21}$	Fe/0.5 M HCl	51
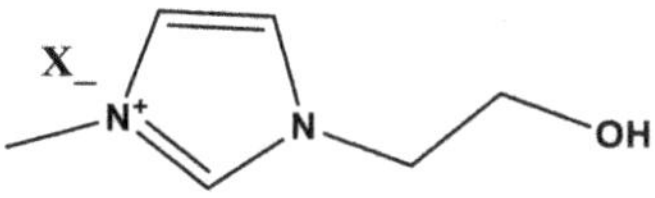 [(HOC2)MIm]PF6: X-=- PF6 [(HOC2)MIm]NTF$_2$: X-=- NTF$_2$	Fe/1 M HCl	53
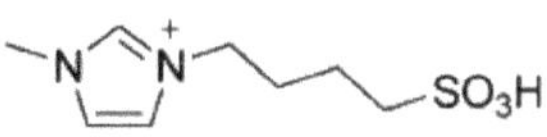 HSO_4^- [BsMIM][HSO$_4$] 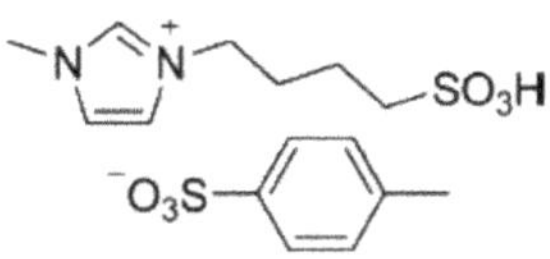 [BsMIM][OTs]	Fe, Ni, and 304 SS/ water	55
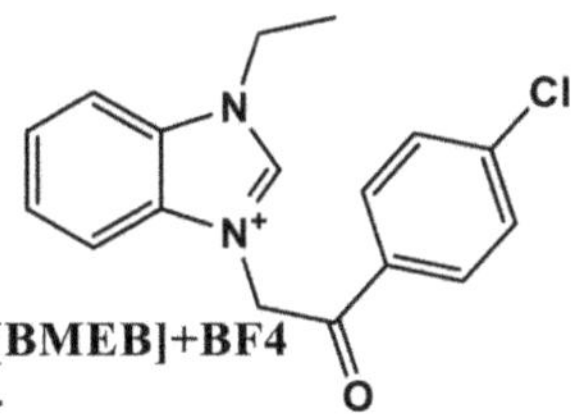 [BMEB]+BF4 -	Fe/1N HCl	57
[bmim][Cl: X= Cl [bmim][Ac: X=-CH_3OO [bmim][CF_3SO_3]: CF_3SO_3 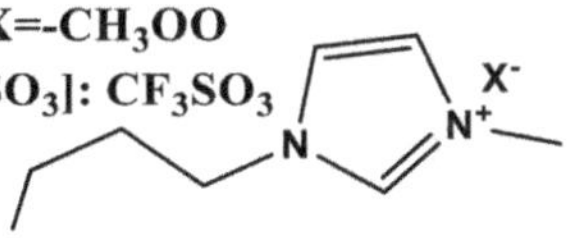	Fe/1 M HCl	59

(*continued*)

Table 4.1 (*continued*)

Chemical str. of ILs	System	Reference
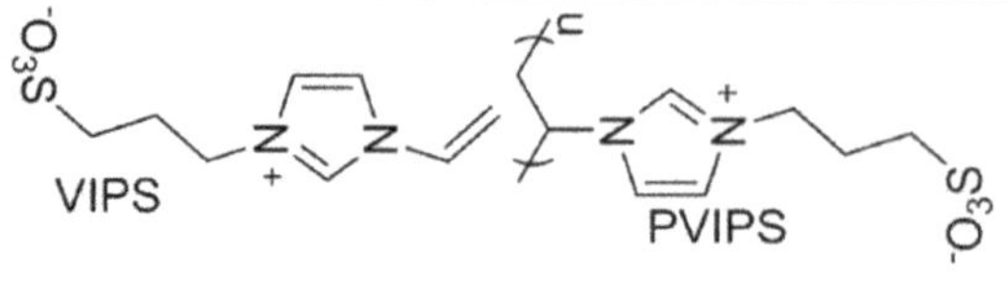	Fe/0.5 M HCl	60
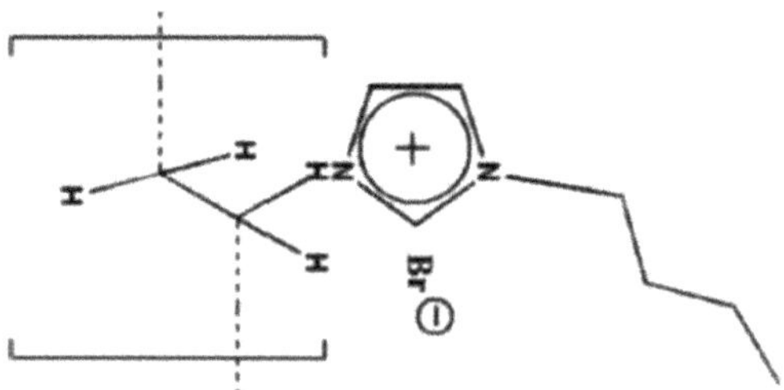	Fe/1 M HCl	62
 **TDTB** 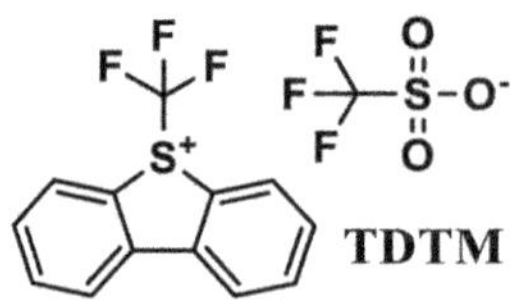 **TDTM** 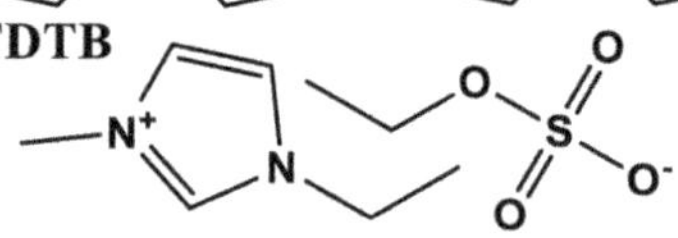 **(EMIM)(ESO4)**	Fe/1 M HCl	64
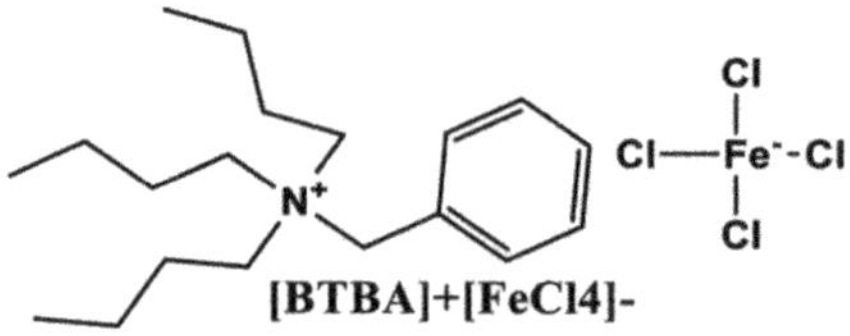 **[BTBA]+[FeCl4]-**	Fe/1N HCl	66
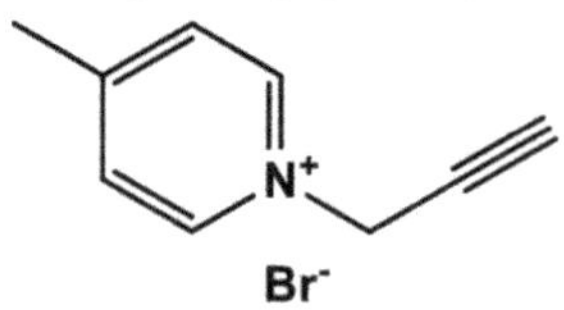	Fe/0.5 M HCl	67

Chemical str. of ILs	System	Reference
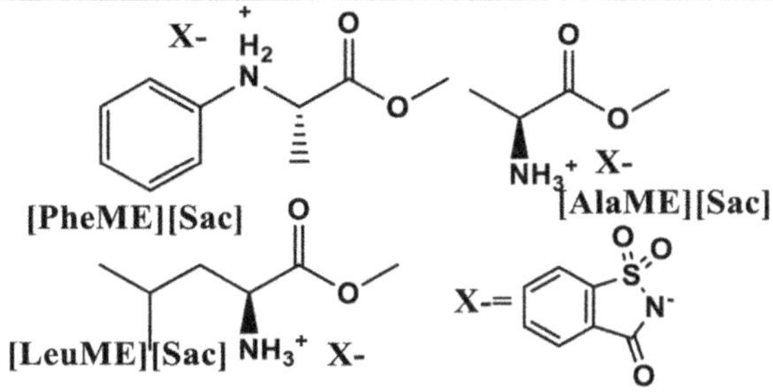	Fe/1 M HCl	61
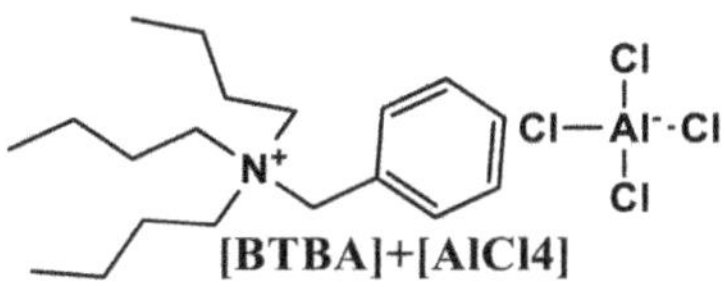	Fe/2N HCl	63
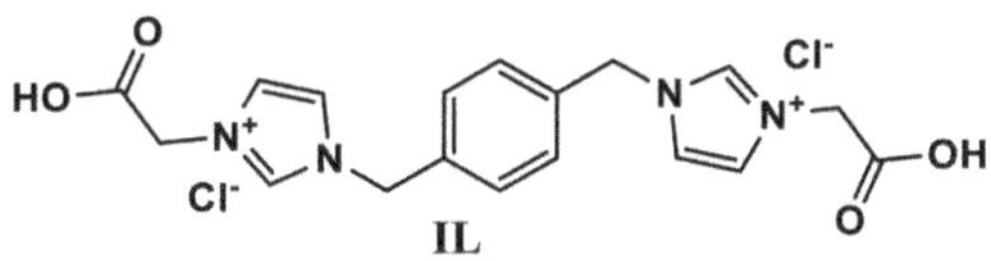	Fe/0.5 M HCl	65
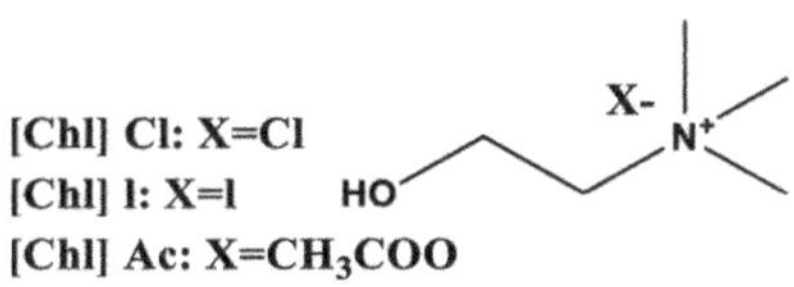	Fe/1 M HCl	16
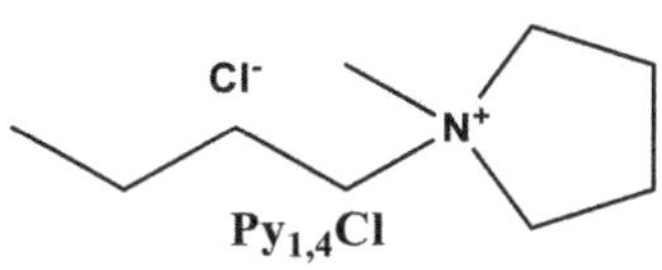	316L SS Fe/1 M HCl	68

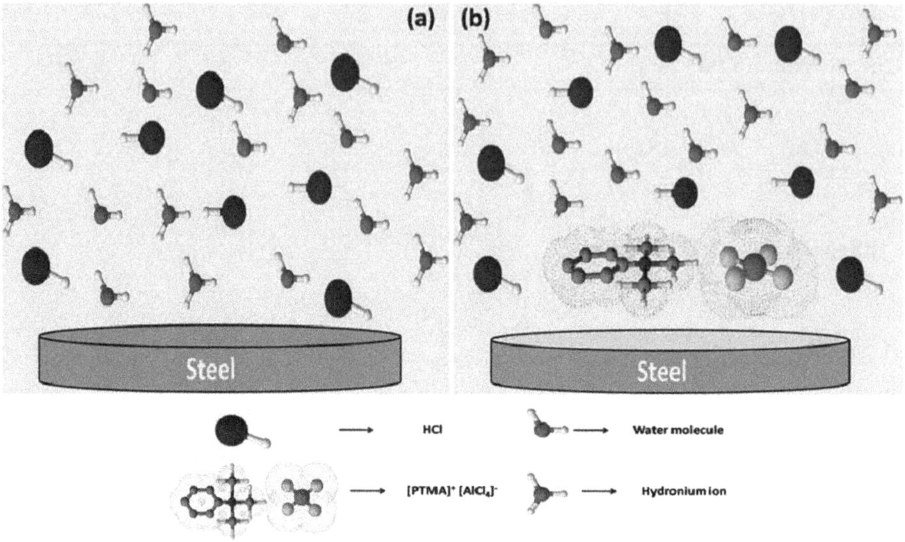

Figure 4.5 Adsorption/interaction of water, HCl and hydronium ions with steel surface in the absence (a) and presence (b) of $[PTMA]^+[AlCl_4]^-$.[24] Reproduced from ref. 24, https://doi.org/10.1016/j.apsadv.2021.100150, under the terms of the CC BY 4.0 licence https://creativecommons.org/licenses/by/4.0/.

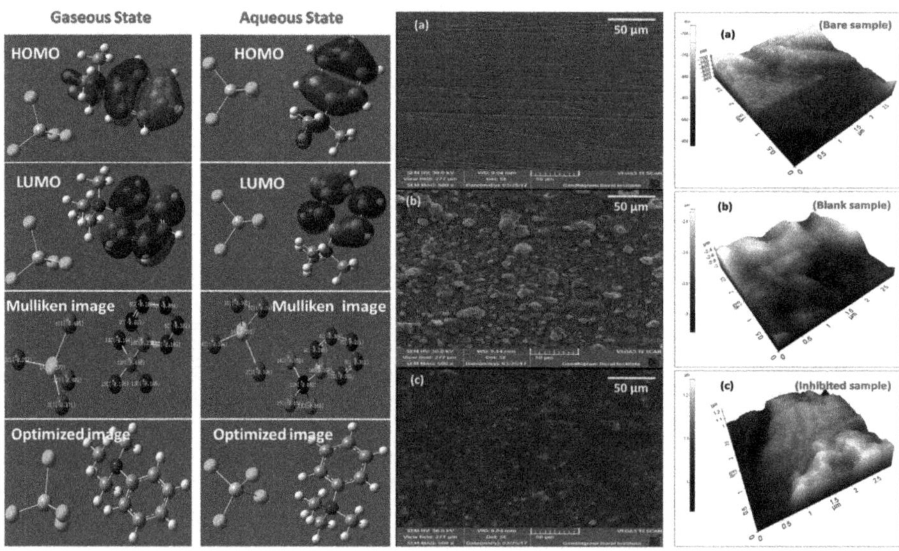

Figure 4.6 Frontier molecular orbital picture (left side) of $[PTMA]^+[AlCl_4]^-$, SEM (middle) and AFM (right side) of (a) polished metal surface, (b) corroded in 1 M HCl without $[PTMA]^+[AlCl_4]^-$ and (c) corroded in 1 M HCl with $[PTMA]^+[AlCl_4]^-$.[24] Reproduced from ref. 24, https://doi.org/10.1016/j.apsadv.2021.100150, under the terms of the CC BY 4.0 licence https://creativecommons.org/licenses/by/4.0/.

and build corrosion protective coverings that isolate the metal from the aggressive environment. However, because of their toxicity and sometimes limited solubility, the use of traditional organic compounds is limited. In view of this, the current employment of ionic liquids as green and sustainable corrosion inhibitors in H_2SO_4 solutions is gaining immense attention.[74–77]

The corrosion inhibition potential of ionic liquids in H_2SO_4 solutions is widely reported.[14,17] A summary of corrosion potentials of ionic liquids in H_2SO_4 and other electrolytes are presented in Tables 4.2 and 4.3, respectively. Because of their strong ability to adsorb onto metal surfaces and extremely high solubility in polar electrolytes, ionic liquids provide excellent anticorrosive activity. Recently, our research team[78] described the inhibition effects of decyltriphenylphosphonium bromide (DTPPB) for mild steel corrosion in 0.5M H_2SO_4 using various methods. Using the potentiodynamic polarization method, it was derived that DTPPB acts as an excellent corrosion inhibitor and its protection efficiency was temperature and concentration dependent. DTPPB exhibits the highest protection effectiveness of 99.86% at 10^{-2} M concentration. Corrosion inhibition efficiency of DTPPB was measured in its concentration range of 10^{-2} to 10^{-5} M. Increases in the DTPPB concentration cause subsequent increases in its protection potential while temperature has adverse effects (Figure 4.7).

Using EIS study, it was observed that DTPPB acts as an interface-type corrosion inhibitor as the presence of the DTPPB increases the value of charge transfer resistance. This observation suggests that DTPPB retards or slows down the charge transfer process. Nyquist curves with and without DTPPB represent a single semicircle indicating that mild steel corrosion with and without ionic liquid involves a single charge transfer mechanism which was also supported by a single maxima in the inhibited and non-inhibited Bode plots. Adsorption and corrosion inhibition behaviour of the DTPPB was also supported by SEM, EDX and AFM analyses, where a significant smoothness in the surface morphologies of protected SEM and AFM images were derived in the presence of DTPPB. Changes in the elemental composition during EDX analyses also confirm this observation. DFT study indicates that DTPPB interacts with the metallic surface using donor–acceptor interactions. Through MD simulations, it was observed that DTPPB interacts with metallic surfaces in such a way that electron rich aromatic rings were directed toward the metal surface, and the hydrophobic hydrocarbon chain was directed toward the solution side (Figure 4.8).

4.4 Summary

Ionic liquids are a green and sustainable substitute for traditional organic corrosion inhibitors. Ionic liquids are associated with various advantages such as low toxicity, low vapour pressure, high solubility, high thermal and

Table 4.2 Nature of metal and H_2SO_4 solutions and chemical structure of ionic liquids used as corrosion inhibitors.

Chemical str. of ILs	System	Reference
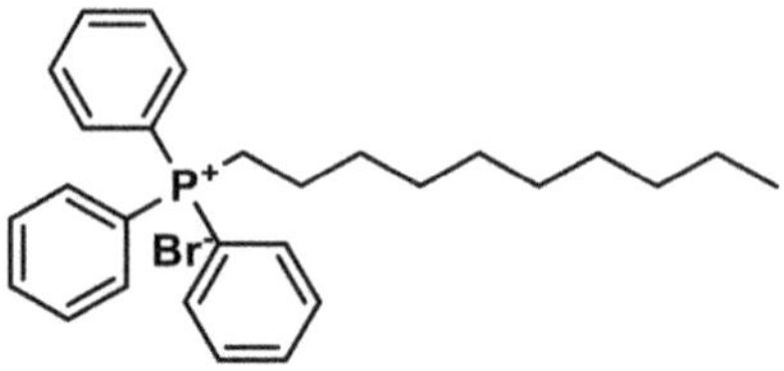	Mild steel/0.5 M H_2SO_4	78
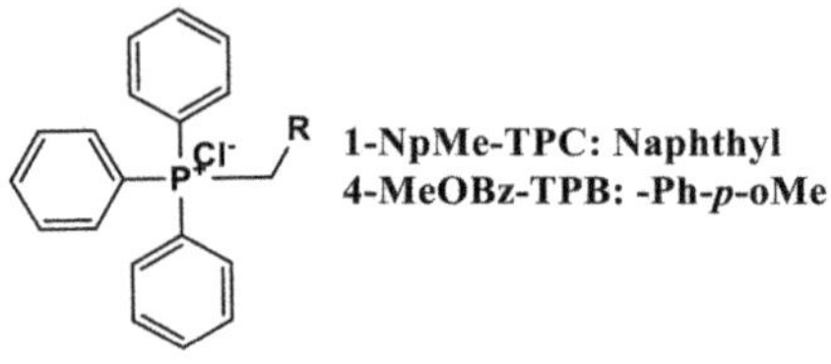	Fe/0.5 M H_2SO_4	80
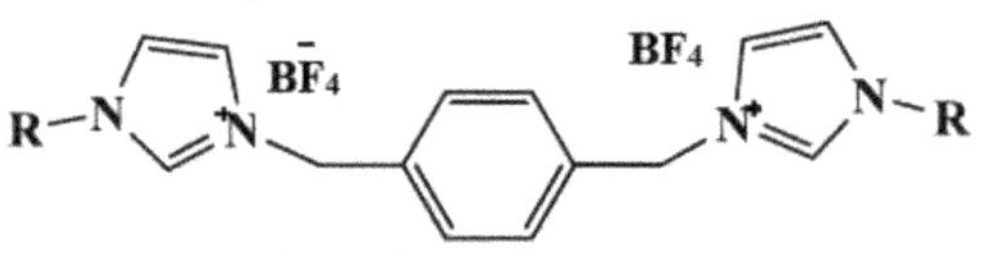 IL1: R=-C_8H_{17}; IL2: R=-$C_{10}H_{21}$ IL3: R=-$C_{12}H_{25}$	304 SS Fe/0.5 M H_2SO_4	82
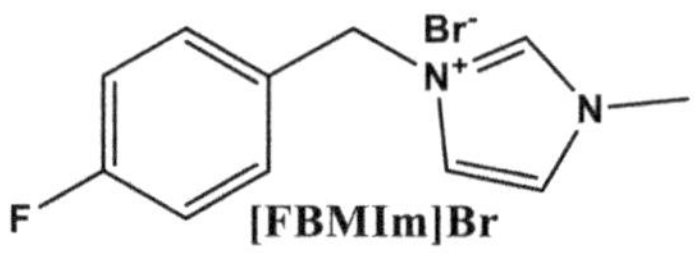	Fe/0.5 M H_2SO_4	84
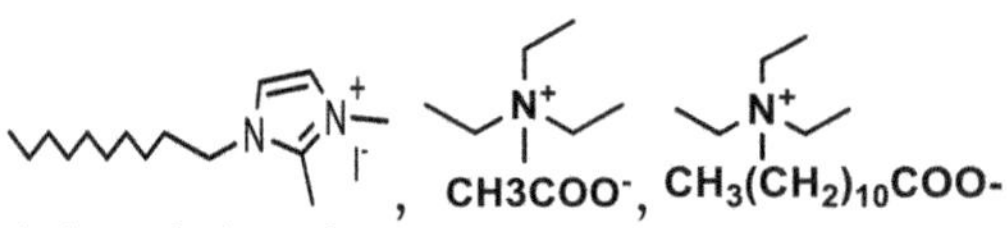 (DDI), (TMA), (TML)	Fe/1 M H_2SO_4	86
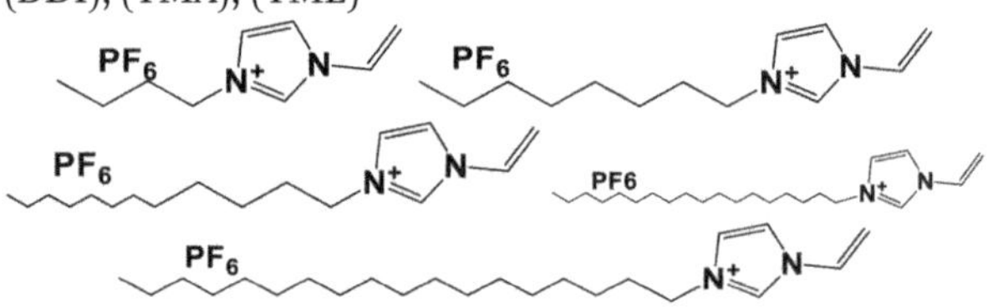 (VImC4PF6) (VImC8PF6) (VImC12PF6) (VImC18PF6) (VImC22PF6)	1 M H_2SO_4	88
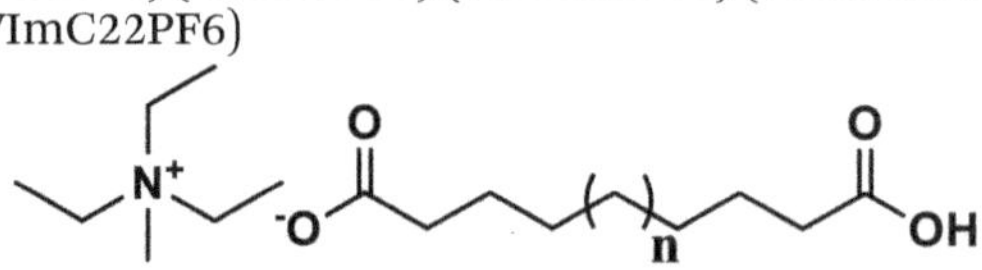 [C_6]: n=1; [C_9]: n=3; [C_{12}]: n=6	API X60 Fe/1 M H_2SO_4	90

Chemical str. of ILs	System	Reference
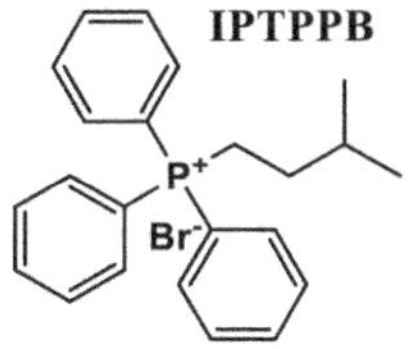	Fe/0.5 M H_2SO_4	79
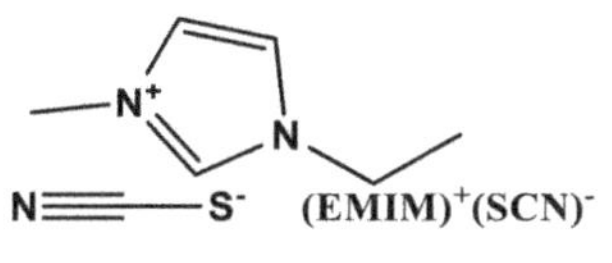	API 5L X52 0.5 M/H_2SO_4 & HCl	81
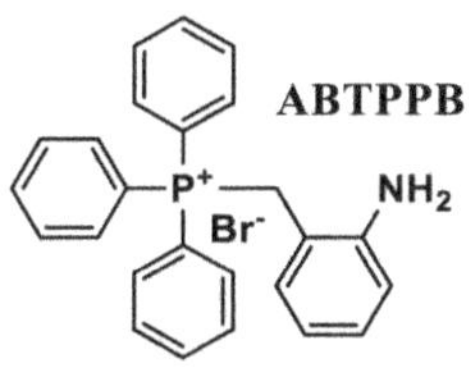	Fe/0.5 M H_2SO_4	83
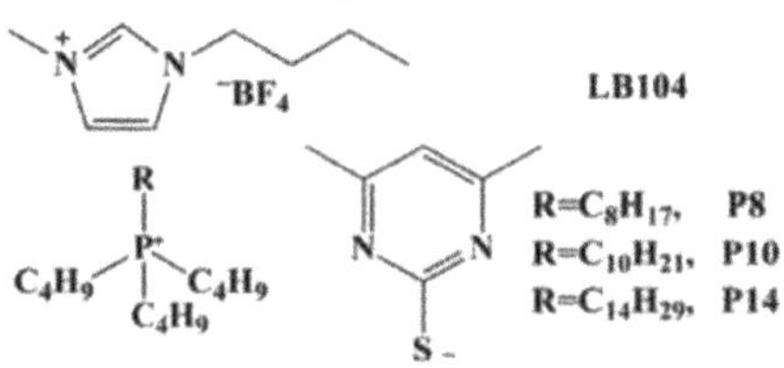	Fe/0.5 M H_2SO_4	85
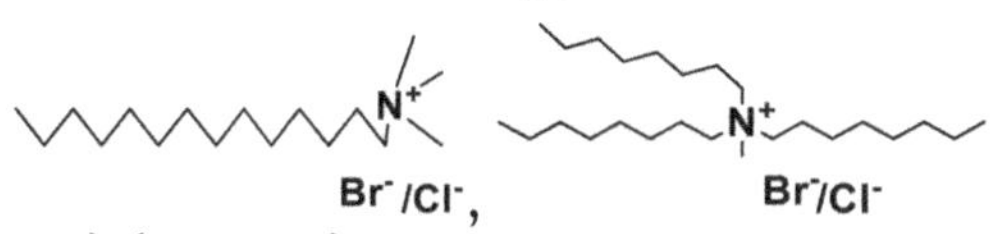 (MTABr), (TOMABr)	Fe/2 M H_2SO_4	87
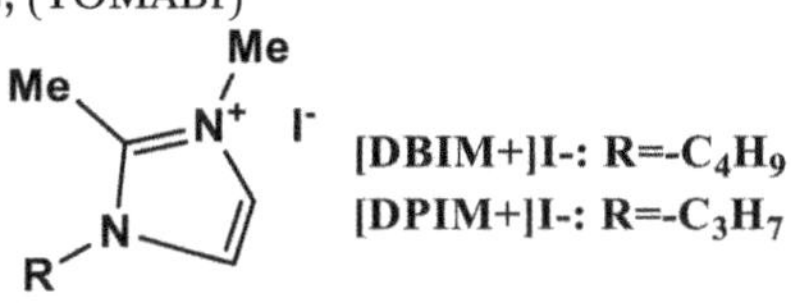	API 5L X52/1 M H_2SO_4	89
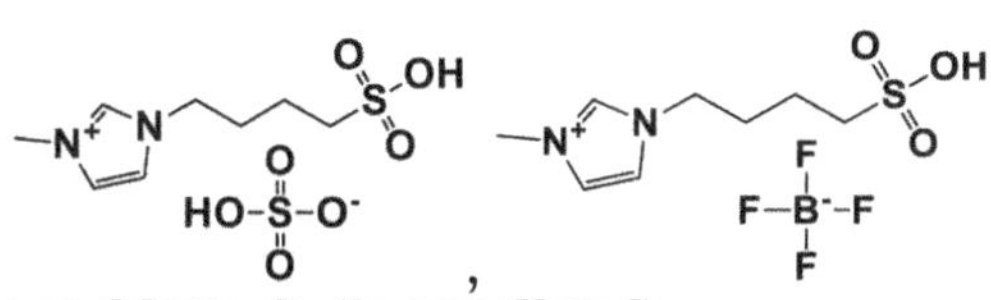 ([BsMIM]-[HSO4]), ([BsMIM][BF4]),	Fe/1 M H_2SO_4	91

(*continued*)

Table 4.2 (*continued*)

Chemical str. of ILs	System	Reference
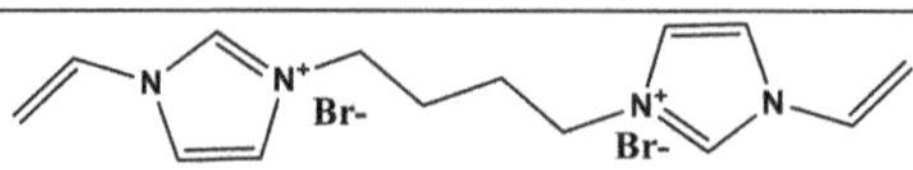	Fe/1 M H_2SO_4	92
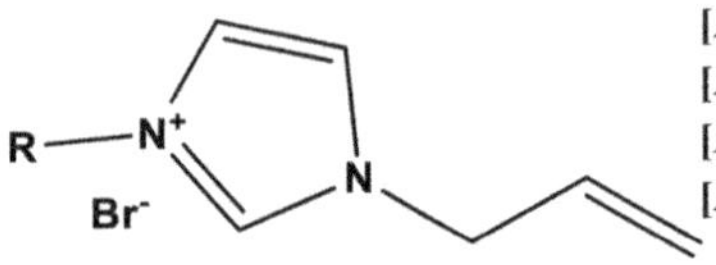 **[AEIM]Br: C_2H_5 [ABIM]Br:C_4H_9 [AHIM]Br: C_6H_{13} [AOIM]Br: C_8H_{17}**	Fe/0.5 M H_2SO_4	74
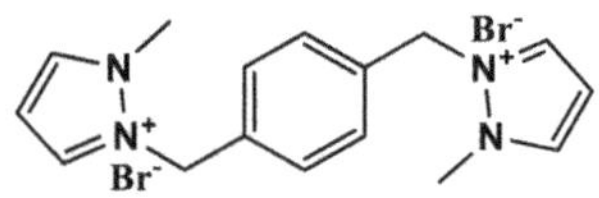 IL	Fe/1 M H_2SO_4	95
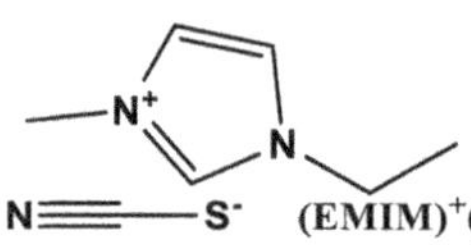 **$(EMIM)^+(SCN)^-$**	API 5L X52 0.5 M/H_2SO_4	97

Table 4.3 Nature of metal and various electrolytes and chemical structure of ionic liquids used as corrosion inhibitors.

Chemical str. of ILs	System	Reference
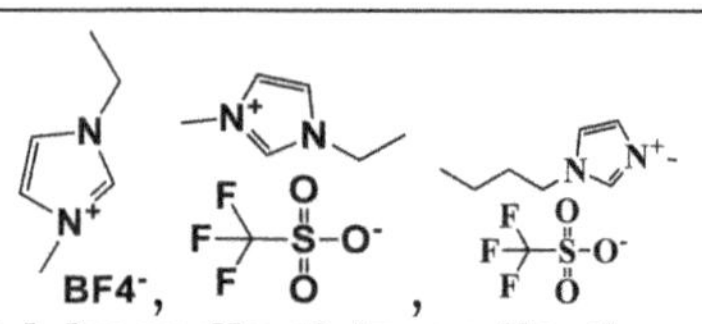 [EMIM][BF4], [BMIM][Otf], [EMIM][Otf]	Fe/CO_2	99
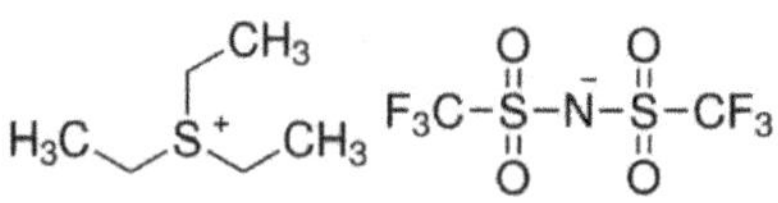 TESFI	304 SS Fe/5% sulfamic acid	101
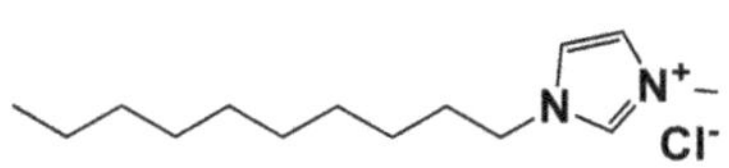 (DMICL)	Fe/NaCl (pH 3.8 & pH 6.8)	102

Chemical str. of ILs	System	Reference
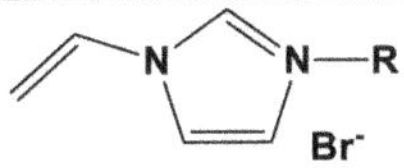 R = IL1:C_4H_9; IL2:C_8H_{17}; IL3:$C_{12}H_{25}$; IL4:$C_{18}H_{37}$; IL5: $C_{22}H_{45}$	Fe/1 M H_2SO_4	93
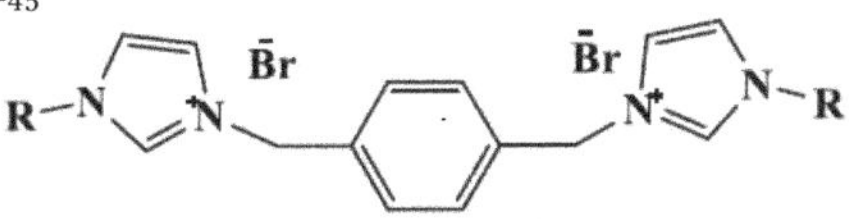 IL1: R=-C_8H_{17}; IL2: R=-$C_{10}H_{21}$ IL3: R=-$C_{12}H_{25}$	304 SS Fe/0.5 M H_2SO_4	94
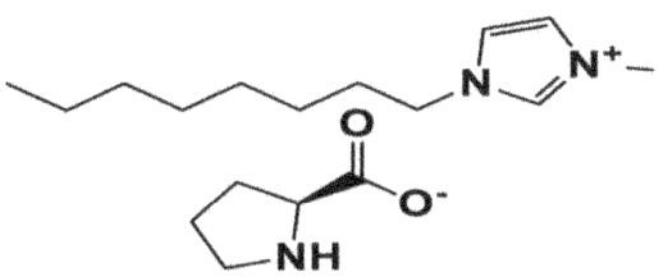	Fe/0.5 M H_2SO_4	96
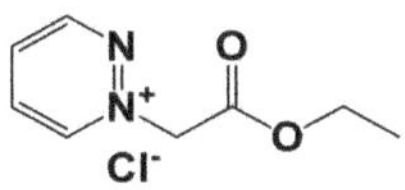 (EOPC)	Fe/0.5 M H_2SO_4	98

Chemical str. of ILs	System	Reference
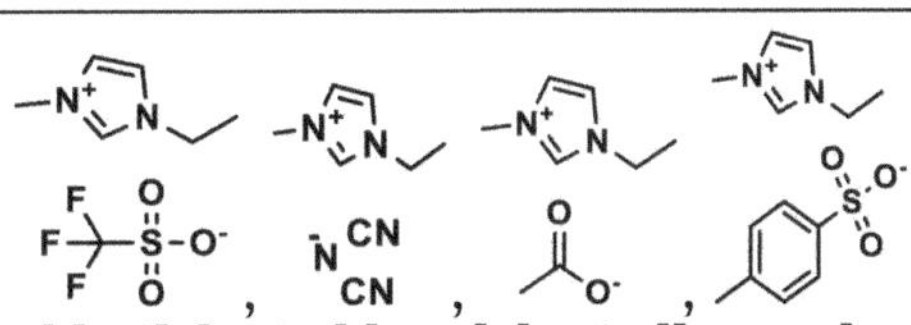 [emim]-[Otf], [emim]-[DCA], [emim][acetate], [emim][tosylate])	Fe/CO_2 Capture system	100
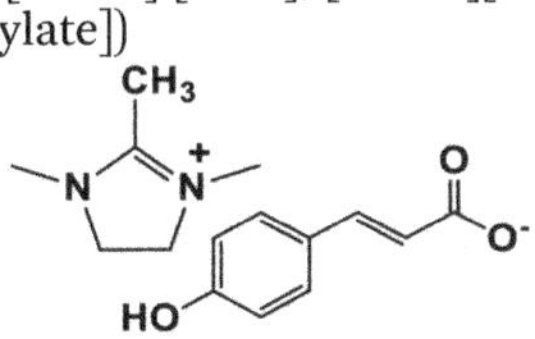	Fe/0.01 M NaCl	91
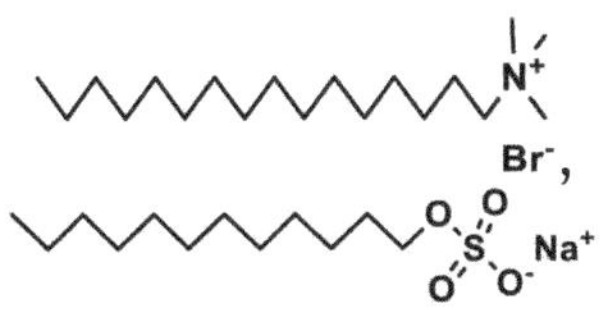 (CTAB) (SDS)	Fe/3.5% NaCl	103

(continued)

Table 4.3 (*continued*)

Chemical str. of ILs	System	Reference
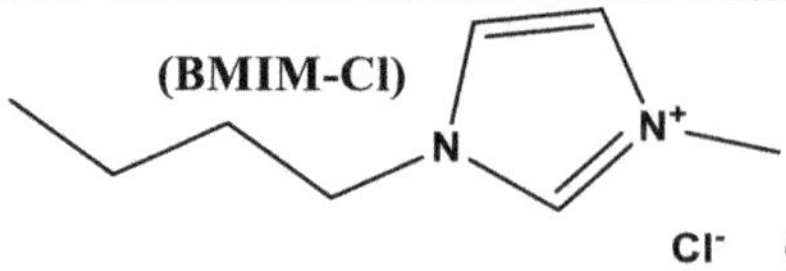 Coating	Fe/3.5% NaCl	104
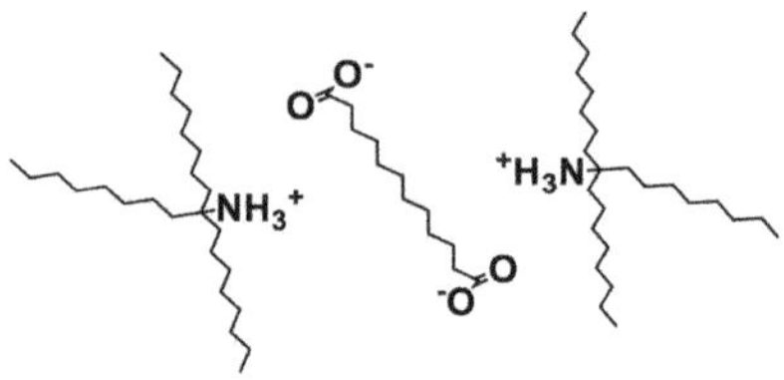	Fe/Production water	106
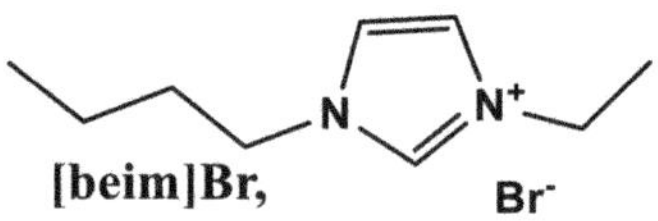	Copper/0.1 M Na_2SO_4	108
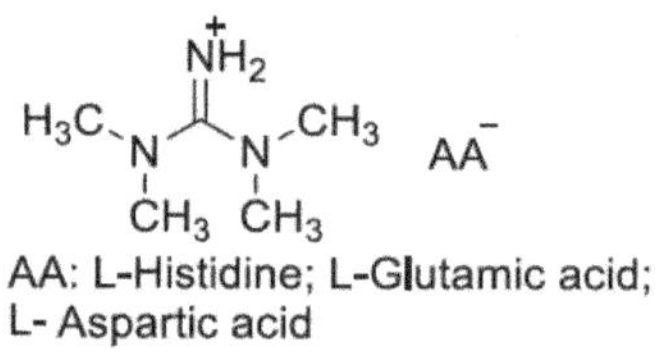	SS Fe/36% HCl + 5% Sb_2O + 4% $SnCl_2$	109
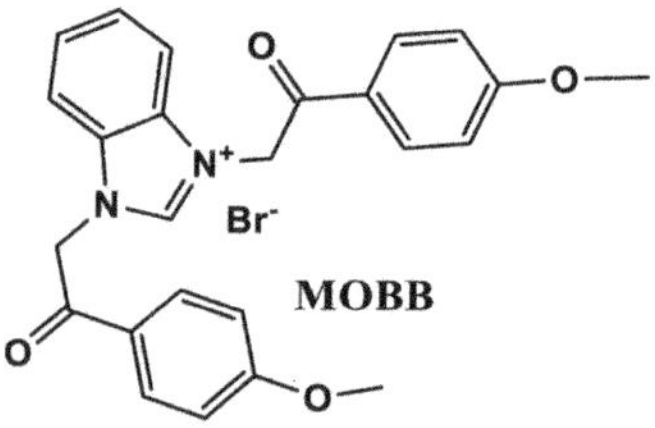	6061 Al-15 vol% SiC (P)/0.1 M HCl & H_2SO_4	111
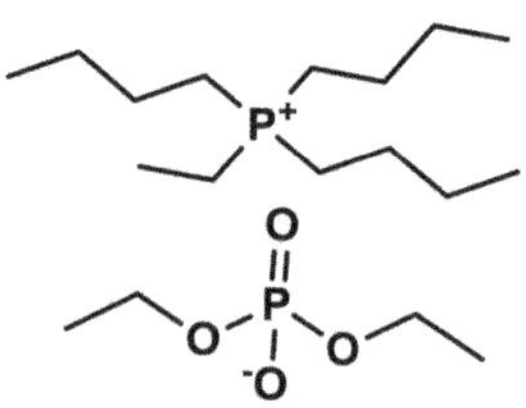	Aluminium alloy AA5052/4% sulfamic acid	113

Chemical str. of ILs	System	Reference
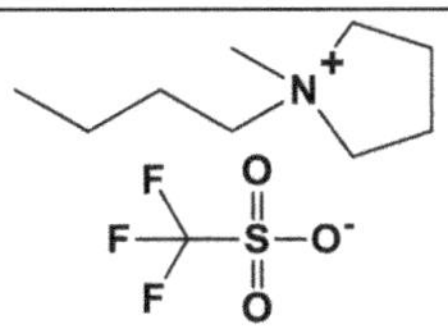	Fe/3.5% NaCl	105
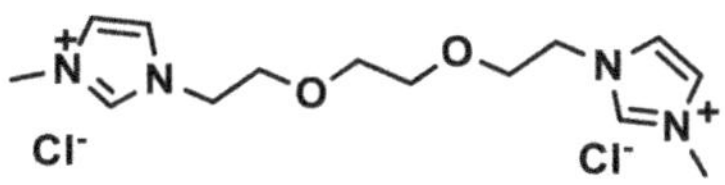 (TSIL)	Fe/open system	107
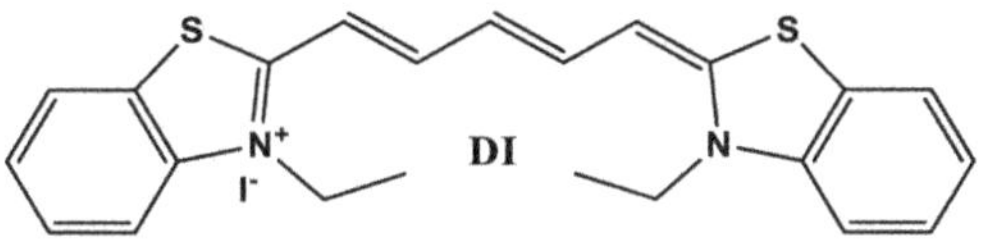	Copper/0.1 M HCl	108
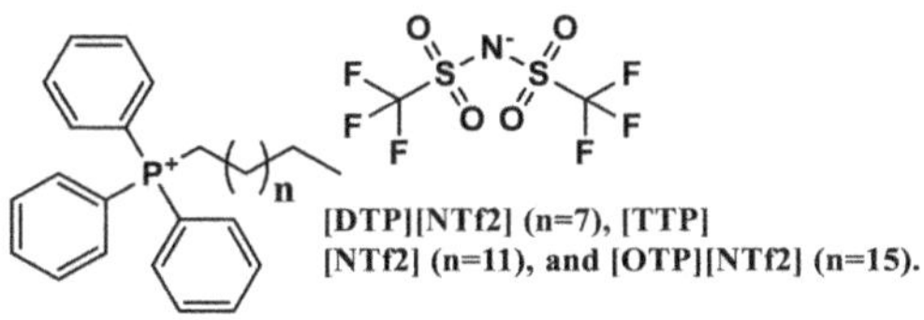	AZ31B Mg/0.05 wt.% NaCl	110
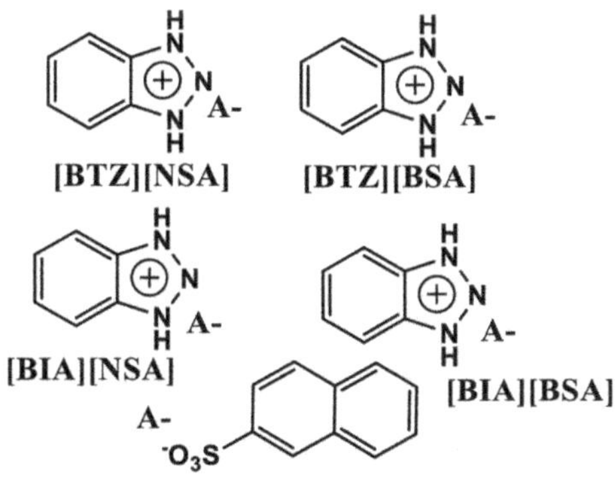	Brass/5 wt% HNO_3	112

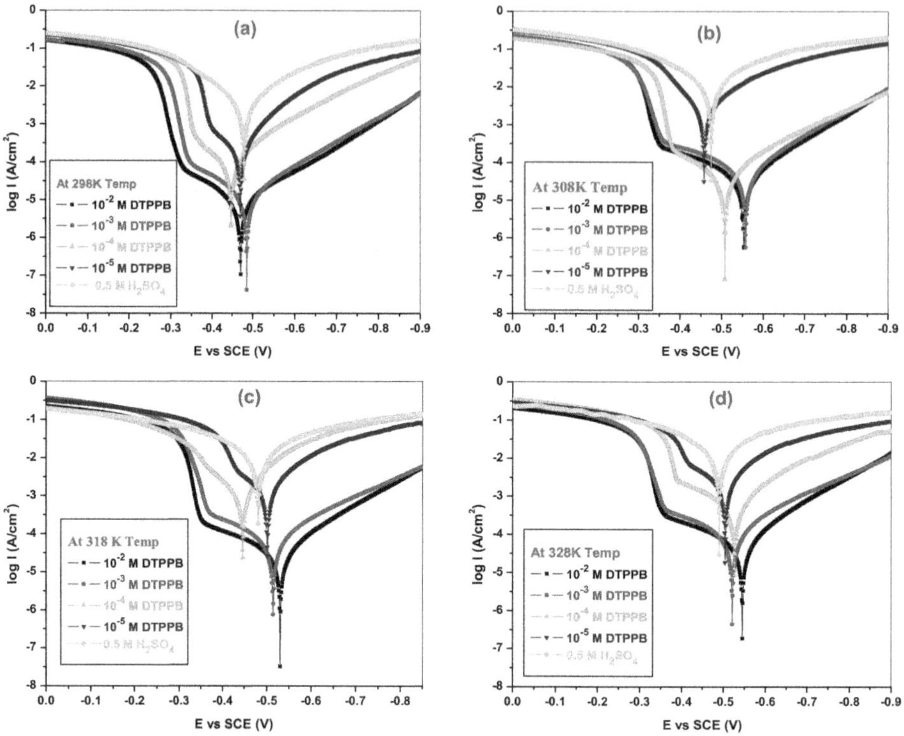

Figure 4.7 Potentiodynamic polarization curves for mild steel corrosion with and without different concentrations of DTPPB in 0.5M H_2SO_4 at (a) 298 K, (b) 308 K, (c) 318 K and (d) 328 K.[78] Reproduced from ref. 78 with permission from Elsevier, Copyright 2021.

chemical stability, and most importantly, being environmentally benign. Various series of ionic liquids, especially imidazolium, pyridinium and tetra-ammonium, are extensively utilized as corrosion inhibitors for different metal/electrolyte systems. These compounds become effective by adsorbing on the metallic surface and forming a protective film. Their adsorption mostly follows the Langmuir adsorption isotherm. Adsorption of ionic liquids on metallic surfaces is supported by SEM, AFM and EDX analyses. Computation analyses were used to demonstrate the corrosion inhibition potential and interaction of ionic liquids with the metallic surface. Ionic liquids act as mixed- and interface-type corrosion inhibitors. They block the active sites through their adsorption and build a corrosion protective film.

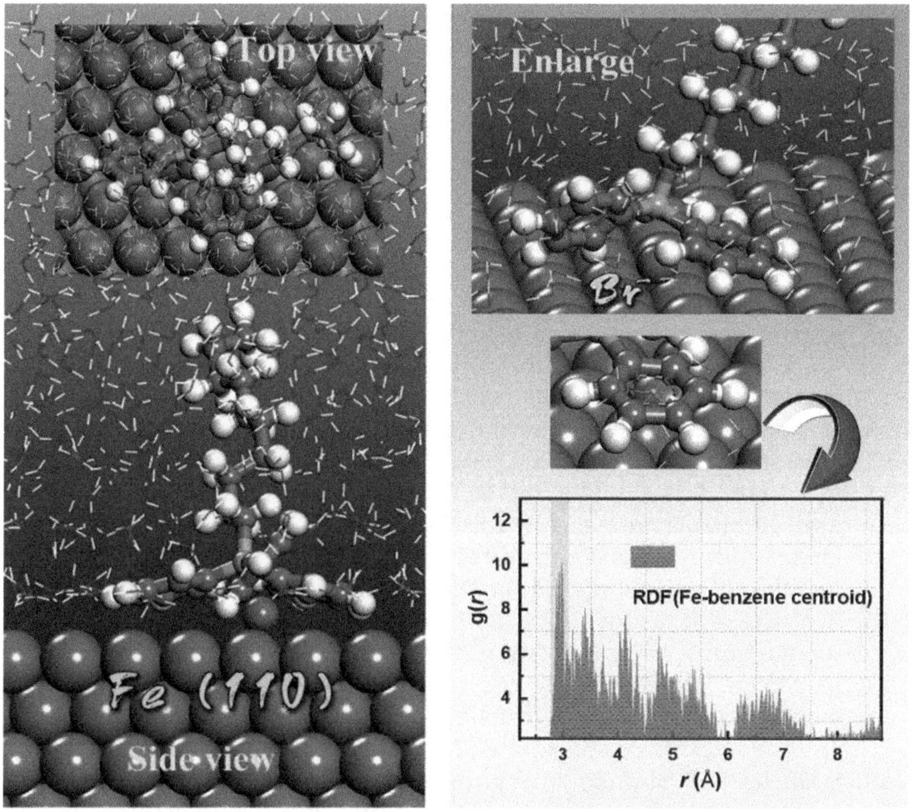

Figure 4.8 Top and side views of DTPPB on a mild steel surface and radial distribution function derived through MD simulation method.[78] Reproduced from ref. 78 with permission from Elsevier, Copyright 2021.

References

1. J. D. Holbrey and K. Seddon, Ionic liquids, *Clean Prod. Proces.*, 1999, **1**, 223–236.
2. R. D. Rogers and K. R. Seddon, Ionic liquids--solvents of the future?, *Science*, 2003, **302**, 792–793.
3. E. W. Castner Jr, C. J. Margulis, M. Maroncelli and J. F. Wishart, Ionic liquids: structure and photochemical reactions, *Annu. Rev. Phys. Chem.*, 2011, **62**, 85–105.
4. J. M. Crosthwaite, M. J. Muldoon, J. K. Dixon, J. L. Anderson and J. F. Brennecke, Phase transition and decomposition temperatures, heat capacities and viscosities of pyridinium ionic liquids, *J. Chem. Thermodyn.*, 2005, **37**, 559–568.

5. N. V. Plechkova and K. R. Seddon, Applications of ionic liquids in the chemical industry, *Chem. Soc. Rev.*, 2008, **37**, 123–150.
6. D. R. MacFarlane, N. Tachikawa, M. Forsyth, J. M. Pringle, P. C. Howlett, G. D. Elliott, J. H. Davis, M. Watanabe, P. Simon and C. A. Angell, Energy applications of ionic liquids, *Energy Environ. Sci.*, 2014, **7**, 232–250.
7. D. D. Patel and J. M. Lee, Applications of ionic liquids, *Chem. Rec.*, 2012, **12**, 329–355.
8. S. Zhu, R. Chen, Y. Wu, Q. Chen, X. Zhang and Z. Yu, A mini-review on greenness of ionic liquids, *Chem. Biochem. Eng. Q.*, 2009, **23**, 207–211.
9. M. L. Dietz and J. A. Dzielawa, Ion-exchange as a mode of cation transfer into room-temperature ionic liquids containing crown ethers: implications for the 'greenness' of ionic liquids as diluents in liquid–liquid extraction, *Chem. Commun.*, 2001, 2124–2125.
10. S. Zhang, N. Sun, X. He, X. Lu and X. Zhang, Physical properties of ionic liquids: database and evaluation, *J. Phys. Chem. Ref. Data*, 2006, **35**, 1475–1517.
11. D. Rooney, J. Jacquemin and R. Gardas, Thermophysical properties of ionic liquids, *Ionic Liq.*, 2009, 185–212.
12. M. A. Malik, M. A. Hashim and F. Nabi, Ionic liquids in supported liquid membrane technology, *Chem. Eng. J.*, 2011, **171**, 242–254.
13. A. J. Greer, J. Jacquemin and C. Hardacre, Industrial Applications of Ionic Liquids, *Molecules*, 2020, **25**, 5207.
14. M. Deyab, Understanding the anti-corrosion mechanism and performance of ionic liquids in desalination, petroleum, pickling, de-scaling, and acid cleaning applications, *J. Mol. Liq.*, 2020, **309**, 113107.
15. S. Gurjar, S. K. Sharma, A. Sharma and S. Ratnani, Performance of imidazolium based ionic liquids as corrosion inhibitors in acidic medium: A review, *Appl. Surf. Sci. Adv.*, 2021, **6**, 100170.
16. C. Verma, I. Obot, I. Bahadur, E.-S. M. Sherif and E. E. Ebenso, Choline based ionic liquids as sustainable corrosion inhibitors on mild steel surface in acidic medium: Gravimetric, electrochemical, surface morphology, DFT and Monte Carlo simulation studies, *Appl. Surf. Sci.*, 2018, **457**, 134–149.
17. C. Verma, S. H. Alrefaee, M. Quraishi, E. E. Ebenso and C. M. Hussain, Recent developments in sustainable corrosion inhibition using ionic liquids: A review, *J. Mol. Liq.*, 2021, **321**, 114484.
18. N. V. Likhanova, M. A. Domínguez-Aguilar, O. Olivares-Xometl, N. Nava-Entzana, E. Arce and H. Dorantes, The effect of ionic liquids with imidazolium and pyridinium cations on the corrosion inhibition of mild steel in acidic environment, *Corros. Sci.*, 2010, **52**, 2088–2097.
19. M. Finšgar and J. Jackson, Application of corrosion inhibitors for steels in acidic media for the oil and gas industry: A review, *Corros. Sci.*, 2014, **86**, 17–41.
20. S. H. Zaferani, M. Sharifi, D. Zaarei and M. R. Shishesaz, Application of eco-friendly products as corrosion inhibitors for metals in acid pickling processes–A review, *J. Environ. Chem. Eng.*, 2013, **1**, 652–657.
21. M. Tourabi, K. Nohair, M. Traisnel, C. Jama and F. Bentiss, Electrochemical and XPS studies of the corrosion inhibition of carbon steel

in hydrochloric acid pickling solutions by 3, 5-bis (2-thienylmethyl)-4-amino-1, 2, 4-triazole, *Corros. Sci.*, 2013, **75**, 123–133.
22. S. Abd El Haleem, S. Abd El Wanees, E. Abd El Aal and A. Farouk, Factors affecting the corrosion behaviour of aluminium in acid solutions. I. Nitrogen and/or sulphur-containing organic compounds as corrosion inhibitors for Al in HCl solutions, *Corros. Sci.*, 2013, **68**, 1–13.
23. E. Abd El Aal, S. Abd El Wanees, A. Farouk and S. Abd El Haleem, Factors affecting the corrosion behaviour of aluminium in acid solutions. II. Inorganic additives as corrosion inhibitors for Al in HCl solutions, *Corros. Sci.*, 2013, **68**, 14–24.
24. P. Kannan, A. Varghese, K. Palanisamy and A. S. Abousalem, Probing the effect of newly synthesized phenyltrimethylammonium tetrachloroaluminate ionic liquid as an inhibitor for carbon steel corrosion, *Appl. Surf. Sci. Adv.*, 2021, **6**, 100150.
25. R. Aslam, M. Mobin, M. Murmu, P. Banerjee and J. Aslam, L-Alanine methyl ester nitrate ionic liquid: synthesis, characterization and anti-corrosive application,*J. Mol. Liq.*, 2021, **334**, 116469.
26. M. Deyab, M. Zaky and M. Nessim, Inhibition of acid corrosion of carbon steel using four imidazolium tetrafluoroborates ionic liquids, *J. Mol. Liq.*, 2017, **229**, 396–404.
27. I. Lozano, E. Mazario, C. Olivares-Xometl, N. Likhanova and P. Herrasti, Corrosion behaviour of API 5LX52 steel in HCl and H_2SO_4 media in the presence of 1, 3-dibencilimidazolio acetate and 1, 3-dibencilimidazolio dodecanoate ionic liquids as inhibitors, *Mater. Chem. Phys.*, 2014, **147**, 191–197.
28. Y. Sasikumar, A. Adekunle, L. O. Olasunkanmi, I. Bahadur, R. Baskar, M. M. Kabanda, I. Obot and E. E. Ebenso, Experimental, quantum chemical and Monte Carlo simulation studies on the corrosion inhibition of some alkyl imidazolium ionic liquids containing tetrafluoroborate anion on mild steel in acidic medium,*J. Mol. Liq.*, 2015, **211**, 105–118.
29. X. Zhou, H. Yang and F. Wang, [BMIM] BF_4 ionic liquids as effective inhibitor for carbon steel in alkaline chloride solution, *Electrochim. Acta*, 2011, **56**, 4268–4275.
30. S. Cao, D. Liu, H. Ding, H. Lu and J. Gui, Towards understanding corrosion inhibition of sulfonate/carboxylate functionalized ionic liquids: An experimental and theoretical study, *J. Colloid Interface Sci.*, 2020, **579**, 315–329.
31. A. Bousskri, A. Anejjar, M. Messali, R. Salghi, O. Benali, Y. Karzazi, S. Jodeh, M. Zougagh, E. E. Ebenso and B. Hammouti, Corrosion inhibition of carbon steel in aggressive acidic media with 1-(2-(4-chlorophenyl)-2-oxoethyl) pyridazinium bromide, *J. Mol. Liq.*, 2015, **211**, 1000–1008.
32. S. M. Tawfik, Ionic liquids based gemini cationic surfactants as corrosion inhibitors for carbon steel in hydrochloric acid solution, *J. Mol. Liq.*, 2016, **216**, 624–635.

33. S. Ullah, M. A. Bustam, A. M. Shariff, G. Gonfa and K. Izzat, Experimental and quantum study of corrosion of A36 mild steel towards 1-butyl-3-methylimidazolium tetrachloroferrate ionic liquid, *Appl. Surf. Sci.*, 2016, **365**, 76–83.
34. A. Yousefi, S. Javadian and J. Neshati, A new approach to study the synergistic inhibition effect of cationic and anionic surfactants on the corrosion of mild steel in HCl solution, *Ind. Eng. Chem. Res.*, 2014, **53**, 5475–5489.
35. P. Kannan, J. Karthikeyan, P. Murugan, T. S. Rao and N. Rajendran, Corrosion inhibition effect of novel methyl benzimidazolium ionic liquid for carbon steel in HCl medium, *J. Mol. Liq.*, 2016, **221**, 368–380.
36. F. A. Azeez, O. A. Al-Rashed and A. A. Nazeer, Controlling of mild-steel corrosion in acidic solution using environmentally friendly ionic liquid inhibitors: effect of alkyl chain, *J. Mol. Liq.*, 2018, **265**, 654–663.
37. N. Subasree and J. A. Selvi, Imidazolium based ionic liquid derivatives; synthesis and evaluation of inhibitory effect on mild steel corrosion in hydrochloric acid solution, *Heliyon*, 2020, **6**, e03498.
38. B. Hammouti, S. Alamry, A. Alzahrani, Z. Moussa, M. Messali and M. Ibrahim, Corrosion inhibition of carbon steel by imidazolium and pyridinium cations ionic liquids in acidic environment, Port, Electrochim, *Acta*, 2011, **29**, 375–389.
39. L. C. Murulana, A. K. Singh, S. K. Shukla, M. M. Kabanda and E. E. Ebenso, Experimental and quantum chemical studies of some bis (trifluoromethyl-sulfonyl) imide imidazolium-based ionic liquids as corrosion inhibitors for mild steel in hydrochloric acid solution, *Ind. Eng. Chem. Res.*, 2012, **51**, 13282–13299.
40. A. Yousefi, S. Javadian, N. Dalir, J. Kakemam and J. Akbari, Imidazolium-based ionic liquids as modulators of corrosion inhibition of SDS on mild steel in hydrochloric acid solutions: experimental and theoretical studies, *RSC Adv.*, 2015, **5**, 11697–11713.
41. S. Yesudass, L. O. Olasunkanmi, I. Bahadur, M. M. Kabanda, I. Obot and E. E. Ebenso, Experimental and theoretical studies on some selected ionic liquids with different cations/anions as corrosion inhibitors for mild steel in acidic medium, *J. Taiwan Inst. Chem. Eng.*, 2016, **64**, 252–268.
42. M. E. Mashuga, L. O. Olasunkanmi, A. S. Adekunle, S. Yesudass, M. M. Kabanda and E. E. Ebenso, Adsorption, thermodynamic and quantum chemical studies of 1-hexyl-3-methylimidazolium based ionic liquids as corrosion inhibitors for mild steel in HCl, *Materials*, 2015, **8**, 3607–3632.
43. M. Ezhilarasi, B. Prabha and T. Santhi, Novel pyrazole based ionic liquid as a corrosion inhibitor for mild steel in acidic media, *Chem. Sci.*, 2015, **4**, 758–767.
44. E. Kowsari, S. Arman, M. Shahini, H. Zandi, A. Ehsani, R. Naderi, A. PourghasemiHanza and M. Mehdipour, In situ synthesis, electrochemical and quantum chemical analysis of an amino acid-derived ionic

liquid inhibitor for corrosion protection of mild steel in 1 M HCl solution, *Corros. Sci.*, 2016, **112**, 73–85.

45. L. Feng, S. Zhang, Y. Qiang, S. Xu, B. Tan and S. Chen, The synergistic corrosion inhibition study of different chain lengths ionic liquids as green inhibitors for X70 steel in acidic medium, *Mater. Chem. Phys.*, 2018, **215**, 229–241.
46. A. Zarrouk, M. Messali, H. Zarrok, R. Salghi, A. Ali, B. Hammouti, S. Al-Deyab and F. Bentiss, Synthesis, characterization and comparative study of new functionalized imidazolium-based ionic liquids derivatives towards corrosion of C38 steel in molar hydrochloric acid, *Int. J. Electrochem. Sci.*, 2012, **7**, 6998–7015.
47. M. Messali and M. Asiri, A green ultrasound-assisted access to some new 1-benzyl-3-(4-phenoxybutyl) imidazolium-based ionic liquids derivatives—Potential corrosion inhibitors of mild steel in acidic environment,*J. Mater. Environ. Sci.*, 2013, **5**, 770–785.
48. P. Arellanes-Lozada, O. Olivares-Xometl, N. V. Likhanova, I. V. Lijanova, J. R. Vargas-García and R. E. Hernández-Ramírez, Adsorption and performance of ammonium-based ionic liquids as corrosion inhibitors of steel,*J. Mol. Liq.*, 2018, **265**, 151–163.
49. A. A. Nkuna, E. D. Akpan, I. Obot, C. Verma, E. E. Ebenso and L. C. Murulana, Impact of selected ionic liquids on corrosion protection of mild steel in acidic medium: Experimental and computational studies, *J. Mol. Liq.*, 2020, **314**, 113609.
50. F. El-Hajjaji, M. Messali, A. Aljuhani, M. Aouad, B. Hammouti, M. Belghiti, D. Chauhan and M. Quraishi, Pyridazinium-based ionic liquids as novel and green corrosion inhibitors of carbon steel in acid medium: electrochemical and molecular dynamics simulation studies, *J. Mol. Liq.*, 2018, **249**, 997–1008.
51. S. Cao, D. Liu, H. Ding, J. Wang, H. Lu and J. Gui, Task-specific ionic liquids as corrosion inhibitors on carbon steel in 0.5 M HCl solution: an experimental and theoretical study, *Corros. Sci.*, 2019, **153**, 301–313.
52. D. Yang, Y. Ye, Y. Su, S. Liu, D. Gong and H. Zhao, Functionalization of citric acid-based carbon dots by imidazole toward novel green corrosion inhibitor for carbon steel, *J. Cleaner Prod.*, 2019, **229**, 180–192.
53. Y. Guo, Z. Chen, Y. Zuo, Y. Chen, W. Yang and B. Xu, Ionic liquids with two typical hydrophobic anions as acidic corrosion inhibitors, *J. Mol. Liq.*, 2018, **269**, 886–895.
54. Y. Guo, B. Xu, Y. Liu, W. Yang, X. Yin, Y. Chen, J. Le and Z. Chen, Corrosion inhibition properties of two imidazolium ionic liquids with hydrophilic tetrafluoroborate and hydrophobic hexafluorophosphate anions in acid medium,*J. Ind. Eng. Chem.*, 2017, **56**, 234–247.
55. Y. Ma, F. Han, Z. Li and C. Xia, Corrosion behavior of metallic materials in acidic-functionalized ionic liquids,*ACS Sustainable Chem. Eng.*, 2016, **4**, 633–639.

56. H. El Sayed, S. Elsaeed, H. Ashour, E. Zaki and H. El Nagy, Novel acrylamide ionic liquids as anti-corrosion for X-65 steel dissolution in acid medium: Adsorption, hydrogen evolution and mechanism, *J. Mol. Struct.*, 2018, **1168**, 106–114.
57. P. Kannan, T. S. Rao and N. Rajendran, Anti-corrosion behavior of benzimidazoliumtetrafluroborate ionic liquid in acid medium using electrochemical noise technique,*J. Mol. Liq.*, 2016, **222**, 586–595.
58. F. El-Hajjaji, E. Ech-chihbi, N. Rezki, F. Benhiba, M. Taleb, D. S. Chauhan and M. Quraishi, Electrochemical and theoretical insights on the adsorption and corrosion inhibition of novel pyridinium-derived ionic liquids for mild steel in 1 M HCl,*J. Mol. Liq.*, 2020, **314**, 113737.
59. C. Verma, L. O. Olasunkanmi, I. Bahadur, H. Lgaz, M. Quraishi, J. Haque, E.-S. M. Sherif and E. E. Ebenso, Experimental, density functional theory and molecular dynamics supported adsorption behavior of environmental benign imidazolium based ionic liquids on mild steel surface in acidic medium,*J. Mol. Liq.*, 2019, **273**, 1–15.
60. J. Wang, D. Liu, S. Cao, S. Pan, H. Luo, T. Wang, H. Ding, B. B. Mamba and J. Gui, Inhibition effect of monomeric/polymerized imidazole zwitterions as corrosion inhibitors for carbon steel in acid medium,*J. Mol. Liq.*, 2020, **312**, 113436.
61. R. Aslam, M. Mobin, I. B. Obot and A. H. Alamri, Ionic liquids derived from α-amino acid ester salts as potent green corrosion inhibitors for mild steel in 1 M HCl,*J. Mol. Liq.*, 2020, **318**, 113982.
62. E. K. Ardakani, E. Kowsari and A. Ehsani, Imidazolium-derived polymeric ionic liquid as a green inhibitor for corrosion inhibition of mild steel in 1.0 M HCl: Experimental and computational study, *Colloids Surf., A*, 2020, **586**, 124195.
63. P. Kannan, A. Varghese, K. Palanisamy and A. S. Abousalem, Evaluating prolonged corrosion inhibition performance of benzyltributylammonium tetrachloroaluminate ionic liquid using electrochemical analysis and Monte Carlo simulation,*J. Mol. Liq.*, 2020, **297**, 111855.
64. A. A. Nkuna, E. D. Akpan, I. Obot, C. Verma, E. E. Ebenso and L. C. Murulana, Impact of selected ionic liquids on corrosion protection of mild steel in acidic medium: Experimental and computational studies, *J. Mol. Liq.*, 2020, **314**, 113609.
65. S. Cao, D. Liu, H. Ding, J. Wang, H. Lu and J. Gui, Corrosion inhibition effects of a novel ionic liquid with and without potassium iodide for carbon steel in 0.5 M HCl solution: An experimental study and theoretical calculation,*J. Mol. Liq.*, 2019, **275**, 729–740.
66. P. Kannan, A. Varghese, A. T. Mathew, K. Palanisamy, A. S. Abousalem, M. Kalaiyarasan and N. Rajendran, Exploring the inhibition performance of tetrachloroferrate ionic liquid in acid environment using scanning electrochemical microscope and theoretical approaches, *Surf. Interfaces*, 2020, **20**, 100594.
67. S. M. Ali, K. M. Emran and M. Messali, Improved protection performance of modified sol-gel coatings with pyridinium-based ionic liquid

for cast iron corrosion in 0.5 M HCL solution, *Prog. Org. Coat.*, 2019, **130**, 226–234.
68. J. H. Ha, J.-H. Cho, J. H. Kim, B. W. Cho and S. H. Oh, 1-Butyl-1-methylpyrrolidinium chloride as an effective corrosion inhibitor for stainless steel current collectors in magnesium chloride complex electrolytes, *J. Power Sources*, 2017, **355**, 90–97.
69. X. Li, S. Deng and H. Fu, Three pyrazine derivatives as corrosion inhibitors for steel in 1.0 M H_2SO_4 solution, *Corros. Sci.*, 2011, **53**, 3241–3247.
70. P. Okafor, V. Osabor and E. Ebenso, Eco-friendly corrosion inhibitors: inhibitive action of ethanol extracts of *Garcinia kola* for the corrosion of mild steel in H_2SO_4 solutions, *Pigm. Resin Technol.*, 2007, **36**(5), 299–305.
71. N. Iroha and A. Hamilton-Amachree, Adsorption and anticorrosion performance of Ocimum Canum extract on mild steel in sulphuric acid pickling environment, *Am. J. Mater. Sci.*, 2018, **8**, 39–44.
72. E.-S. M. Sherif, A. T. Abbas, H. Halfa and A. El-Shamy, Corrosion of high strength steel in concentrated sulfuric acid pickling solutions and its inhibition by 3-amino-5-mercapto-1, 2, 3-triazole, *Int. J. Electrochem. Sci.*, 2015, **10**, 1777–1791.
73. M. Goyal, S. Kumar, I. Bahadur, C. Verma and E. E. Ebenso, Organic corrosion inhibitors for industrial cleaning of ferrous and non-ferrous metals in acidic solutions: A review, *J. Mol. Liq.*, 2018, **256**, 565–573.
74. Y. Qiang, S. Zhang, L. Guo, X. Zheng, B. Xiang and S. Chen, Experimental and theoretical studies of four allyl imidazolium-based ionic liquids as green inhibitors for copper corrosion in sulfuric acid, *Corros. Sci.*, 2017, **119**, 68–78.
75. X. Zeng, X. Zheng, L. Guo, Q. Xu, H. Huang and B. Tan, Three imidazole ionic liquids as green and eco-friendly corrosion inhibitors for mild steel in sulfuric acid medium, *J. Mol. Liq.*, 2021, **324**, 115063.
76. G. Tian and K. Yuan, Performance and mechanism of alkylimidazolium ionic liquids as corrosion inhibitors for copper in sulfuric acid solution, *Molecules*, 2021, **26**, 4910.
77. B. Tan, B. Xiang, S. Zhang, Y. Qiang, L. Xu, S. Chen and J. He, Papaya leaves extract as a novel eco-friendly corrosion inhibitor for Cu in H_2SO_4 medium, *J. Colloid Interface Sci.*, 2021, **582**, 918–931.
78. M. Goyal, H. Vashisht, S. H. Alrefaee, R. Jain, S. Kumar, S. Kaya, L. Guo and C. Verma, Decyltriphenylphosphonium bromide containing hydrophobic alkyl-chain as a potential corrosion inhibitor for mild steel in sulfuric acid: Theoretical and experimental studies, *J. Mol. Liq.*, 2021, **336**, 116166.
79. M. Goyal, H. Vashisht, A. Kumar, S. Kumar, I. Bahadur, F. Benhiba and A. Zarrouk, Isopentyltriphenylphosphonium bromideionic liquid as a newly effective corrosion inhibitor on metal-electrolyte interface in acidic medium: Experimental, surface morphological (SEM-EDX & AFM) and computational analysis, *J. Mol. Liq.*, 2020, **316**, 113838.

80. M. Goyal, S. Kumar, I. Bahadur, E. E. Ebenso, H. Lgaz and I.-M. Chung, Interfacial adsorption behavior of quaternary phosphonium based ionic liquids on metal-electrolyte interface: Electrochemical, surface characterization and computational approaches, *J. Mol. Liq.*, 2020, **298**, 111995.
81. M. Corrales-Luna, T. Le Manh, M. Romero-Romo, M. Palomar-Pardavé and E. M. Arce-Estrada, 1-Ethyl 3-methylimidazolium thiocyanate ionic liquid as corrosion inhibitor of API 5L X52 steel in H_2SO_4 and HCl media, *Corros. Sci.*, 2019, **153**, 85–99.
82. M. Zaky, M. Nessim and M. Deyab, Synthesis of new ionic liquids based on dicationic imidazolium and their anti-corrosion performances, *J. Mol. Liq.*, 2019, **290**, 111230.
83. M. Goyal, H. Vashisht, S. Kumar and I. Bahadur, Anti-corrosion performance of eco-friendly inhibitor (2-aminobenzyl) triphenylphosphonium bromide ionic liquid on mild steel in 0.5 M sulfuric acid, *J. Mol. Liq.*, 2018, **261**, 162–173.
84. P. D. Pancharatna, S. Lata and G. Singh, Imidazolium based ionic liquid as an efficient and green corrosion constraint for mild steel at acidic pH levels, *J. Mol. Liq.*, 2019, **278**, 467–476.
85. Y. Li, S. Zhang, Q. Ding, B. Qin and L. Hu, Versatile 4, 6-dimethyl-2-mercaptopyrimidine based ionic liquids as high-performance corrosion inhibitors and lubricants, *J. Mol. Liq.*, 2019, **284**, 577–585.
86. O. Olivares-Xometl, C. López-Aguilar, P. Herrastí-González, N. V. Likhanova, I. Lijanova, R. Martínez-Palou and J. A. Rivera-Márquez, Adsorption and corrosion inhibition performance by three new ionic liquids on API 5L X52 steel surface in acid media, *Ind. Eng. Chem. Res.*, 2014, **53**, 9534–9543.
87. R. Fuchs-Godec, The Erosion–Corrosion inhibition of AISI 431 martensitic stainless steel in 2.0 M H_2SO_4 solution using N-alkyl quaternary ammonium salts as inhibitors, *Ind. Eng. Chem. Res.*, 2010, **49**, 6407–6415.
88. N. V. Likhanova, O. Olivares-Xometl, D. Guzmán-Lucero, M. A. Domínguez-Aguilar, N. Nava, M. Corrales-Luna and M. C. Mendoza, Corrosion inhibition of carbon steel in acidic environment by imidazolium ionic liquids containing vinyl-hexafluorophosphate as anion, *Int. J. Electrochem. Sci.*, 2011, **6**, 4514–4536.
89. P. Arellanes-Lozada, V. Díaz-Jiménez, H. Hernández-Cocoletzi, N. Nava, O. Olivares-Xometl and N. V. Likhanova, Corrosion inhibition properties of iodide ionic liquids for API 5L X52 steel in acid medium, *Corros. Sci.*, 2020, **175**, 108888.
90. O. Olivares-Xometl, I. V. Lijanova, N. V. Likhanova, P. Arellanes-Lozada, H. Hernández-Cocoletzi and J. Arriola-Morales, Theoretical and experimental study of the anion carboxylate in quaternary-ammonium-derived ionic liquids for inhibiting the corrosion of API X60 steel in 1 M H_2SO_4, *J. Mol. Liq.*, 2020, 114075.
91. A. L. Chong, J. I. Mardel, D. R. MacFarlane, M. Forsyth and A. E. Somers, Synergistic corrosion inhibition of mild steel in aqueous chloride

solutions by an imidazolinium carboxylate salt, *ACS Sustainable Chem. Eng.*, 2016, **4**, 1746–1755.
92. A. Jannat, R. Naderi, E. Kowsari, H. Zandi, M. Saybani, R. Safari and A. Ehsani, Electrochemical techniques and quantum chemical analysis as tools to study effect of a dicationic ionic liquid on steel behavior in H_2SO_4, *J. Taiwan Inst. Chem. Eng.*, 2019, **99**, 18–28.
93. D. Guzmán-Lucero, O. Olivares-Xometl, R. Martínez-Palou, N. V. Likhanova, M. A. Domínguez-Aguilar and V. Garibay-Febles, Synthesis of selected vinylimidazolium ionic liquids and their effectiveness as corrosion inhibitors for carbon steel in aqueous sulfuric acid, *Ind. Eng. Chem. Res.*, 2011, **50**, 7129–7140.
94. M. Nessim, M. Zaky and M. Deyab, Three new gemini ionic liquids: synthesis, characterizations and anticorrosion applications, *J. Mol. Liq.*, 2018, **266**, 703–710.
95. A. P. Hanza, R. Naderi, E. Kowsari and M. Sayebani, Corrosion behavior of mild steel in H_2SO_4 solution with 1, 4-di [1′-methylene-3′-methyl imidazolium bromide]-benzene as an ionic liquid, *Corros. Sci.*, 2016, **107**, 96–106.
96. X. Zheng, S. Zhang, M. Gong and W. Li, Experimental and theoretical study on the corrosion inhibition of mild steel by 1-octyl-3-methylimidazolium L-prolinate in sulfuric acid solution, *Ind. Eng. Chem. Res.*, 2014, **53**, 16349–16358.
97. M. C. Luna, T. Le Manh, R. C. Sierra, J. M. Flores, L. L. Rojas and E. A. Estrada, Study of corrosion behavior of API 5L X52 steel in sulfuric acid in the presence of ionic liquid 1-ethyl 3-methylimidazolium thiocyanate as corrosion inhibitor, *J. Mol. Liq.*, 2019, **289**, 111106.
98. O. Benali, O. Cherkaoui and A. Lallam, Adsorption and corrosion inhibition of new synthesized pyridazinium-based ionic liquid on carbon steel in 0.5 M H_2SO_4, *J. Mater. Environ. Sci.*, 2015, **6**, 598–606.
99. A. Acidi, M. Hasib-ur-Rahman, F. Larachi and A. Abbaci, Ionic liquids [EMIM][BF_4], [EMIM][Otf] and [BMIM][Otf] as corrosion inhibitors for CO_2 capture applications, *Korean J. Chem. Eng.*, 2014, **31**, 1043–1048.
100. M. Hasib-ur-Rahman and F. Larachi, Prospects of using room-temperature ionic liquids as corrosion inhibitors in aqueous ethanolamine-based CO_2 capture solvents, *Ind. Eng. Chem. Res.*, 2013, **52**, 17682–17685.
101. M. Deyab, Sulfonium-based ionic liquid as an anticorrosive agent for thermal desalination units, *J. Mol. Liq.*, 2019, **296**, 111742.
102. D. Yang, M. Zhang, J. Zheng and H. Castaneda, Corrosion inhibition of mild steel by an imidazolium ionic liquid compound: the effect of pH and surface pre-corrosion, *RSC Adv.*, 2015, **5**, 95160–95170.
103. A. Yousefi, S. Javadian, M. Sharifi, N. Dalir and A. Motaee, An experimental and theoretical study of biodegradable gemini surfactants and surfactant/carbon nanotubes (CNTs) mixtures as new corrosion inhibitor, *J. Bio- Tribo-Corros.*, 2019, **5**, 82.
104. A. K. Dermani, E. Kowsari, B. Ramezanzadeh and R. Amini, Utilizing imidazole based ionic liquid as an environmentally friendly process for enhancement of the epoxy coating/graphene oxide composite corrosion resistance, *J. Ind. Eng. Chem.*, 2019, **79**, 353–363.

105. A. El-Shamy, K. Zakaria, M. Abbas and S. Z. El Abedin, Anti-bacterial and anti-corrosion effects of the ionic liquid 1-butyl-1-methylpyrrolidinium trifluoromethylsulfonate, *J. Mol. Liq.*, 2015, **211**, 363–369.
106. M. Ontiveros-Rosales, O. Olivares-Xometl, N. V. Likhanova, I. V. Lijanova, D. Guzman-Lucero and M. D. C. Mendoza-Herrera, Use of the ionic liquid trioctylmethyl ammonium dodecanedioate as a corrosion inhibitor of steel in production water, *Res. Chem. Intermed.*, 2017, **43**, 641–660.
107. E. Kowsari, M. Payami, R. Amini, B. Ramezanzadeh and M. Javanbakht, Task-specific ionic liquid as a new green inhibitor of mild steel corrosion, *Appl. Surf. Sci.*, 2014, **289**, 478–486.
108. G. Vastag, A. Shaban, M. Vraneš, A. Tot, S. Belić and S. Gadžurić, Influence of the N-3 alkyl chain length on improving inhibition properties of imidazolium-based ionic liquids on copper corrosion, *J. Mol. Liq.*, 2018, **264**, 526–533.
109. A. M. Sadanandan, P. K. Khatri, R. Saxena and S. L. Jain, Guanidine based amino acid derived task specific ionic liquids as noncorrosive lubricant additives for tribological performance, *J. Mol. Liq.*, 2020, **313**, 113527.
110. H. Su, L. Wang, Y. Wu, Y. Zhang and J. Zhang, Insight into inhibition behavior of novel ionic liquids for magnesium alloy in NaCl solution: Experimental and theoretical investigation, *Corros. Sci.*, 2020, **165**, 108410.
111. S. K. Shetty and A. N. Shetty, Eco-friendly benzimidazolium based ionic liquid as a corrosion inhibitor for aluminum alloy composite in acidic media, *J. Mol. Liq.*, 2017, **225**, 426–438.
112. P. Hu, Z. Wu, J. Wang, Y. Huang, Q. Liu and S.-F. Zhou, Corrosion inhibiting performance and mechanism of protic ionic liquids as green brass inhibitors in nitric acid, *Green Energy Environ.*, 2020, **5**, 214–222.
113. M. Deyab and Q. Mohsen, Impact of phosphonium-based ionic liquid on the corrosion control of aluminum alloy AA5052 in MED desalination plants during acid cleaning process, *J. Mol. Liq.*, 2021, **334**, 116121.

CHAPTER 5

Synthetic Green Corrosion Inhibitors

5.1 Introduction

5.1.1 Corrosion and Corrosion Inhibitors

Corrosion is a natural process that converts metals and alloys into their chemically more stable forms, including carbonates, oxides, hydroxides or sulfides.[1,2] Corrosion involves the slow but gradual decomposition of metallic structures through electrochemical and/or chemical reactions with their environment.[3–5] It is a highly damaging phenomenon; therefore various methods have been developed for its prevention and control depending upon the nature of metal and electrolyte. Some of the common corrosion prevention methods include alloying or dealloying, painting, coating, galvanization, cathodic protection and the use of corrosion inhibitors.[6,7] In the aqueous electrolytes, the use of corrosion inhibitors is one of the most valuable and admired methods of corrosion anticipation and management.[8,9] Previously, inorganic compounds such as chromates, silicates and molybdates were developed and used as effective corrosion inhibitors. Nevertheless, because of their toxicity and non-environmentally friendly nature, the anticorrosive use of these materials is highly restricted. The increasing ecological awareness throughout the world does not authorize the exploitation of these poisonous materials and they were replaced by relatively greener organic substituents.[10,11]

Literature study demonstrates that organic compounds having electron donors *i.e.* electron rich sites including functional groups (–OH, $-NH_2$, $-NR_2$, –NHR, –OR, –COOH, $-CONH_2$, –COOEt, –O–, –S–, –S–S– *etc.*) and multiple

Green Chemistry Series No. 73
Green Corrosion Inhibition: Fundamentals, Design, Synthesis and Applications
By Chandrabhan Verma, Mumtaz A. Quraishi and Dheeraj Singh Chauhan

Published by the Royal Society of Chemistry, www.rsc.org

bonds (>C=C<, >C=N–, >C=S, –C≡N, –C≡C–, >C=O *etc.*) behave as effective anticorrosive species.[12,13] It is well documented that the organic compounds become effective by adsorbing on the metallic surface either through chemisorption, or physisorption or mixed, *i.e.* physiochemisorption, mechanisms. Nevertheless, physiochemisorption, with the standard Gibbs free energy (ΔG°) values of -20 kJ mol^{-1} to -40 kJ mol^{-1}, most commonly describes the adsorption behaviour of organic corrosion inhibitors.[14] Subsequent to getting adsorbed, these compounds form a defensive shelter which isolates metallic structures and avoids corrosion. Noticeably, not all organic compounds serve as effective corrosion inhibitors; some of them are highly effective and some of them are very poor in corrosion inhibition performance. Based on their anticorrosive activity, organic compounds may be divided into poor, moderate and effective types having percentage inhibition efficiencies (%IE) of <40%, 41–71% and >71%, respectively. The %IE of organic compounds significantly depends upon their electronic structures.[15] An organic compound having a greater number of polar functional groups is expected to be a more effective corrosion inhibitor compared to an organic compound having fewer functional groups. Obviously, increases in the number of multiple bonds and aromatic rings exert a positive effect on the %IE of organic compounds. Nevertheless, different substituents affect the corrosion inhibition potential of organic compounds differently. Literature investigation reveals that electron donating/releasing functional groups exert more prominent effects compared to electron withdrawing functional groups. The effect of substituents on the corrosion inhibition effect of organic corrosion inhibitors can be assessed using Hammett substituent constant (σ) or Taft substituent constants (σ^*) for aromatic and linear compounds, respectively. Different forms of Hammett equation are presented below:[16]

$$\log\frac{K_{\mathrm{R}}}{K_{\mathrm{H}}} = \rho\sigma \tag{5.1}$$

$$\log\frac{1-\%IE_{\mathrm{R}}}{1-\%IE_{\mathrm{H}}} = \rho\sigma \tag{5.2}$$

$$\log\frac{IE_{\mathrm{R}}}{IE_{\mathrm{H}}} = \log\frac{C_{\mathrm{rH}}}{C_{\mathrm{rR}}} = \rho\sigma - \log\frac{\theta_{\mathrm{R}}}{\theta_{\mathrm{H}}} \tag{5.3}$$

All physical symbols have their usual meaning as described before.[16,17] Generally, the substituents with a more negative value of σ^* (or σ) are joined with a greater electron donating ability and therefore higher %IE compared to the substituents with relatively more positive values of σ^* (or σ). The Hammett substituent constant values of some common functional groups can be found elsewhere.

Our research team worked on the effect of hydroxyl (–OH) as electron donating and nitro ($-NO_2$) as electron withdrawing substituents on the corrosion inhibition potential of synthetic organic compounds.[18,19] In these studies, we observed that compared to non-substituted compounds, –OH and $-NO_2$ substituents exert positive and negative effects, respectively. During the investigation of the %IE of the 2-amino-4-arylquinoline-3-carbonitriles (AACs) for mild steel/1 M HCl system, we observed that inhibition efficiencies of these compounds followed the sequence: AAC-3 (–OH) > AAC-2 (–H) > AAC-1 (–NO_2).[18] These compounds behave as mixed-type corrosion inhibitors and their adsorption on the metallic surface follows the Langmuir adsorption isotherm model. Density functional theory (DFT) based study shows that the –OH group increases the contribution of HOMO and/or LUMO whereas the –NO_2 group minimizes the HOMO and/or LUMO contributions (as compared to the non-substituted compound; AAC-1). The optimized HOMO and LUMO molecular orbitals of AACs are presented in Figure 5.1.

A similar observation was derived while studying the corrosion inhibition effects of 2,4-Diamino-5-(phenylthio)-5*H*-chromeno [2,3-*b*]pyridine-3-carbonitriles (DHPCs) for mild steel/1 M HCl system.[19] Various experimental, surface and computational investigations were adopted to prove the inhibition potential of the investigated DHPCs. Through experimental and computational studies it was derived that the %IE of DHPCs followed the order: DHPC-1 (–NO_2) < DHPC-2 (–H) < DHPC-3 (–OH). All studied DHPCs behaved as mixed-type corrosion inhibitors, and they become effective by inhibiting anodic and cathodic Tafel reactions. Their adsorption on the metallic surface followed the Langmuir adsorption isotherm model. DFT study indicated that the contributions of both HOMO and LUMO was greatly reduced by –NO_2 compared to the non-substituted and hydroxyl substituted DHPC molecules.

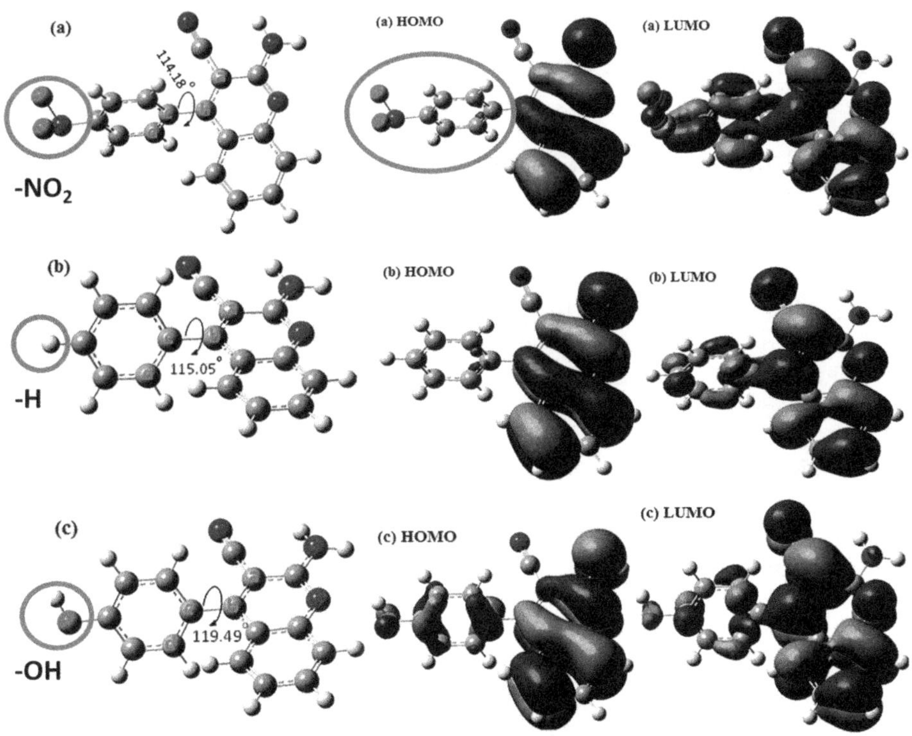

Figure 5.1 Optimized, HOMO and LUMO molecular orbital picture of (a) AAC-1 (–NO_2), (b) AAC-2 (–H) and (c) AAC-3 (–OH).[18] Reproduced from ref. 18 with permission from the Royal Society of Chemistry.

While demonstrating the corrosion inhibition effect of 5-Arylpyrimido-[4,5-*b*]quinoline-diones (APQDs), it was also derived that their inhibition efficiency also increases on increasing the number of electron releasing –OH groups. The greater the number of –OH groups, the greater the %IE. The %IE of the studied APQDs molecules followed the sequence: APQD-4 (2(–OH)) > APQD-3 (–OH) > APQD-2 (–H) > APQD-1 (–NO_2).[20] It can be seen from the frontier molecular orbital picture that the contributions of HOMO and LUMO were minimum in APQD-1 whereas the HOMO and LUMO contributions were maximum in APQD-4 (Figure 5.2). Through molecular dynamic simulations (MDS), it was derived that in the presence of the electron withdrawing –NO_2 group, a major fraction of the APQD-1 molecule acquired a vertical orientation and provided a minimum surface coverage and protection. On the other hand, in the absence of –NO_2 and the presence of electron releasing –OH group(s), APQD molecules acquire a flat orientation and provide maximum surface coverage and protection (Figure 5.3). Therefore, it can be concluded that substituents play a significant role while determining the anticorrosive effect of synthetic corrosion inhibitors. They affect electron sharing capability as well as the orientation of the corrosion inhibitor molecules. Organic compounds derived from different synthetic methods are extensively used as corrosion inhibitors for different metals and alloys in numerous electrolytes.

5.2 Synthetic Green Corrosion Inhibitors: Literature Survey

5.2.1 Green Corrosion Inhibitors Derived from Multicomponent Reactions (MCRs)

Multicomponent reactions (MCRs) are characterized by the combination of more than two reactants to offer exceedingly selective products that preserve the majority of the atoms of the starting materials.[21,22] Historically, MCRs have been documented for over 150 years. The condensation of ammonia (NH_3), benzaldehyde (Ph-CHO) and hydrogen cyanide (HCN) to form aminocyanide is documented as the first MCR. The synthesis of aminocyanide was documented by Laurent and Gerhardt (1838).[23] Literature investigation shows that different series of heterocyclic compounds, especially imidazole, quinoline, indole, pyridine, isoquinoline *etc.*, are synthesized using MCRs. Some well-known MCRs include Kabachnik–Fields, alkyne trimerization and polymerization, Grieco three-component, Strecker synthesis, Biginelli, Gewald, Bucherer–Bergs, the Hantzsch reaction, the Mannich reaction, Passerini, Petasis, Ugi synthesis, Pauson–Khand and Asinger synthesis, *etc.* (Figure 5.4).[24] MCRs are associated with the following advantages (Figure 5.4):[21,22,25]

(i) Ease of handling and experimental simplicity
(ii) Rapid access to large libraries of heterocyclics

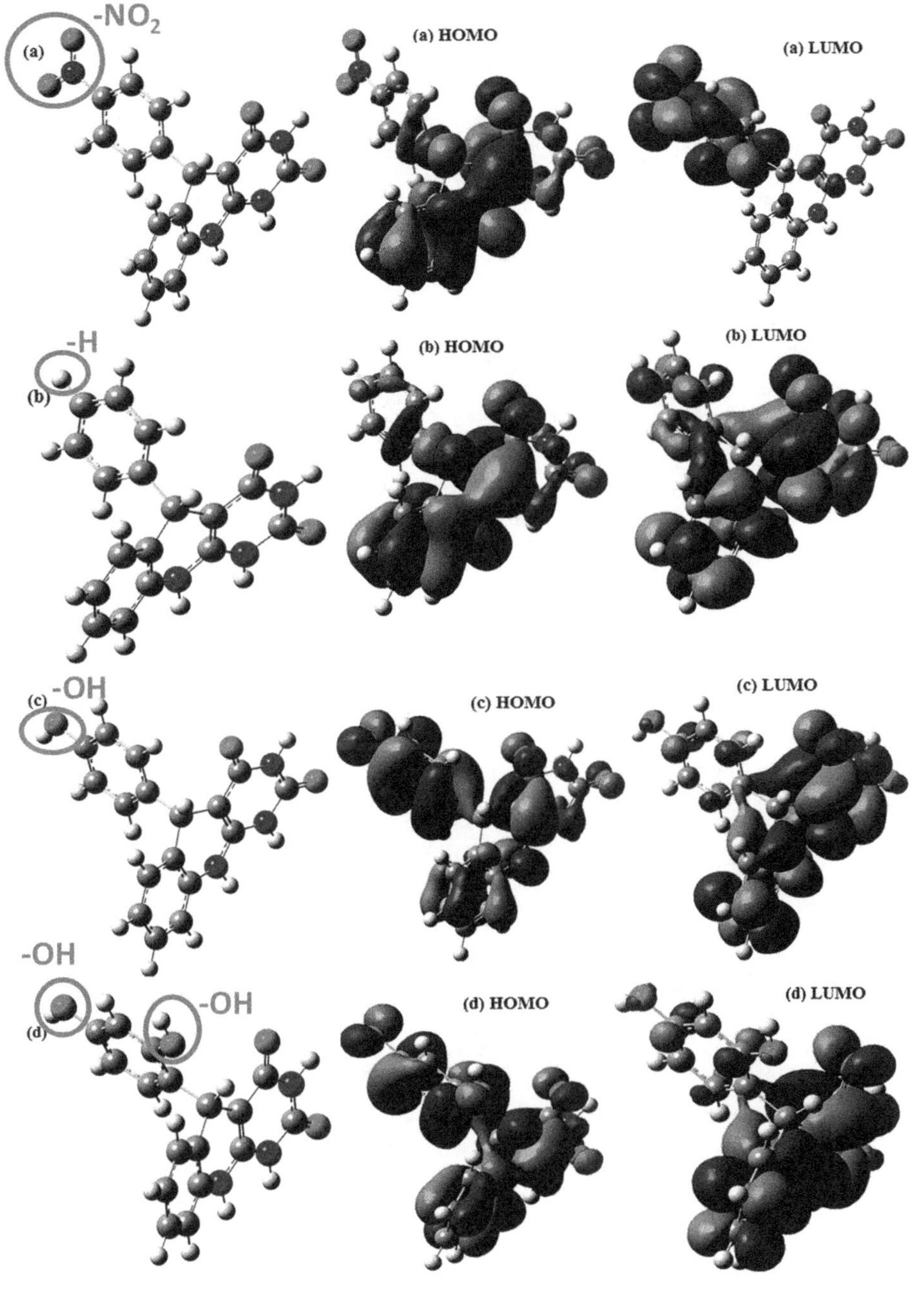

Figure 5.2 Optimized, HOMO and LUMO frontier molecular orbital picture of (a) APQD-1 (–NO_2), (b) APQD-2 (–H), (c) APQD-3 (–OH) and (d) APQD-4 (2(–OH)).[20] Reproduced from ref. 20 with permission from the Royal Society of Chemistry.

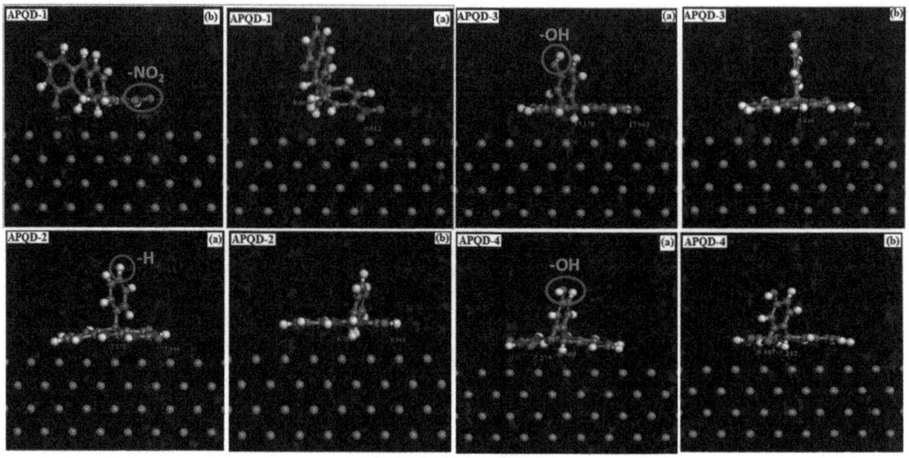

Figure 5.3 Most stable orientations of (a) APQD-1, (b) APQD-2, (c) APQD-3 and (d) APQD-4 derived from molecular dynamic simulations.[20] Reproduced from ref. 20 with permission from the Royal Society of Chemistry.

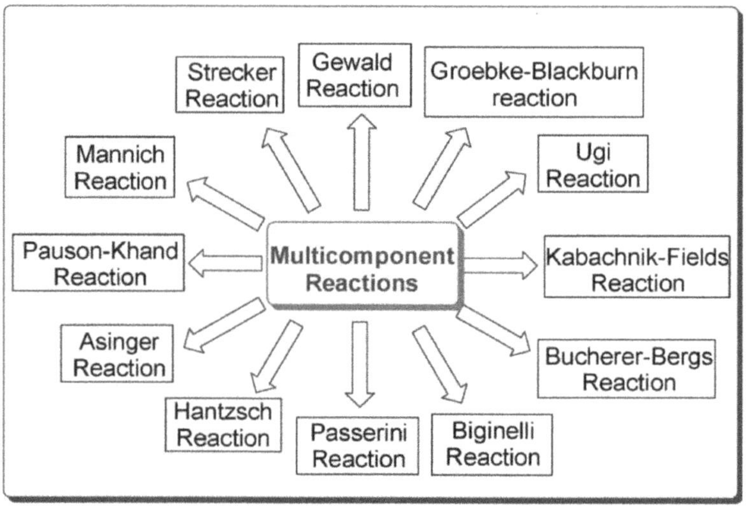

Figure 5.4 Different MCRs.[24] Reproduced from ref. 24 with permission from the Royal Society of Chemistry.

(iii) Eco-friendly (low waste production)
(iv) High synthetic yield and atom economy
(v) Reduced number of purification and work-up steps
(vi) Exceptionally high chemo- and regio-selectivity
(vii) Solvent-friendly *i.e.* MCRs can easily be assessed in common solvents, including alcohols, water and amines, *etc.*

(viii) Useful for the synthesis of simple to complex organic/heterocyclic molecules
(ix) Utilize cheap and readily available starting materials
(x) Facile automation
(xi) Cost effective and resource and time saving

Literature study suggests that organic compounds, particularly heterocyclic compounds, derived from MCRs are extensively used as corrosion inhibitors for different metal/electrolyte systems.[26,27] These compounds assessed from MCRs become effective by adsorbing on the metallic surface, mostly following the Langmuir adsorption isotherm model. Through their adsorption, compounds derived through MCRs block the active sites responsible for corrosion. Potentiodynamic polarization investigation shows that these compounds retard both anodic and cathodic reactions and act as mixed-type corrosion inhibitors, though some anodic or cathodic predominance has also been observed in a few cases. The adsorption mechanism of corrosion inhibition using these compounds is greatly demonstrated using electrochemical impedance spectroscopy (EIS) study where a significant increase in the charge transfer resistance is observed in the presence of corrosion inhibitors. Obviously, after adsorption, the corrosion inhibitors impose a barrier for the charge transfer process. Various surface investigations including scanning electron microscopy (SEM), energy dispersive X-ray (EDX), atomic force microscopy (AFM), X-ray diffraction (XRD) and X-ray photoelectron spectroscopy (XPS) *etc.* are also widely used to support the adsorption of these compounds on the metallic surface.

A summary of some major reports on heterocyclics derived from MCRs as corrosion inhibitors is in Table 5.1. It can be seen that in these reports the effects of substituents are greatly studied. Generally, the presence of electron donating substituents increases the corrosion inhibition effectiveness and electron withdrawing substituents exert adverse effects. Using weight loss, electrochemical, surface and DFT analyses, our research team demonstrate that %IE of 2-amino-4-arylquinoline-3-carbonitriles (AACs) for mild steel corrosion in acidic hydrochloric acid solution followed the order: AAC-3 (–OH) > AAC-2 (–H) > AAC-1 (–NO_2).[18] This observation is also demonstrated in other studies. Weight loss and electrochemical studies indicate that protection effectiveness of AACs was concentration and temperature dependent. Increases in the AACs concentrations causes subsequent increases in the %IE, until the optimum concentration is observed. However, temperature exerts the opposite effect, *i.e.* an increase in temperature causes an increase in the corrosion rate. AACs become effectual by adsorbing at the interface of mild steel and 1 M HCl. SEM and AFM studies indicated that the presence of the AACs in the corrosive media causes a significant decrease in the surface roughness, which is attributed to their adsorption and the formation of a corrosion protective film. As discussed above, as derived through DFT analyses, in AAC-3, the presence of –OH increases HOMO and LUMO contributions and in AAC-1 the presence of –NO_2 decreases the HOMO and LUMO contributions.

Table 5.1 Summary of some major reports on heterocyclic compounds derived from one step multicomponent reactions.

Synthetic scheme(s)	%IE	Optimum Conc.	Electrolyte/ Metal/alloys	Reference
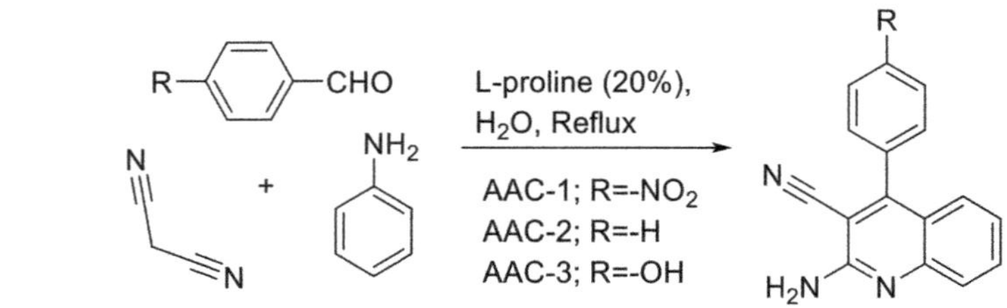	96.95% (AAC-3)	50 ppm	1 M HCl/Mild steel	18
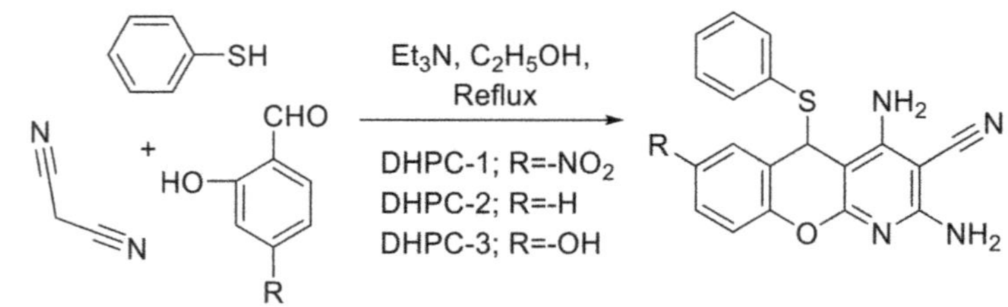	95.69% (DHPC-3)	10.22×10^{-5} M	1 M HCl/Mild steel	19
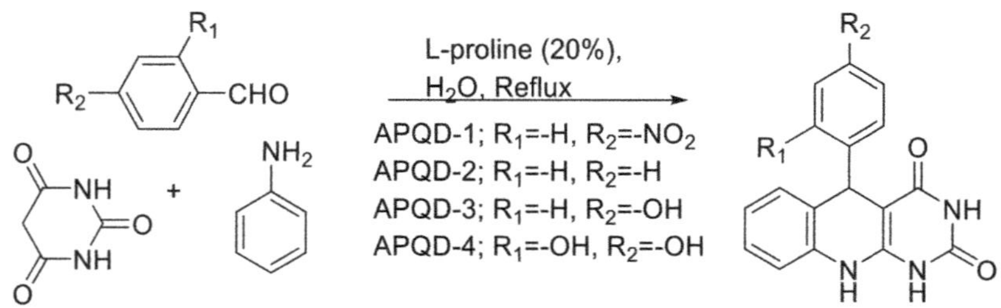	98.30% (APQD-4)	20 mg L^{-1}	1 M HCl/Mild steel	20

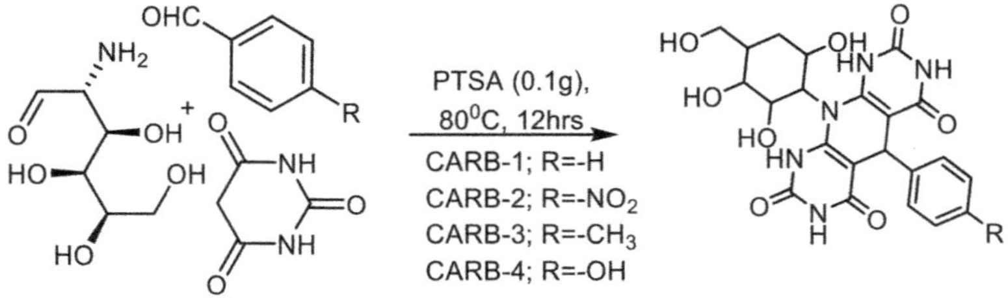	96.52% (CARB-4)	7.41 × 10−5 M	1 M HCl/Mild steel	13
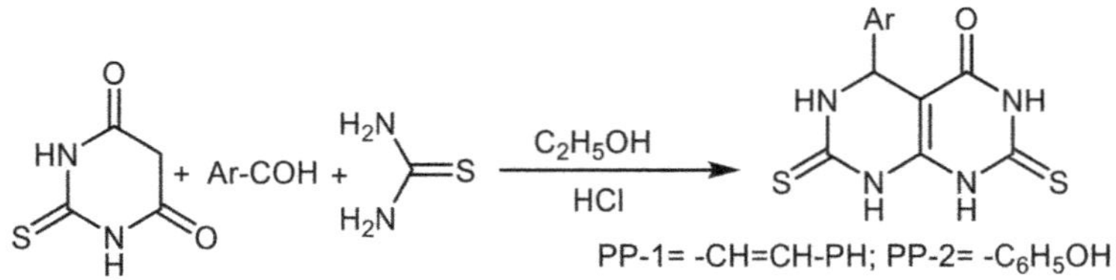	89.1%/(PP-1)	(PP-1)%IE	15% HCl/N80	28
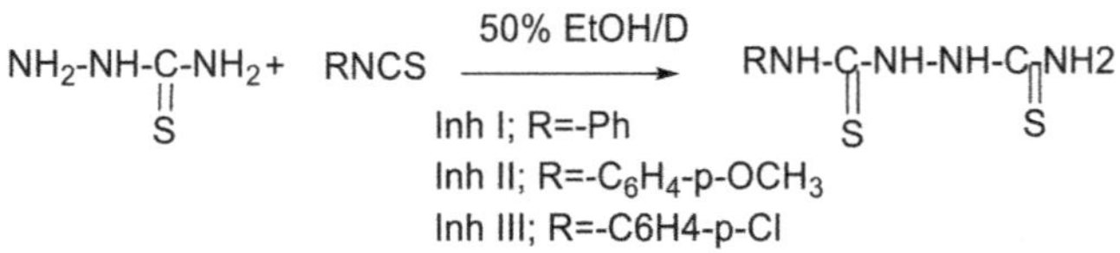	95.6% (BMBD)	200 ppm	3.5% NaCl & 1 M HCl/ Copper	29
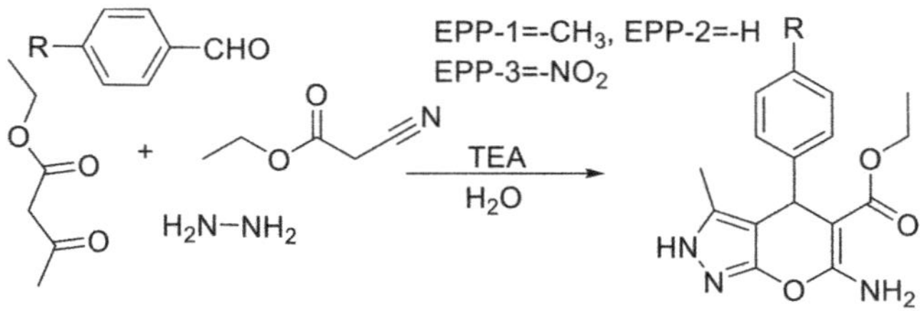	98.8% (EPP-1)	100 mg L^{-1}	1 M HCl/Mild steel	30

(*continued*)

Table 5.1 (*continued*)

Synthetic scheme(s)	%IE	Optimum Conc.	Electrolyte/ Metal/alloys	Reference
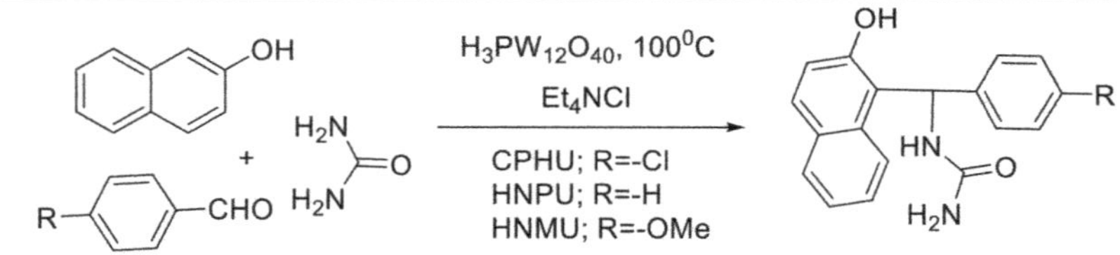	90% (CPHU)	80 ppm	0.5 M HCl/ Mild steel	31
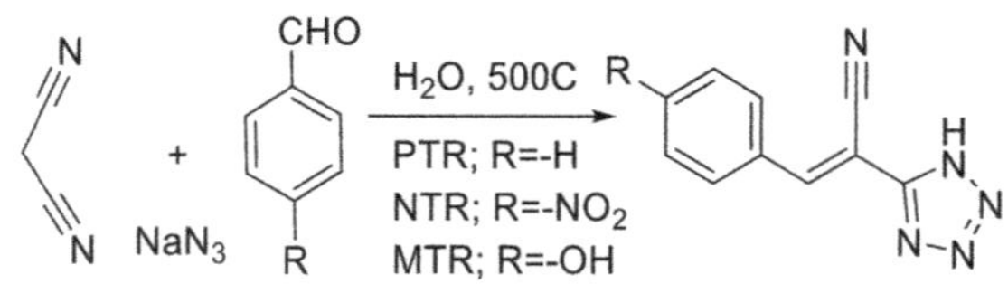	98.69% (HTA)	40 mg L^{-1}	1 M HCl/Mild steel	32
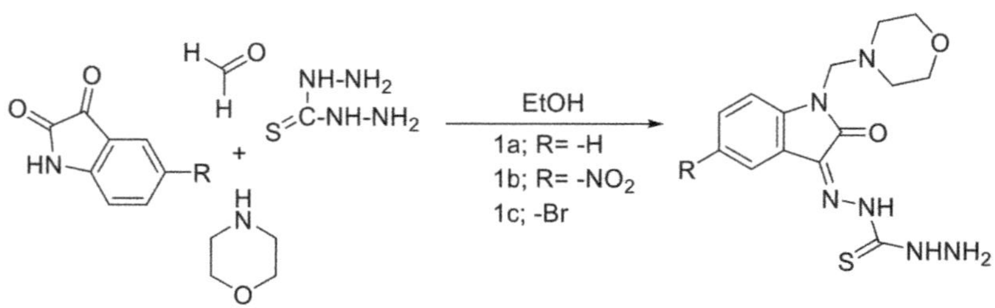	94.67% (5-BMOH)	200 ppm	1 M HCl/Mild steel	33
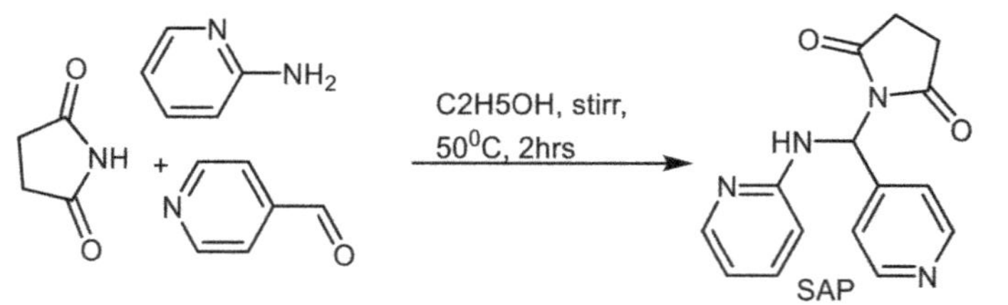	94.5% (SAP)	200 ppm	1 M HCl/Mild steel	34

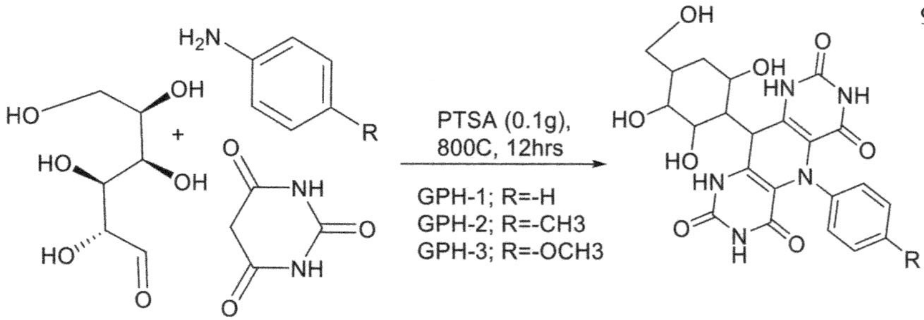	97.82% (GPH-3)	10.15×10^{-5} M	1 M HCl/Mild steel	35
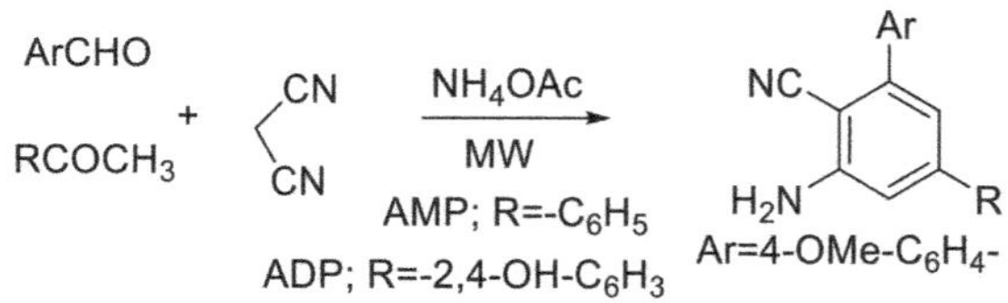	90.3% (ADP)	200 ppm	15% HCl/N80	36
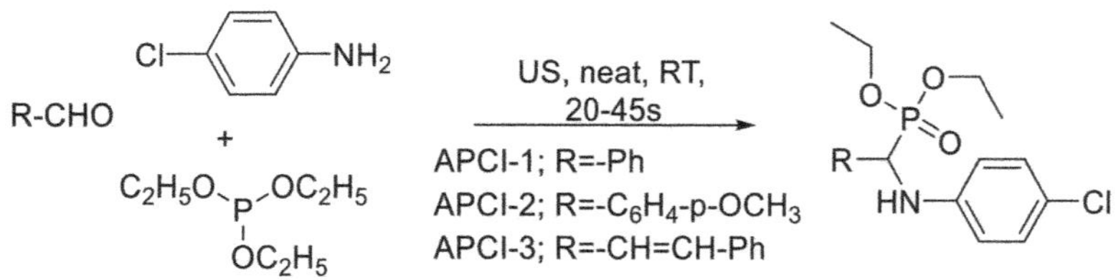	96.90% (APCl-3)	564×10^{-6} M	1 M HCl/Mild steel	37
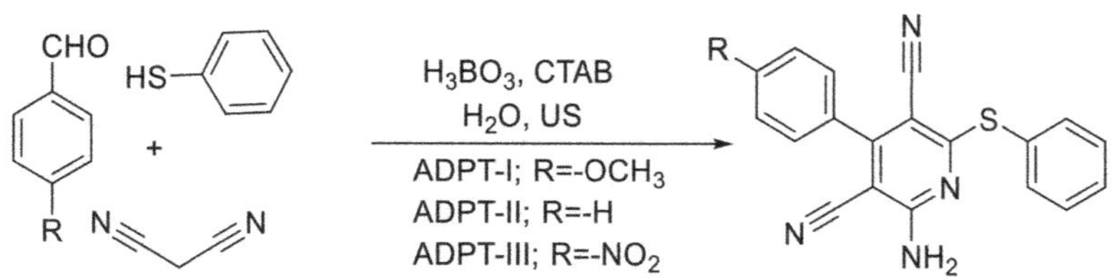	97.6% (ADTP-1	1.22 mmol L^{-1})	1 M HCl/Mild steel	38

(*continued*)

Table 5.1 (*continued*)

Synthetic scheme(s)	%IE	Optimum Conc.	Electrolyte/ Metal/alloys	Reference
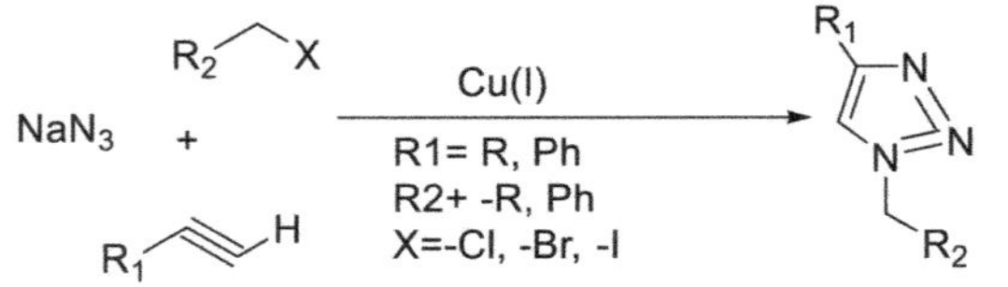	95.9%	10 ppm	1 M HCl/API 5 L X52 steel	39
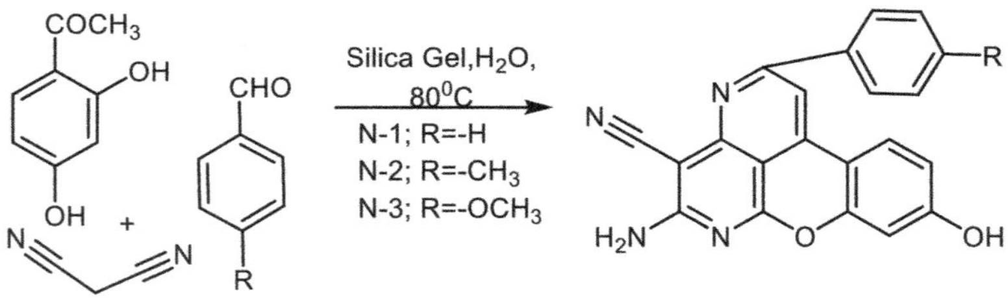	98.09% (N-3)	6.54×10^{-5} M	1 M HCl/Mild steel	40
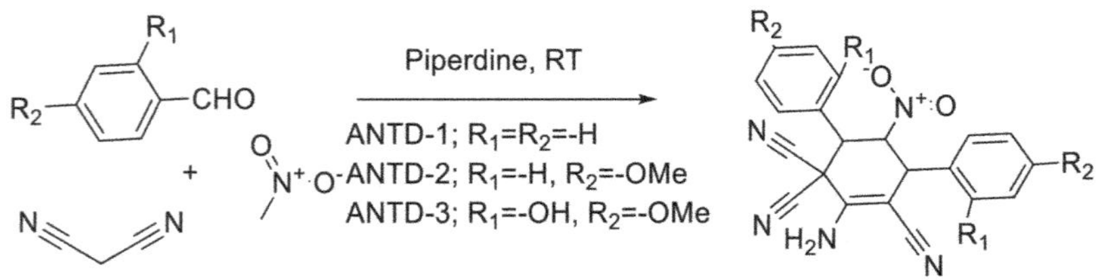	98.96% (ANDT-3)	25 mg L^{-1}	1 M HCl/Mild steel	14
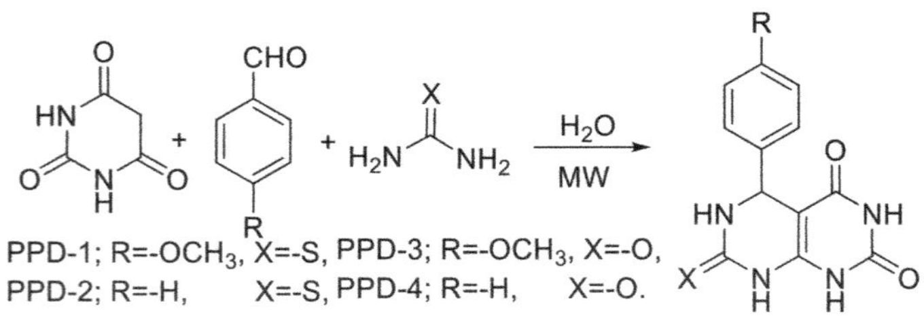	97.1% (PDP-1)	400 ppm	1 M HCl/Mild steel	41

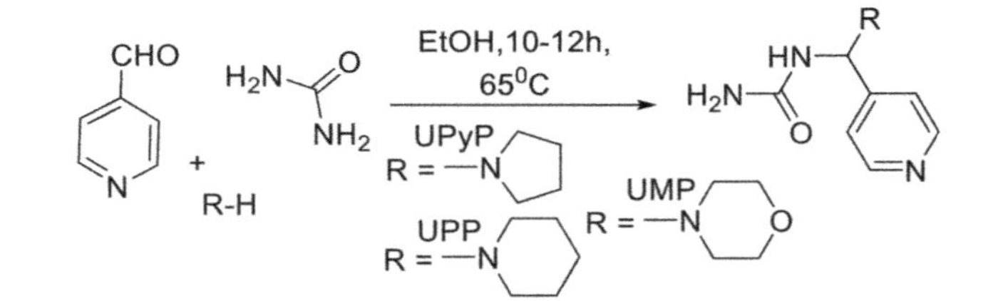	86% (UPP)	200 ppm	1 M HCl/Mild steel	42
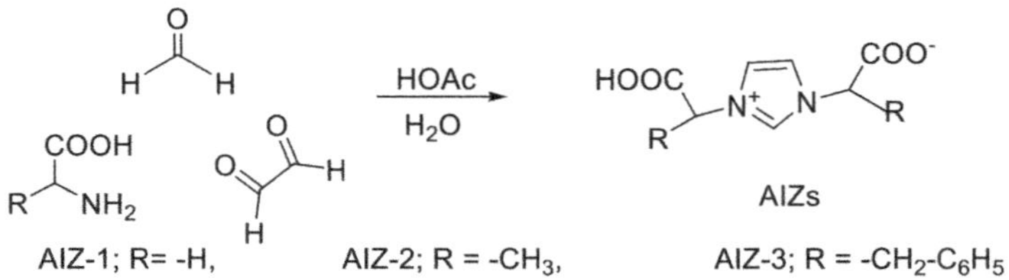	96.08% (AIZ-3)	0.55 mM/200 ppm	1 M HCl/Mild steel	43
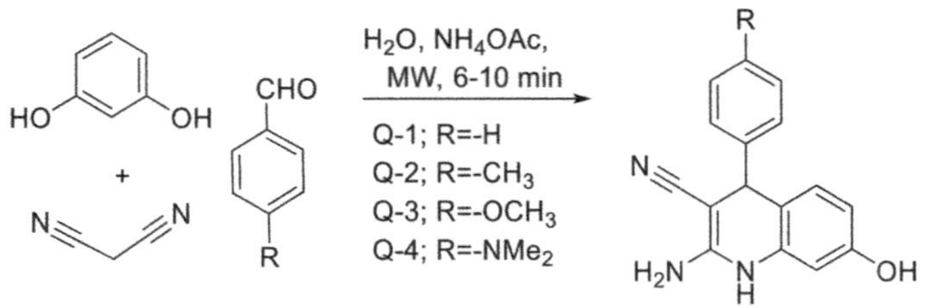	98.09% (Q-4)	150 mg L^{-1}	1 M HCl/Mild steel	44
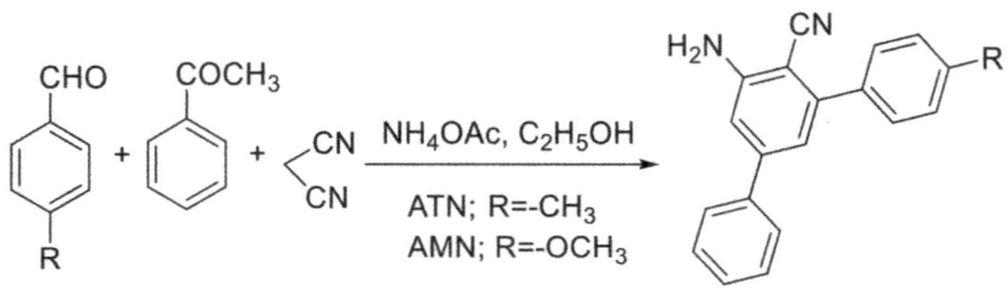	97.14% (AMN)	0.33 mM	1 M HCl/Mild steel	45

(continued)

Table 5.1 *(continued)*

Synthetic scheme(s)	%IE	Optimum Conc.	Electrolyte/ Metal/alloys	Reference
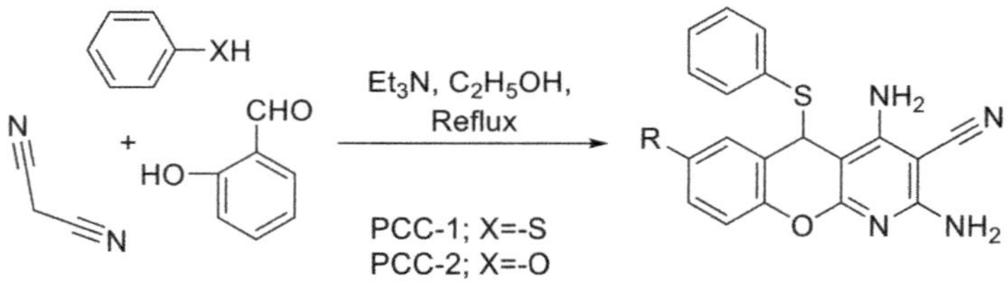	92.4 (PPC-1)	200 mg L^{-1}	15% HCl/N-80 steel	46
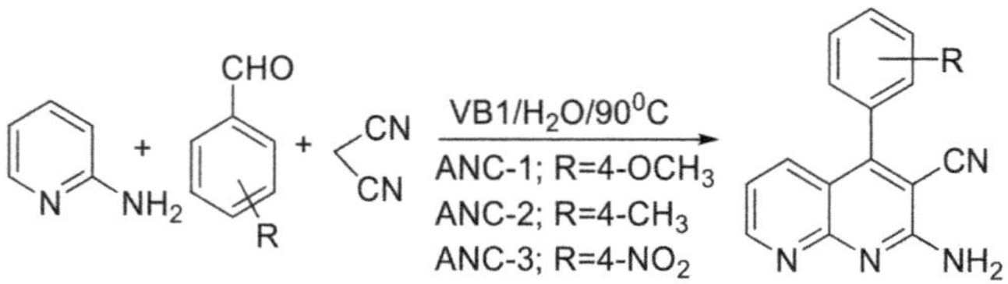	93% (ANC-1)	200 ppm	1 M HCl/Mild steel	47
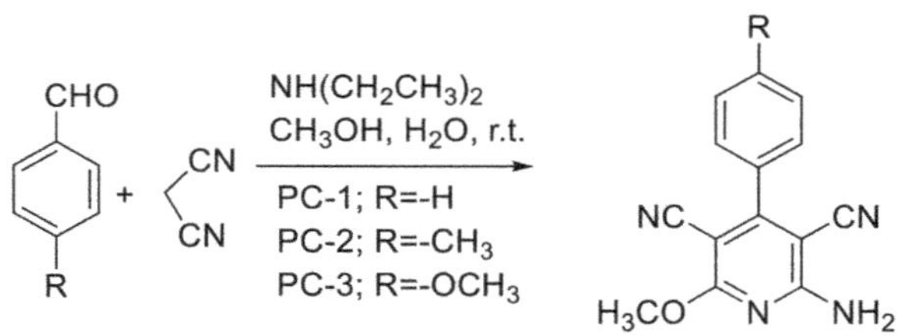	97.4% (PC-1)	400 mg L^{-1}	1 M HCl/Mild steel	48

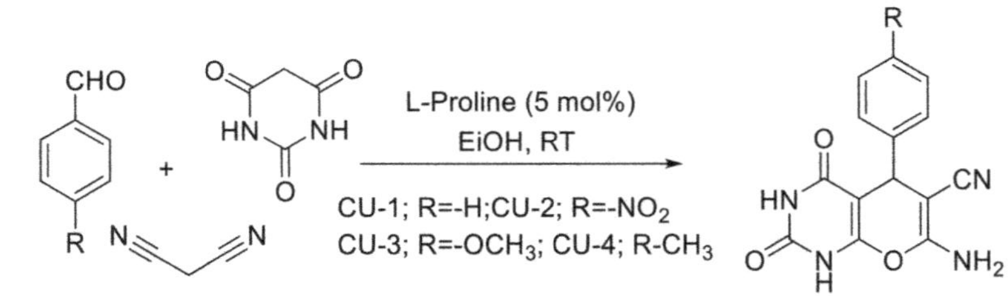	96.1% (CU-3)	450 ppm/14.4 × 10^{-4} M	1 M HCl/Mild steel	49
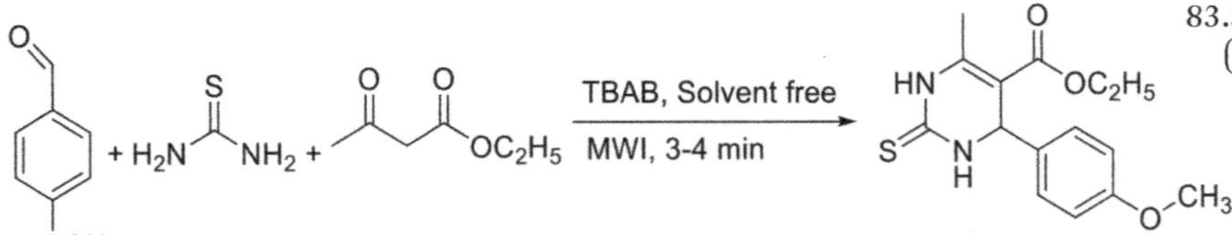 INH	83.33% (INH)	500 ppm	1 M HCl/Al	50

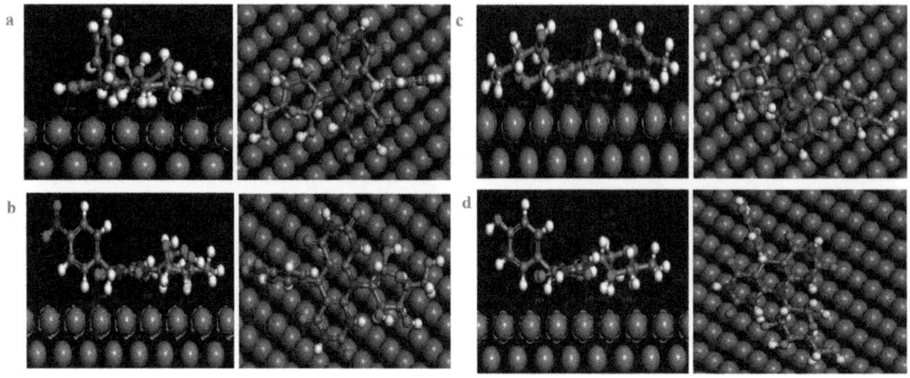

Figure 5.5 Top and side views of most stable configurations of (a) CARB-1, (b) CARB-2, (c) CARB-3 and (d) CARB-4 derived from MC simulations.[13] Reproduced from ref. 13 with permission from American Chemical Society, Copyright 2016.

Nevertheless, this is not true in all cases, as sometimes the presence of an electron withdrawing $-NO_2$ substituent enhances the %IE compared to structurally similar compounds without any substituents. For example, during studying the anticorrosive effect of glucosamine-based, pyrimidine-fused heterocycles (CARBs) for mild steel in 1 M HCl, our team observed that CARBs containing both electron donating (–OH and –Me) and withdrawing ($-NO_2$) substituents exhibit better anticorrosive effects compared to the compounds without any substituents.[13] In this study, the authors proposed that the substituents effect was observed in the medium sized molecules. In either too big or too small molecules, the substituents effect is not observed and other factors such as solubility and the number of adsorption sites come into play. Weight loss, electrochemical, surface, DFT and MC (Monte Carlo) simulations studies suggest the following order of CARBs %IE: CARB-1 (–H) < CARB ($-NO_2$) < CARB-3 ($-CH_3$) < CARB-4 (–OH). From this order of %IE it can be concluded that both electron donating and withdrawing groups increase the %IE. However, this increase is more pronounced in the presence of electron donating groups. These compounds act as mixed-type corrosion inhibitors and their presence increases the values of charge transfer resistance. DFT study suggests that $-NO_2$ groups decrease the HOMO/LUMO contribution compared to $-CH_3$ and –OH groups. MC simulations indicate that the major portion of CARB-1 acquires a vertical orientation with the plane of the metal surface (Figure 5.5). Therefore, it provides little surface coverage and protection compared to the other CARB molecules.

5.3 Green Corrosion Inhibitors Derived from Microwave and Ultrasound

Conventional hot plate heating for chemical transformations is associated with various shortcomings, including slow and non-uniform heating, less efficiency, low selectivity, long duration, tedious work-up, utilization of toxic solvents for purification and low synthetic yield.[51] The increasing ecological awareness and strict environmental rules do not permit the use of conventional hot plate heating for chemical transformation. Therefore, recently some new heating approaches such as microwave (MW) and ultrasound (US) have been developed for the activation of reactant molecules.[52,53] In the last 35 years, MW and US based nonconventional heating has been widely used for various chemical transformations.[52,53] Numerous well-known chemical reactions are successfully catalysed using MW and US heating (Figure 5.6). Nonconventional heating is associated with plenty of benefits, including the following:[54–56]

(i) Rapid and instantaneous heating and activation of reactant molecules
(ii) Uniform heating and activation
(iii) Extremely high reaction rate (sometimes reaction rate increases up to 1000 fold)
(iv) High synthetic yield and atom economy
(v) Minimized waste production
(vi) Facile automation
(vii) High selectivity and purity of developed compounds
(viii) Low operating cost and time saving
(ix) Avoids unwanted heating of the equipment
(x) Avoids undesired heating of the reaction vessel
(xi) Minimized environmental heating
(xii) Reduction in number of purification and work-up steps

Outcomes of the literature investigation show that organic compounds, principally heterocyclic compounds derived from microwave and ultrasound heating, are extensively used as corrosion inhibitors for different metals and alloys in various electrolytes.[57,58] A more detailed description of

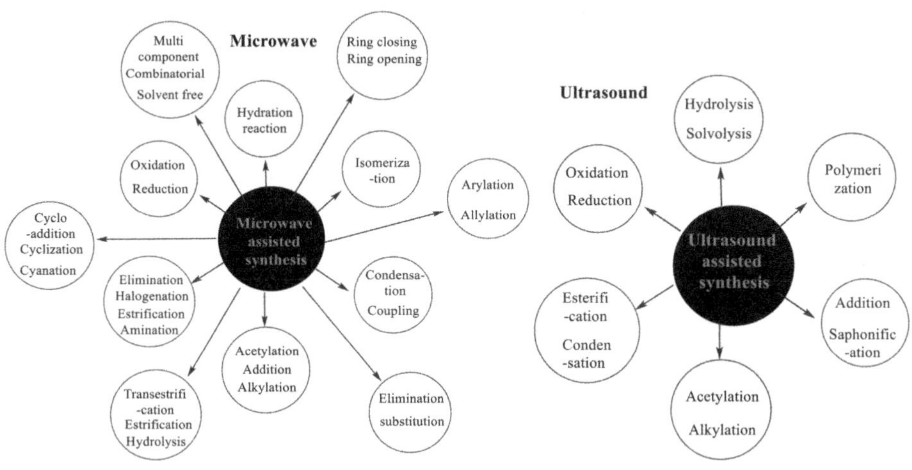

Figure 5.6 Some representative examples of microwave (left-hand-side) and ultrasound (right-hand-side) catalysed reactions.

the corrosion inhibition potential of compounds synthesized from MW and US irradiation can be found elsewhere in our previous reports.[11,59] Similar to traditional organic corrosion inhibitors, compounds synthesized from US and MW behave as mixed-type corrosion inhibitors and they become effective by blocking the active sites through their adsorption. They retard both anodic and cathodic Tafel reactions after getting adsorbed. EIS study indicates that these compounds adsorb at the interface of metal and electrolyte and increase the value of charge transfer resistance. Adsorption behaviour of corrosion inhibition using organic compounds developed through MW and US irradiations is highly reinforced with surface analyses including SEM, AFM, EDX, XPS and XRD *etc.* DFT study shows that these compounds interact with the metallic surface using the charge sharing mechanism and their %IE is substituents dependent. Obviously, electron releasing groups such as $-NH_2$, –NHMe, $-NMe_2$, –OMe and –OH have positive effects on the corrosion inhibition effectiveness, whereas electron withdrawing substituents such as –CN, –COOH and $-NO_2$ have the opposite effect.

5.4 Green Corrosion Inhibitors Derived Using Green Solvents

Solvents are indispensable element of synthetic chemistry.[60] In pharmaceutical and fine-chemical production based industries, solvents are used in large amounts.[61,62] Solvents are essential candidates for chemical industries. Most of the common and known solvents used in synthetic chemistry are toxic and non-ecofriendly and cause various toxicity related problems to the environment and living organisms after their discharge. The physical properties of some common solvents are presented in Table 5.2. The growing demands of green chemistry retard the utilization of such toxic solvents and therefore, recently, green solvents have been developed and used as solvents for the synthesis of industrially and biologically useful organic compounds.[63–65] These solvents are called eco-friendly, green or environmentally sustainable solvents. Obviously, these solvents have either no or very little adverse effect on living organisms and to the environment. Solvents are used in huge quantities, especially in pharmaceutical and fine-chemical production. The following six points can be made behind the selection of green solvents:

(a) Application of bio-solvents
(b) Replacement of toxic solvents with less hazardous solvents
(c) Replacement of organic solvents with supercritical fluids
(d) Replacement of organic solvents with ionic liquids
(e) Replacement of organic solvents with water

In view of the above, traditional hydrocarbon, oxygenated and hydrogenated solvents can be replaced by relatively more eco-friendly solvents such

Table 5.2 Physical properties of some common solvents.

Solvents	bp °C	Density	Dipole moment	Water solubility	mp °C	Relative polarity	Dielectric constant	Viscosity (1/1000 Pa s.)
Acetonitrile	81.6	0.786	3.5	M	−46	0.460	37.5	0.34
Diethylamine	56.3	0.706	1.2	M	−48	0.145	3.8	0.32
2-Aminoethanol	170.9	1.018	2.4	M	10.5	0.651	37.7	20.8
2-Butanol	99.5	0.808	1.7	18.1	−114.7	0.506	17.3	3.1
Benzonitrile	205	0.996	4.1	0.2	−13	0.333	26	1.27
t-Butyl alcohol	82.2	0.786	1.7	M	25.5	0.389	12.4	3.35
Chloroform	61.2	1.498	1.0	0.8	−63.5	0.259	4.8	0.54
Aniline	184.4	1.022	1.6	3.4	−6.0	0.420	6.8	3.8
Acetone	56.2	0.786	2.85	M	−94.3	0.355	21	0.30
Dimethoxyethane	85	0.868	1.7	M	−58	0.231	7.3	1.1
Iso-butanol	107.9	0.803	1.8	8.5	−108.2	0.552	17.9	6.68
Di-*n*-butylphthalate	340	1.049		0.0011	−35	0.272		16.6
Cyclohexane	80.7	0.779	0	0.005	−16.4	0.006	2	0.89
Diglyme	162	0.945	1.9	M	−64	0.244	7.23	1.88
Benzyl alcohol	205.4	1.042	1.7	3.5	−15.3	0.608	13	5.47
1,2-Dichloroethane	83.5	1.235	1.8	0.87	−35.4	0.327	10.4	0.78
Carbon tetrachloride	76.7	1.594	0	0.08	−22.4	0.052	2.3	0.90
Diethylene glycol	245	1.118	2.3	M	−10	0.713	31.8	30.2
N,*N*-dimethylaniline	194.2	0.956	4.40	0.14	2,4	0.179	1.61	1.40
Ehylene glycol	197	1.115	2.3	M	−13	0.790	37.7	16.1
Ethyl acetoacetate	180.4	1.028		2.9	−80	0.577		
Anisole	153.7	0.996	1.4	0.10	−37.5	0.198	4.3	0.98
Pyridine	115.5	0.982	2.2	M	−42	0.302	13	0.88
Acetic acid	118	1.049	1.68	M	16.6	0.648	6.2	1.12
Hexane	69	0.655	0	0.0014	−95	0.009	1.9	0.29
1,1-Dichloroethane	57.3	1.176	1.8	0.5	−97	0.269	10	0.84
Dimethyl phthalate	283.8	1.190		0.43	1	0.309		

(*continued*)

Table 5.2 (*continued*)

Solvents	bp °C	Density	Dipole moment	Water solubility	mp °C	Relative polarity	Dielectric constant	Viscosity (1/1000 Pa s.)
1-Heptanol	176.4	0.819	1.7	0.17	−35	0.549	12	6.0
Cyclohexanone	155.6	0.948	2.9	2.3	−16.4	0.281	15	2.00
Pentane	36.1	0.626	0	0.004	−129.7	0.009	1.84	0.23
Cyclohexanol	161.1	0.962	1.9	4.2	25.2	0.509	15	
2-Pentanone	102.3	0.809	2.7	4.3	−76.9	0.321	15.4	0.50
Carbon disulfide	46.3	1.263	0	0.2	−111.6	0.065	2.6	0.36
Acetyl acetone	140.4	0.975	3.0	16	−23	0.571	23	0.6
2-Butanone	79.6	0.805	2.8	25.6	−86.3	0.327	18.6	0.41
Chlorobenzene	132	1.106	1.54	0.05	−45.6	0.188	5.7	0.75
1-Butanol	117.6	0.81	1.7	7.7	−89.5	0.586	17.5	2.59
Benzene	80.1	0.879	0	0.18	5.1	0.111	2.3	0.60
2-Pentanol	119.0	0.810	1.7	4.5	−50	0.488	13.7	3.5
Ethanol	78.5	0.789	1.7	M	−114.1	0.654	24	1.08
3-Pentanol	115.3	0.821	1.7	5.1	−8	0.463	13.3	
Toluene	110.6	0.867	0.36	0.05	−93	0.099	2.4	0.55
Dimethylformamide	153	0.944	3.8	M	−61	0.386	37	0.80
1-Hexanol	158	0.814	1.60	0.59	−46.7	0.559	12.5	0.59
Water	100.00	0.998	1.85	M	0.00	1.000	80.1	0.89
Methylene chloride	39.8	1.326	1.6	1.32	−96.7	0.309	9.0	0.42
3-Pentanone	101.7	0.814	2.8	3.4	−39.8	0.265	17.0	0.44
1-Octanol	194.4	0.827	1.7	0.096	−15	0.537	10.3	7.4
Methanol	64.6	0.791	1.6	M	−98	0.762	33	0.54
Water, heavy	101.3	1.107	1.84	M	4	0.991	787.3	1.10
1-Propanol	97	0.803	1.68	M	−126	0.617	22	1.95
p-Xylene	138.3	0.861	0	0.02	13.3	0.074	2.27	0.65
Heptanes	98	0.684	0	0.0003	−90.6	0.012	1.9	0.39
Ethyl benzoate	213	1.047	2.0	0.07	−34.6	0.228	6.0	
Methyl *t*-butyl ether	55.2	0.741	1.4	4.8	−109	0.124	2.6	0.36

Glycerin	290	1.261	2.7	M	17.8	0.812	42.5	934
Dimethyl sulfoxide	189	1.092	3.9	M	18.4	0.444	46.7	2.00
Ether	34.6	0.713	1.25	7.5	−116.3	0.117	4.3	0.22
Ethyl acetate	77	0.894	1.78	8.7	−83.6	0.228	6.0	0.43
Tetrahydrofuran	66	0.886	1.63	30 or M (7)	−108.4	0.207	7,5	0.46
Dioxane	101.1	1.033	0.4	M	11.8	0.164	2	1.18
Methyl acetate	56.9	0.933	1.7	24.4	−98.1	0.253	6.68	0.36
2-Propanol	82.4	0.785	1.66	M	−88.5	0.546	19	2.07
1-Pentanol	138.0	0.814	1.7	2.2	−78.2	0.568	14	3.5

as water, polyethylene glycol (PEG), supercritical CO_2 and ionic liquids. It is important to mention that every solvent has its own benefits and shortcoming as the "best solvent is no solvent". Nevertheless, they are highly toxic, therefore their recent application is highly restricted owing to increasing demands for green solvents. Literature study suggests water, supercritical CO_2 and ionic liquids are extensively used as solvents for a variety of chemical transformations.[66–70]

Water, ionic liquids, PEG and supercritical CO_2 are associated with various advantages over the traditional organic solvents that make them ideal candidates to be used as corrosion inhibitors. The advantages and shortcomings of some major green solvents are tabulated in Table 5.3. Compounds synthesized in these solvents are extensively used as corrosion inhibitors for different metals and alloys in various electrolytic media.[71] They act as mixed-type corrosion inhibitors and they retard both anodic as well cathodic reactions through their adsorption at the active sites. This type of adsorption results in the subsequent increase in the value of charge transfer resistance. Adsorption of these compounds on a metallic surface is greatly supported by surface analyses especially through SEM, EDX, AFM, XPS and XRD techniques. More so, ILs and PEG (polyethylene glycol) themselves are used as effective corrosion inhibitors for different metal/electrolyte systems. Obviously, they acquire very high protection efficiencies compared to traditional organic compounds. A detailed description of the corrosion inhibition potential of ionic liquids and PEG can be found elsewhere in our previous reports.[72,73] The relatively high corrosion inhibition potential of PEG is attributed to its polymeric nature, while the extremely high %IE of ionic liquids is attributed to their high chelating ability and high solubility in polar electrolytes.

5.5 Synthetic Green Corrosion Inhibitors: Miscellaneous

Along with compounds synthesized using one step multicomponent reactions (MCRs), microwave and ultrasound heating and using green solvents, the organic/heterocyclic compounds derived from previously existing green chemicals such as carbohydrates and amino acids can also be regarded as green corrosion inhibitors.[12,13] Literature study suggests that heterocyclic compounds derived from carbohydrates and amino acids are widely used as corrosion inhibitors for various systems.[74–77] These compounds act as mixed- and interface-type corrosion inhibitors and their presence increases the value of charge transfer resistance and decreases the value of current density. They become effective by adsorbing on the metallic surface using their electron rich sites, called adsorption sites. The adsorption mechanism of these compounds on the metallic surface is greatly studied by surface investigations. Computational demonstrations have been used to describe the nature and bonding between metal surfaces and inhibitor molecules as well as the

Table 5.3 Properties, advantages and shortcomings of some major green solvents.

Solvent	Properties	Advantages	Shortcomings
Water	Highest density: 1000 kg m^{-3} (4 °C). MP: 0 °C (273.15 K or 32 °F), BP: 99.98 °C (373.13 K or 211.73 °F) pK_a (acidity) value: 15.74 & Dipole moment (1.85 D)	(i) Excellent solubilizing capacity (ii) High dielectric constant (iii) Elevated heat capacity (iv) High dipole moment (v) High liquidus array (0–100 °C) (vi) High heat of vaporization (vii) High adhesive & cohesive forces (viii) Liquid is denser than solid form (ix) Non-flammable (x) Nontoxic (xi) Cost effective & commercially available (xii) Bio-renewable (xiii) CO_2 dissolving capability (xiv) Unique redox potential	(i) Energy intensive (ii) A lot of chemicals are water sensitive (iii) Steam is difficult to treat (iv) Solvent use is limited to polar compounds (v) Poor solvent for organics (vi) It may change the acidity and basicity of the re-agents
Supercritical CO_2	Critical temperature: 31 °C, 87.8 °F, 304.13 K Critical pressure: 73.8 bar, 1070 psi, 72.8 atm., 7.3773 MPa	(i) Relatively cheaper (ii) Nontoxic and eco-friendly (iii) Ease recycling and recovery (iv) Gases (*e.g.* H_2) solubilizing tendency (v) Well-known liquidus range (vi) Non-flammable (vii) Low viscosity (viii) Fast diffusion	(i) Expensive instrumenta-tion (ii) Liquid rage only at high partial pressure (iii) Relatively poor solvent (iv) Susceptible to reacting with strong nucleophiles (amines)

(*continued*)

Table 5.3 (*continued*)

Solvent	Properties	Advantages	Shortcomings
Ionic liquids	Liquid below 100 °C	(i) Very low vapour pressure (ii) Ease synthesis (iii) ILs can act as reagents, solvents and catalysts (iv) Designer chemicals *i.e.* their properties can be tuned (v) High chemical stability (vi) High liquidus range (vii) High solubilizing capability (viii) Recyclability (ix) High thermal stability (x) Soluble in organic and aqueous solvents (xi) Behave as strong ligands	(i) Synthesis and purification are expensive and toxic (ii) Non-biodegradable (iii) Purification requires toxic organic solvents. (iv) Synthesis of ILs mainly utilizes toxic haloalkanes

orientation of the inhibitor molecules on the metallic surface. Literature outcomes suggest that these compounds mainly acquire a flat orientation and thereby cover the larger metallic surface area and provide excellent surface protection.

5.6 Summary

From the ongoing discussion, it can be seen that there are numerous forms of synthetic organic compounds developed and implemented as green or environmentally sustainable corrosion inhibitors. Some of the common protocols for the synthesis of green corrosion inhibitors include one step multicomponent reactions (MCRs) and nonconventional microwave (MW) and ultrasound (US) heating. Besides these, compounds derived from previously existing environmentally sustainable chemicals, including carbohydrates and amino acids are also treated as green compounds to be implemented as green corrosion inhibitors. Water, ionic liquids (ILs), supercritical CO_2 and polyethylene are treated as green solvents because of their various beneficial properties. Therefore, organic compounds synthesized in these solvents can also be used as environmentally sustainable corrosion inhibitors. In view of the greenness of ILs and PEG, they are also utilized as green corrosion inhibitors in various metal/alloy-electrolyte systems. Obviously, these compounds become effectual by building a protective covering at the metal-electrolyte interfaces through their adsorption. Adsorption of these compounds mostly follows the physiochemisorption mode, which is consistent with ΔG° values of -20 kJ mol^{-1} to -40 kJ mol^{-1}, though in some cases pure physisorption or chemisorption have also been observed. Electrochemically, these compounds behave as mixed- and interface-type corrosion inhibitors as, through their adsorption, they block the active sites and impose a barrier to the charge transfer process. Their adsorption on the metallic surface is reinforced with numerous surface investigations, including SEM, XRD, XPS, EDX and AFM *etc.* The mechanism of interactions of these compounds with the metallic surface is greatly studied by computational studies mainly through DFT, MDS and MCS methods.

References

1. M. Kendig and R. Buchheit, Corrosion inhibition of aluminum and aluminum alloys by soluble chromates, chromate coatings, and chromate-free coatings, *Corrosion*, 2003, **59**, 379–400.
2. L. L. Shreir, *Corrosion: Metal/environment Reactions*, Newnes, 2013.
3. J. Kruger, Electrochemistry of corrosion, *Electrochemistry Encyclopedia*, 2001.
4. E. E. Stansbury and R. A. Buchanan, *Fundamentals of Electrochemical Corrosion*, ASM International, 2000.

5. R. G. Kelly, J. R. Scully, D. Shoesmith and R. G. Buchheit, *Electrochemical Techniques in Corrosion Science and Engineering*, CRC Press, 2002.
6. R. S. Razavi, *Recent Researches in Corrosion Evaluation and Protection*, BoD–Books on Demand, 2012.
7. M. Aliofkhazraei, *Developments in Corrosion Protection*, BoD–Books on Demand, 2014.
8. M. Goyal, S. Kumar, I. Bahadur, C. Verma and E. E. Ebenso, Organic corrosion inhibitors for industrial cleaning of ferrous and non-ferrous metals in acidic solutions: A review, *J. Mol. Liq.*, 2018, **256**, 565–573.
9. B. Sanyal, Organic compounds as corrosion inhibitors in different environments—a review, *Prog. Org. Coat.*, 1981, **9**, 165–236.
10. N. Chaubey, A. Qurashi, D. S. Chauhan and M. Quraishi, Frontiers and advances in green and sustainable inhibitors for corrosion applications: A critical review, *J. Mol. Liq.*, 2021, **321**, 114385.
11. C. Verma, E. E. Ebenso, M. Quraishi and C. M. Hussain, Recent developments in sustainable corrosion inhibitors: design, performance and industrial scale applications, *Mater. Adv.*, 2021, **2**, 3806–3850.
12. S. A. Umoren and U. M. Eduok, Application of carbohydrate polymers as corrosion inhibitors for metal substrates in different media: a review, *Carbohydr. Polym.*, 2016, **140**, 314–341.
13. C. Verma, L. O. Olasunkanmi, E. E. Ebenso, M. A. Quraishi and I. B. Obot, Adsorption behavior of glucosamine-based, pyrimidine-fused heterocycles as green corrosion inhibitors for mild steel: experimental and theoretical studies, *J. Phys. Chem. C*, 2016, **120**, 11598–11611.
14. C. Verma, M. Quraishi and A. Singh, 2-Amino-5-nitro-4, 6-diarylcyclohex-1-ene-1, 3, 3-tricarbonitriles as new and effective corrosion inhibitors for mild steel in 1 M HCl: Experimental and theoretical studies, *J. Mol. Liq.*, 2015, **212**, 804–812.
15. C. Verma, M. Quraishi and K. Y. Rhee, Electronic effect *vs.* molecular size effect: Experimental and computational based designing of potential corrosion inhibitors, *Chem. Eng. J.*, 2021, **430**, 132645.
16. C. Verma, L. Olasunkanmi, E. E. Ebenso and M. Quraishi, Substituents effect on corrosion inhibition performance of organic compounds in aggressive ionic solutions: a review, *J. Mol. Liq.*, 2018, **251**, 100–118.
17. C. Verma, E. E. Ebenso and M. Quraishi, Molecular structural aspects of organic corrosion inhibitors: influence of–CN and–NO_2 substituents on designing of potential corrosion: inhibitors for aqueous media, *J. Mol. Liq.*, 2020, **316**, 113874.
18. C. Verma, M. Quraishi, L. Olasunkanmi and E. E. Ebenso, L-Proline-promoted synthesis of 2-amino-4-arylquinoline-3-carbonitriles as sustainable corrosion inhibitors for mild steel in 1 M HCl: experimental and computational studies, *RSC Adv.*, 2015, **5**, 85417–85430.
19. C. Verma, L. O. Olasunkanmi, I. Obot, E. E. Ebenso and M. Quraishi, 2, 4-Diamino-5-(phenylthio)-5 H-chromeno [2, 3-b] pyridine-3-carbonitriles as green and effective corrosion inhibitors: gravimetric, electrochemical, surface morphology and theoretical studies, *RSC Adv.*, 2016, **6**, 53933–53948.

20. C. Verma, L. Olasunkanmi, I. Obot, E. E. Ebenso and M. Quraishi, 5-Arylpyrimido-[4, 5-*b*] quinoline-diones as new and sustainable corrosion inhibitors for mild steel in 1 M HCl: a combined experimental and theoretical approach, *RSC Adv.*, 2016, **6**, 15639–15654.
21. J. Zhu and H. Bienaymé, *Multicomponent Reactions*, John Wiley & Sons, 2006.
22. B. H. Rotstein, S. Zaretsky, V. Rai and A. K. Yudin, Small heterocycles in multicomponent reactions, *Chem. Rev.*, 2014, **114**, 8323–8359.
23. I. Ugi and S. Heck, The multicomponent reactions and their libraries for natural and preparative chemistry, *Comb. Chem. High Throughput Screening*, 2001, **4**, 1–34.
24. M. S. Singh and S. Chowdhury, Recent developments in solvent-free multicomponent reactions: a perfect synergy for eco-compatible organic synthesis, *RSC Adv.*, 2012, **2**, 4547–4592.
25. R. C. Cioc, E. Ruijter and R. V. Orru, Multicomponent reactions: advanced tools for sustainable organic synthesis, *Green Chem.*, 2014, **16**, 2958–2975.
26. C. Verma, J. Haque, M. Quraishi and E. E. Ebenso, Aqueous phase environmental friendly organic corrosion inhibitors derived from one step multicomponent reactions: a review, *J. Mol. Liq.*, 2019, **275**, 18–40.
27. C. Verma and M. A. Quraishi, *Green Corrosion Inhibitors Derived from Synthesis: Progress and Future Directions, Sustainable Corrosion Inhibitors II: Synthesis, Design, and Practical Applications*, ACS Publications, 2021, pp. 121–147.
28. J. Haque, K. Ansari, V. Srivastava, M. Quraishi and I. Obot, Pyrimidine derivatives as novel acidizing corrosion inhibitors for N80 steel useful for petroleum industry: a combined experimental and theoretical approach, *J. Ind. Eng. Chem.*, 2017, **49**, 176–188.
29. S. Kumar, D. Sharma, P. Yadav and M. Yadav, Experimental and quantum chemical studies on corrosion inhibition effect of synthesized organic compounds on N80 steel in hydrochloric acid, *Ind. Eng. Chem. Res.*, 2013, **52**, 14019–14029.
30. M. Yadav, L. Gope, N. Kumari and P. Yadav, Corrosion inhibition performance of pyranopyrazole derivatives for mild steel in HCl solution: Gravimetric, electrochemical and DFT studies, *J. Mol. Liq.*, 2016, **216**, 78–86.
31. M. Bahrami, S. Hosseini and P. Pilvar, Experimental and theoretical investigation of organic compounds as inhibitors for mild steel corrosion in sulfuric acid medium, *Corros. Sci.*, 2010, **52**, 2793–2803.
32. C. Verma, M. Quraishi and A. Singh, 5-Substituted 1*H*-tetrazoles as effective corrosion inhibitors for mild steel in 1 M hydrochloric acid, *J. Taibah Univ. Sci.*, 2016, **10**, 718–733.
33. A. K. Singh and M. Quraishi, Inhibiting effects of 5-substituted isatin-based Mannich bases on the corrosion of mild steel in hydrochloric acid solution, *J. Appl. Electrochem.*, 2010, **40**, 1293–1306.
34. M. Jeeva, M. susai Boobalan and G. V. Prabhu, Adsorption and anticorrosion behavior of 1-((pyridin-2-ylamino)(pyridin-4-yl) methyl) pyrrolidine-2, 5-dione on mild steel surface in hydrochloric acid solution, *Res. Chem. Intermed.*, 2018, **44**, 425–454.

35. C. Verma, M. A. Quraishi, K. Kluza, M. Makowska-Janusik, L. O. Olasunkanmi and E. E. Ebenso, Corrosion inhibition of mild steel in 1 M HCl by D-glucose derivatives of dihydropyrido [2, 3-*d*: 6, 5-*d*′] dipyrimidine-2, 4, 6, 8 (1*H*, 3*H*, 5*H*, 7*H*)-tetraone, *Sci. Rep.*, 2017, **7**, 44432.
36. K. Ansari, M. Quraishi and A. Singh, Pyridine derivatives as corrosion inhibitors for N80 steel in 15% HCl: Electrochemical, surface and quantum chemical studies, *Measurement*, 2015, **76**, 136–147.
37. N. K. Gupta, C. Verma, R. Salghi, H. Lgaz, A. Mukherjee and M. Quraishi, New phosphonate based corrosion inhibitors for mild steel in hydrochloric acid useful for industrial pickling processes: experimental and theoretical approach, *New J. Chem.*, 2017, **41**, 13114–13129.
38. M. A. Quraishi, 2-Amino-3, 5-dicarbonitrile-6-thio-pyridines: new and effective corrosion inhibitors for mild steel in 1 M HCl, *Ind. Eng. Chem. Res.*, 2014, **53**, 2851–2859.
39. R. González-Olvera, V. Román-Rodríguez, G. E. Negrón-Silva, A. Espinoza-Vázquez, F. J. Rodríguez-Gómez and R. Santillan, Multicomponent synthesis and evaluation of new 1, 2, 3-triazole derivatives of dihydropyrimidinones as acidic corrosion inhibitors for steel, *Molecules*, 2016, **21**, 250.
40. P. Singh, E. E. Ebenso, L. O. Olasunkanmi, I. Obot and M. Quraishi, Electrochemical, theoretical, and surface morphological studies of corrosion inhibition effect of green naphthyridine derivatives on mild steel in hydrochloric acid,*J. Phys. Chem. C*, 2016, **120**, 3408–3419.
41. K. Ansari, Sudheer, A. Singh and M. Quraishi, Some pyrimidine derivatives as corrosion inhibitor for mild steel in hydrochloric acid,*J. Dispersion Sci. Technol.*, 2015, **36**, 908–917.
42. M. Jeeva, G. V. Prabhu, M. S. Boobalan and C. M. Rajesh, Interactions and inhibition effect of urea-derived Mannich bases on a mild steel surface in HCl,*J. Phys. Chem. C*, 2015, **119**, 22025–22043.
43. V. Srivastava, J. Haque, C. Verma, P. Singh, H. Lgaz, R. Salghi and M. Quraishi, Amino acid based imidazolium zwitterions as novel and green corrosion inhibitors for mild steel: Experimental, DFT and MD studies, *J. Mol. Liq.*, 2017, **244**, 340–352.
44. P. Singh, V. Srivastava and M. Quraishi, Novel quinoline derivatives as green corrosion inhibitors for mild steel in acidic medium: electrochemical, SEM, AFM, and XPS studies,*J. Mol. Liq.*, 2016, **216**, 164–173.
45. P. Singh, M. Makowska-Janusik, P. Slovensky and M. Quraishi, Nicotinonitriles as green corrosion inhibitors for mild steel in hydrochloric acid: Electrochemical, computational and surface morphological studies, *J. Mol. Liq.*, 2016, **220**, 71–81.
46. K. Ansari, M. Quraishi and A. Singh, Chromenopyridin derivatives as environmentally benign corrosion inhibitors for N80 steel in 15% HCl,*J. Assoc. Arab Univ. Basic Appl. Sci.*, 2017, **22**, 45–54.
47. K. Ansari and M. Quraishi, Experimental and computational studies of naphthyridine derivatives as corrosion inhibitor for N80 steel in 15% hydrochloric acid, *Phys. E*, 2015, **69**, 322–331.

48. K. Ansari, M. Quraishi and A. Singh, Corrosion inhibition of mild steel in hydrochloric acid by some pyridine derivatives: an experimental and quantum chemical study, *J. Ind. Eng. Chem.*, 2015, **25**, 89–98.
49. D. K. Yadav and M. A. Quraishi, Application of some condensed uracils as corrosion inhibitors for mild steel: gravimetric, electrochemical, surface morphological, UV–visible, and theoretical investigations, *Ind. Eng. Chem. Res.*, 2012, **51**, 14966–14979.
50. R. Korde, C. B. Verma, E. Ebenso and M. Quraishi, Electrochemical and thermo dynamical investigation of 5-ethyl 4-(4-methoxyphenyl)-6-methyl-2-thioxo-1, 2, 3, 4 tetrahydropyrimidine-5-carboxylate on corrosion inhibition behavior of aluminium in 1 M hydrochloric acid medium, *Int. J. Electrochem. Sci.*, 2015, **10**, 1081–1093.
51. C. O. Kappe, Controlled microwave heating in modern organic synthesis, *Angew. Chem., Int. Ed.*, 2004, **43**, 6250–6284.
52. R. N. Baig and R. S. Varma, Alternative energy input: mechanochemical, microwave and ultrasound-assisted organic synthesis, *Chem. Soc. Rev.*, 2012, **41**, 1559–1584.
53. A. Bruckmann, A. Krebs and C. Bolm, Organocatalytic reactions: effects of ball milling, microwave and ultrasound irradiation, *Green Chem.*, 2008, **10**, 1131–1141.
54. M. N. Nadagouda, T. F. Speth and R. S. Varma, Microwave-assisted green synthesis of silver nanostructures, *Acc. Chem. Res.*, 2011, **44**, 469–478.
55. N. Kaval, W. Dehaen, P. Mátyus and E. Van der Eycken, Convenient and rapid microwave-assisted synthesis of pyrido-fused ring systems applying the tert-amino effect, *Green Chem.*, 2004, **6**, 125–127.
56. A. M. Sarotti, R. A. Spanevello and A. G. Suárez, An efficient microwave-assisted green transformation of cellulose into levoglucosenone. Advantages of the use of an experimental design approach, *Green Chem.*, 2007, **9**, 1137–1140.
57. A. K. Singh, S. Thakur, B. Pani, E. E. Ebenso, M. A. Quraishi and A. K. Pandey, 2-Hydroxy-*N′*-((Thiophene-2-yl) methylene) benzohydrazide: ultrasound-assisted synthesis and corrosion inhibition study, *ACS Omega*, 2018, **3**, 4695–4705.
58. H. I. Beltrán, R. Esquivel, A. Sosa-Sánchez, J. L. Sosa-Sánchez, H. Höpfl, V. Barba, N. Farfán, M. Galicia García, O. Olivares-Xometl and L. S. Zamudio-Rivera, Microwave assisted stereoselective synthesis of cis-substituted tinIV phthalocyanine dicarboxylates. Application as corrosion inhibitors, *Inorg. Chem.*, 2004, **43**, 3555–3557.
59. C. Verma, M. Quraishi and E. E. Ebenso, Microwave and ultrasound irradiations for the synthesis of environmentally sustainable corrosion inhibitors: An overview, *Sustainable Chem. Pharm.*, 2018, **10**, 134–147.
60. J. A. Riddick, W. B. Bunger and T. K. Sakano, *Organic Solvents: Physical Properties and Methods of Purification*, 1986.
61. K. Grodowska and A. Parczewski, Organic solvents in the pharmaceutical industry, Acta Poloniae Pharmaceutica, *Drug Res.*, 2010, **67**, 3–12.
62. D. R. Lide, *Handbook of Organic Solvents*, CRC Press, 1994.

63. P. G. Jessop, Searching for green solvents, *Green Chem.*, 2011, **13**, 1391–1398.
64. M. Cvjetko Bubalo, S. Vidović, I. Radojčić Redovniković and S. Jokić, Green solvents for green technologies, *J. Chem. Technol. Biotechnol.*, 2015, **90**, 1631–1639.
65. W. M. Nelson, *Green Solvents for Chemistry: Perspectives and Practice*, Oxford University Press, 2003.
66. T. Welton, Ionic liquids in green chemistry, *Green Chem.*, 2011, **13**, 225.
67. S. Mallakpour and M. Dinari, Ionic liquids as green solvents: progress and prospects, *Green Solvents II*, Springer, 2012, pp. 1–32.
68. D. Lozowski, Supercritical CO_2: a green solvent, *Chem. Eng.*, 2010, **117**, 15.
69. S. P. Nalawade, F. Picchioni and L. Janssen, Supercritical carbon dioxide as a green solvent for processing polymer melts: Processing aspects and applications, *Prog. Polym. Sci.*, 2006, **31**, 19–43.
70. U. Ejaz and M. Sohail, Ionic liquids: green solvent for biomass pretreatment, *Nanotechnology-Based Industrial Applications of Ionic Liquids*, Springer, 2020, pp. 27–36.
71. V. S. Sastri, *Green Corrosion Inhibitors: Theory and Practice*, John Wiley & Sons, 2012.
72. C. Verma, E. E. Ebenso and M. Quraishi, Ionic liquids as green and sustainable corrosion inhibitors for metals and alloys: an overview, *J. Mol. Liq.*, 2017, **233**, 403–414.
73. V. Srivastava, D. S. Chauhan, P. G. Joshi, V. Maruthapandian, A. A. Sorour and M. A. Quraishi, PEG-functionalized chitosan: A biological macromolecule as a novel corrosion inhibitor, *ChemistrySelect*, 2018, **3**, 1990–1998.
74. M. A. Quraishi, D. S. Chauhan and V. S. Saji, Heterocyclic biomolecules as green corrosion inhibitors, *J. Mol. Liq.*, 2021, **341**, 117265.
75. M. Yadav, T. Sarkar and I. Obot, Carbohydrate compounds as green corrosion inhibitors: electrochemical, XPS, DFT and molecular dynamics simulation studies, *RSC Adv.*, 2016, **6**, 110053–110069.
76. L. Hamadi, S. Mansouri, K. Oulmi and A. Kareche, The use of amino acids as corrosion inhibitors for metals: a review, *Egypt. J. Pet.*, 2018, **27**, 1157–1165.
77. B. El Ibrahimi, A. Jmiai, L. Bazzi and S. El Issami, Amino acids and their derivatives as corrosion inhibitors for metals and alloys, *Arabian J. Chem.*, 2020, **13**, 740–771.

CHAPTER 6

Natural Polymers as Green Corrosion Inhibitors

6.1 Introduction

The area of corrosion inhibitor research is finding new pathways in the realms of environment-friendly inhibitor development.[1–3] This is partly because of the strict environmental regulations and the high cost and tedious synthetic procedures required to use the conventional corrosion inhibitors used in industries. Hence, new categories of green and sustainable inhibitors from the classes of natural extracts, drugs, ionic liquids, *etc.*, have come forward.[4–6] A major requirement from the inhibitors is the thorough coverage of the metallic surface area. Accordingly, polymeric molecules having a large size are being explored. The polymeric molecules from the classes of carbohydrates and proteins are of prime research interest. Among these, the major categories are chitosan, cellulose, exudate gums, and polyamino acid-based corrosion inhibitors.[7,8]

This chapter presents a comprehensive overview of naturally occurring polymers derived from polysaccharides and proteins. Furthermore, the single-step chemical modifications, nanocomposite formations, copolymer formation, *etc.*, methods to improve the polymeric materials are described to improve the corrosion inhibition performance.

6.2 Significance of Natural Polymers as Corrosion Inhibitors

Natural polymers can be derived from naturally occurring resources, produced by living organisms, and chemically produced from biological materials.[9] The global market share of biopolymers can be expected to grow to 10

Green Chemistry Series No. 73
Green Corrosion Inhibition: Fundamentals, Design, Synthesis and Applications
By Chandrabhan Verma, Mumtaz A. Quraishi and Dheeraj Singh Chauhan

Published by the Royal Society of Chemistry, www.rsc.org

billion USD by 2021, at a rate of approximately 17%.[10] This can be ascribed to their inherent biodegradable nature that can be applied for specific application purposes combined with excellent structural attributes. These polymers present a significant advantage over plastics due to their natural origin. Further attributes include biocompatibility, ease of extraction, cheaper availability, production scale-up, barrier action against moisture/gases, and low toxicity. The application of naturally occurring polymers, in general, is following the provisos of green nature and environmental regulations. Therefore, these polymeric materials have found extensive research interest in the domains of corrosion protection due to the afforded greater coverage of metallic substrate and the presence of an abundance of adsorption sites. A number of research reports are available in the literature mentioning the application of polymers as anticorrosion coatings and as inhibitors of corrosion. In general, natural polymeric corrosion inhibitors refer to the inhibitors developed from polymers from the polysaccharides and proteins categories. A beneficial attribute of applying these polymers as corrosion inhibitors are their natural origin, due to which these do not create any environmental concerns. Moreover, having a large structure, these molecules contain an abundance of heteroatoms and functional groups, and have a large molecular weight, which can provide substantial metallic surface coverage. To further enhance the applicability of these polymers, researchers have focused on improving the solubility and processability of these polymers, employing various techniques of chemical functionalization, grafting and copolymer formation, synthesis of polymeric nanoparticles, composites, nanocomposites, *etc.* The following sections provide a detailed account of the natural polymeric corrosion inhibitors. Figure 6.1 lists some of the natural polymers that have been explored as corrosion inhibitors.[10]

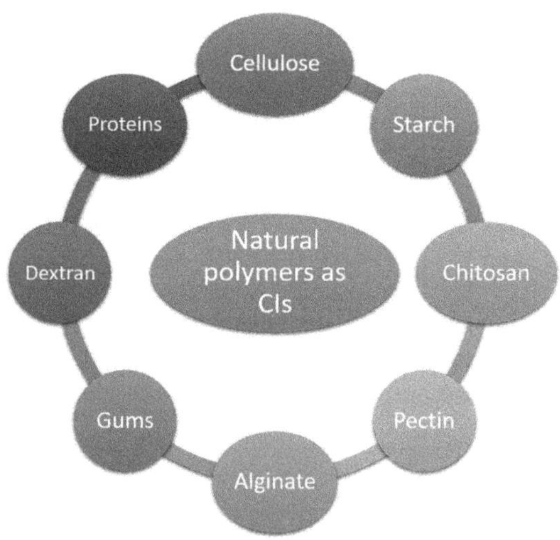

Figure 6.1 Natural polymers reported as corrosion inhibitors.

6.3 Natural Polymers as Corrosion Inhibitors

Carbohydrate polymers abundantly occur in nature, and are well known for their biocompatibility, biodegradability, wide applicability, cost-effectiveness, and greenness.[7] In addition, several polypeptides (proteins) are also studied as corrosion inhibitors. The following section provides a brief account of some of the major carbohydrate and protein-based corrosion inhibitors.

6.3.1 Chitosan

Chitosan (CS) (Figure 6.2) is the *N*-deacetylated derivative of chitin which is a naturally occurring polysaccharide found in the crustacean shells or fungi cell walls. CS exhibits excellent biological properties such as biodegradability, immunogenicity, antibacterial behaviour, and wound-healing activity.[11–13] This biopolymer finds extensive application in cosmetics, artificial skin, photography, food and nutrition, ophthalmology, and wastewater treatment.[14–18] The presence of many free $-NH_2$ and $-OH$ groups on this natural polymer endows its anticorrosion properties and the capability of complexation with metal ions.[6] For a mild steel substrate, it was examined in 0.1 M HCl, wherein the inhibition efficiency increased with increasing inhibitor dosage, and a maximum efficiency of 68% was attained.[19] CS shows a high affinity for metal complexation.[20,21] Its complexation with copper follows two mechanisms: (a) the "pendant model" and (b) the "bridge model". According to the "bridge model", the metallic ions show bonding to several free $-NH_2$ of the same or to different chains by molecular complexation, whereas in the "pendant model", the metal ion is closer to an $-NH_2$ group forming a pendant manner.[22,23] CS was evaluated as an inhibitor for the copper surface in 0.5 M HCl.[24] SEM and FTIR analyses provided evidence for the adsorption of the corrosion inhibitor on the metallic surface. CS was also studied in sulfide-containing synthetic seawater medium.[25] At 800 mg L^{-1}, CS revealed a physical mode of adsorption and high efficiency of 89%. A long immersion time study of 90 days showed strong adsorption and the formation of a stable

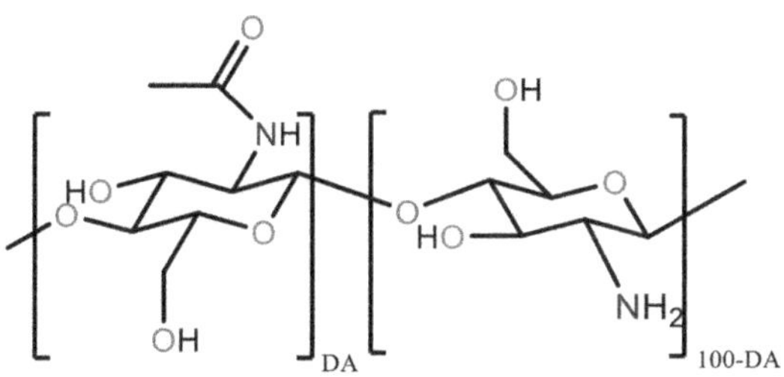

Figure 6.2 Molecular structure of chitosan (CS) monomeric unit, where the DA represents the degree of deacetylation.[27] Reproduced from ref. 27 with permission from the Royal Society of Chemistry.

inhibitor film over the Cu surface. Considering the possible application in the sugar industry, CS was studied for mild steel surface in 1 M sulphamic acid solution using gravimetric as well as electroanalytical techniques.[26] At a dose of 200 mg L^{-1}, CS showed an efficiency of 73.8% which increased to 90% upon addition of KI as a synergistic agent. Surface investigation *via* SEM and AFM supported the formation of inhibitor film on the metal surface.

6.3.2 Gums

Exudate gums are derived from plants as exudates that ooze out of the bark as a result of an injury according to the protection mechanism. The natural exudates of different plants and trees have wide applicability in food and non-food industries. Guar gum (GG) (Figure 6.3) consists of a straight chain of D-mannopyranose units linked together by β-(1→4) linkage with side-branching units of a single D-galactopyranose unit linked to every other mannose by α-(1→6) links. This polysaccharide was evaluated for the protection of carbon steel in 1 M H_2SO_4 media using weight loss and polarization studies.[28] High efficiency of 93.2% was achieved at a concentration of 1000 mg L^{-1} following an immersion time of 24 h. The inhibitor adsorption was in accordance with Langmuir isotherm. Over the surface of pure aluminium in 1 M HCl medium, GG exhibited a high efficiency of 83.19% at 0.4 g L^{-1}.[29] It was also studied for carbon steel surface in 2 M H_3PO_4.[30] Remarkable adsorption and corrosion inhibition behaviour was observed in the range of 0.1 to 1.0 g L^{-1} at 298 to 328 K.

Gum Arabic or Gum acacia (GA) is a complex arabinogalactan derived from acacia trees. This polysaccharide shows high solubility in water. It mainly consists of highly branched polysaccharides (Figure 6.4) (typically 42% galactosyl, 27% arabinosyl, 15% rhamnosyl, 14.5% glucuronosyl, and 1.5% 4-*O*-methyl-glucuronosyl) and a protein-polysaccharide complex as a minor component.[32] It was studied for mild steel in 1 M HCl, which revealed a 95%

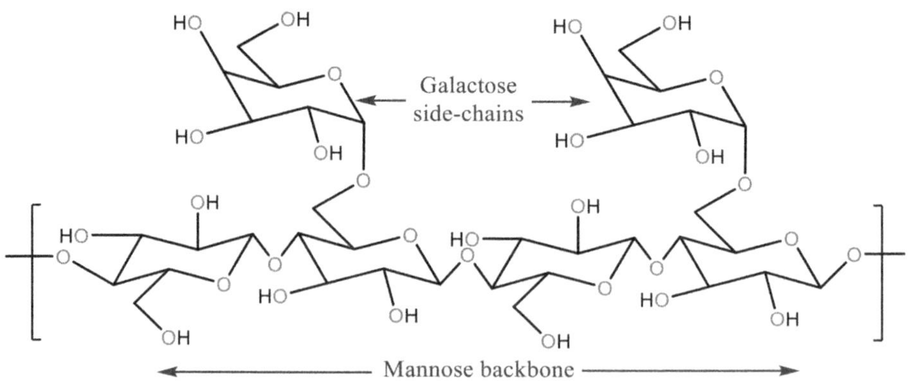

Figure 6.3 Structure of guar gum (GG).[31] Reproduced from ref. 31 with permission from Elsevier, Copyright 2016.

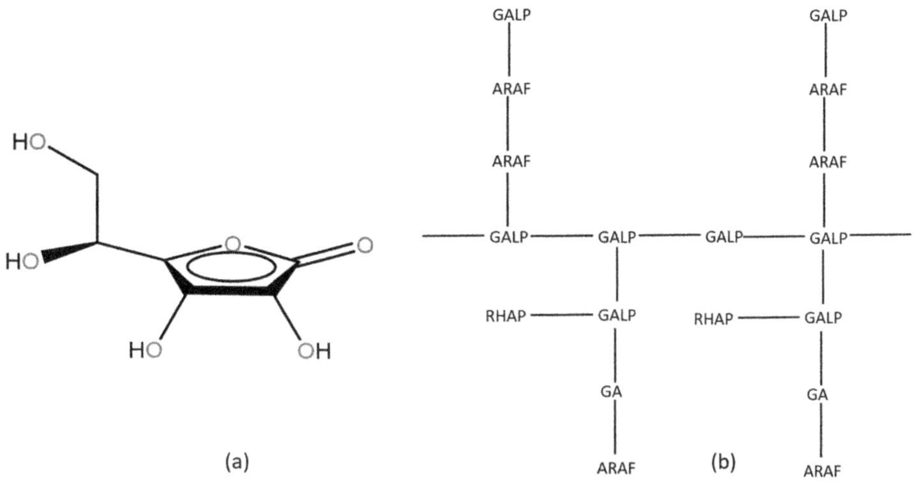

Figure 6.4 Structure of monosaccharides (a) Segment of Gum Arabic (GA) molecule. (b) The polysaccharide backbone is composed of D-Galactopyranose (GALP), with linked branches of L-Arabofuranose (ARAF), L-Rhamnopyranose (RHAP), and D-Glucuronic Acid (GA).[37] Reproduced from ref. 37, http://www.electrochemsci.org/papers/vol7/71009303.pdf, under the terms of the CC BY 4.0 licence https://creativecommons.org/licenses/by/4.0/.

inhibition performance at 1 g L^{-1}.[33] Surface examination *via* AFM and XPS investigations indicated inhibitor adsorption and the formation of a protective film on a metallic substrate that was fortified using DFT-based computations. In synthetic brine, GA was also evaluated for carbon steel surface *via* electrochemical measurements, wherein a high efficiency of 94% at 5 g L^{-1} concentration was attained.[34] Molecular simulation investigation on the Fe (111) surface provided further evidence for the adsorption of high molecular weight carbohydrate species. The same polysaccharide was also studied for N80 steel in a CO_2-saturated 0.5 M KCl medium *via* weight loss tests and electrochemical investigations.[35] At 500 mg L^{-1}, 70.27% protection efficiency was afforded from the EIS analyses, whereas the PDP measurements suggested mixed-type inhibition behaviour. FTIR and the SEM-EDX investigations showed further evidence for inhibitor adsorption and the formation of a protective film. In NaOH solution, on the surface of aluminium, GA was studied in the range of 0.1–0.5 g L^{-1} concentration, and high efficiency was observed, supported by kinetic and thermodynamic analyses.[36]

Xanthan gum (XG) (Figure 6.5) is a polysaccharide produced from the fermentation of glucose and sucrose. It is quite common in salad dressings and sauces. For mild steel in 1 M HCl, XG was evaluated along with some synergistic surfactant additives.[38] The introduction of trace amounts of sodium dodecyl sulfate (SDS), cetyl pyridinium chloride (CPC), and Triton X-100 (TX), increased the inhibition performance of XG at a dose of 1000 mg L^{-1}. XG was also evaluated for carbon steel in 0.1% NaCl solution as an inhibitor using

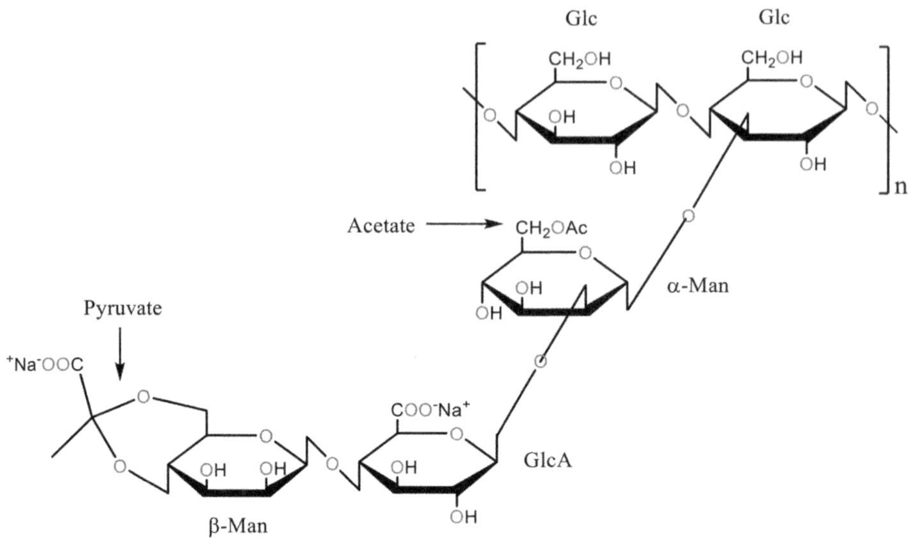

Figure 6.5 Chemical structure of xanthan gum.[63–65] Reproduced from ref. 63 with permission from Elsevier, Copyright 2007.

electrochemical and surface studies.[39] The introduction of biogenic trehalose lipid surfactants (TLS) significantly improved the inhibition performance. The same inhibitor was also evaluated in 0.5 M HCl for corrosion protection of aluminium surfaces.[40] Mixed-type inhibition behaviour with cathodic predominance was noted. Some of the different polysaccharides that have been used as corrosion inhibitors for metals and alloys are listed in Table 6.1.

6.3.3 Starch

Starch or amylum can be described as a polymeric carbohydrate consisting of numerous glucose units joined by glycosidic bonds. Most green plants produce this polysaccharide for the purpose of energy storage. Starch consists of two structurally varying components: amylose in 15–20% and amylopectin in 80–85% (Figure 6.6). There are few reports available on starch as a corrosion inhibitor, although chemically modified starch is reported as a corrosion inhibitor as described subsequently. Millet starch was extracted from millet grains and used as an inhibitor for mild steel corrosion in 0.5 M H_2SO_4. At a dose of 1.4 g L^{-1}, a high inhibition efficiency of 80.44% was afforded, which improved further to 85.44% with the introduction of 0.4 g L^{-1} of KI in the test solution.[42] Kinetic-thermodynamic parameters indicated the adsorption mode *via* the physisorption mechanism. In another study, amylose was evaluated for mild steel in 0.1 M H_2SO_4, affording an efficiency of 64.78% at 500 mg L^{-1}.[41] The addition of cetrimonium bromide (CTAB) surfactant as a synergistic agent provided further improvement in the protection performance. For aluminium alloy AA6061 in seawater solution, tapioca

Table 6.1 Performance of different polysaccharides as corrosion inhibitors.

Name	Metal surface/medium	Inhibition efficiency/ concentration	Reference
Chitosan (CS)	Mild steel/0.1 M HCl	68%/4 μM	19
CS	Cu/0.5 M HCl	93%/8 μM	24
CS	Cu/Sulfide-containing synthetic seawater	89%/800 mg L^{-1}	25
CS + KI	Mild steel/1 M sulphamic acid	90%/200 mg L^{-1}	26
Guar gum (GG)	Carbon steel/1 M H_2SO_4	93.2%/1 g L^{-1}	28
GG	Al/1 M HCl	83.19%/400 mg L^{-1}	29
GG	Carbon steel/2 M H_3PO_4	1 g L^{-1}/95.8%	30
Gum Arabic (GA)	Mild steel/1 M HCl	95%/1 g L^{-1}	33
GA	Carbon steel 1018/Oilfield water	94.2%/5 g L^{-1}	34
GA	N80 steel/CO_2-saturated 0.5 M KCl	70.27%/500 mg L^{-1}	35
GA	Al alloy 1060/1 M NaOH	74%/500 mg L^{-1}	36
Xanthan gum (XG)	Mild steel/1 M HCl	74.24%/1 g L^{-1}	38
XG + SDS (5 mg L^{-1})		83.17%/1 g L^{-1}	
XG + CPC (5 mg L^{-1})		75.19%/1 g L^{-1}	
XG + TX (5 mg L^{-1})		82.31%/1 g L^{-1}	
XG	St37 steel/0.1% NaCl	87.34%/2 g L^{-1}	39
XG + TLS (500 mg L^{-1})		96.97%/2 g L^{-1}	
XG	Al/0.5 M HCl	65%/5 g L^{-1}	40
Starch	Mild steel/0.1 M H_2SO_4	66.21%/300 mg L^{-1}	41
Millet starch	Mild steel/0.5 M H_2SO_4	80.44%/1.4 g L^{-1}	42
Millet starch + KI (400 mg L^{-1})		85.44%/1.4 g L^{-1}	
Amylose	Mild steel/0.1 M H_2SO_4	64.78%/500 mg L^{-1}	41
Tapioca starch	AA6061 alloy/seawater	96%/1 g L^{-1}	43
Pectin	X60 steel/0.5 M HCl	78.7%/1000 mg L^{-1}	44
Pectin	Mild steel/1 M HCl	94.2%/2 g L^{-1}	45
Pectin	Al/2 M HCl	92.1%/2 g L^{-1}	46
Pectin from tomato peel waste	Tin/2% NaCl, 1% acetic acid and 0.5% citric acid	72.98%/4 g L^{-1}	47
Pectin + CeO_2 Nanoparticles (5 mM)	X60 steel/0.5 M HCl	91.3%/500 mg L^{-1}	48

(continued)

Table 6.1 (*continued*)

Name	Metal surface/medium	Inhibition efficiency/ concentration	Reference
Carboxymethylcellulose (CMC)	API 5L X60 pipeline steel/CO_2 saturated 3.5% NaCl	39%/100 mg L^{-1}	49
CMC	Mild steel/2 M H_2SO_4	60.8%/500 mg L^{-1}	50
Hydroxyethyl cellulose (HEC)	Zn/26% NH_4Cl	92.07%/300 mg L^{-1}	51
HEC	Mild steel/0.5 M H_2SO_4	93.61%/2 g L^{-1}	52
	Al/0.5 M H_2SO_4	64.18%/2 g L^{-1}	
HEC	Mg alloy AZ31/3.5% NaCl	64.13%/2 g L^{-1}	53
Hydroxypropyl cellulose (HPC)	Al/0.5 M HCl	80.2%/5 g L^{-1}	54
	Al/2 M H_2SO_4	65.3%/5 g L^{-1}	
Sodium alginate (SA)	Mg alloy AZ31/3.5% NaCl	58.17%/2 g L^{-1}	53
SA	Mg alloy AZ91DF/3.5% NaCl	86.62%/500 mg L^{-1}	55
SA	Cu/1 M HCl	83%/0.1 mg L^{-1}	56
SA + chondroitin sulfate (400 mg L^{-1})	Mild steel/1 M HCl	96.88%/400 mg L^{-1}	57
SA	Carbon steel/0.5 M HCl	80%/300 mg L^{-1}	58
SA	API X60 steel/3.5% NaCl	87.23%/1 g L^{-1}	59
SA	Mg alloy AZ31 in NaCl	90%/500 mg L^{-1}	60
Dextran (DEX)	Al alloy 6061/1 M HCl	74.6%/0.4 g L^{-1}	61
DEX	6061Al—15%$_{(v)}$ $SiC_{(P)}$ composite/0.25 M HCl	80%/0.4 g L^{-1}	62

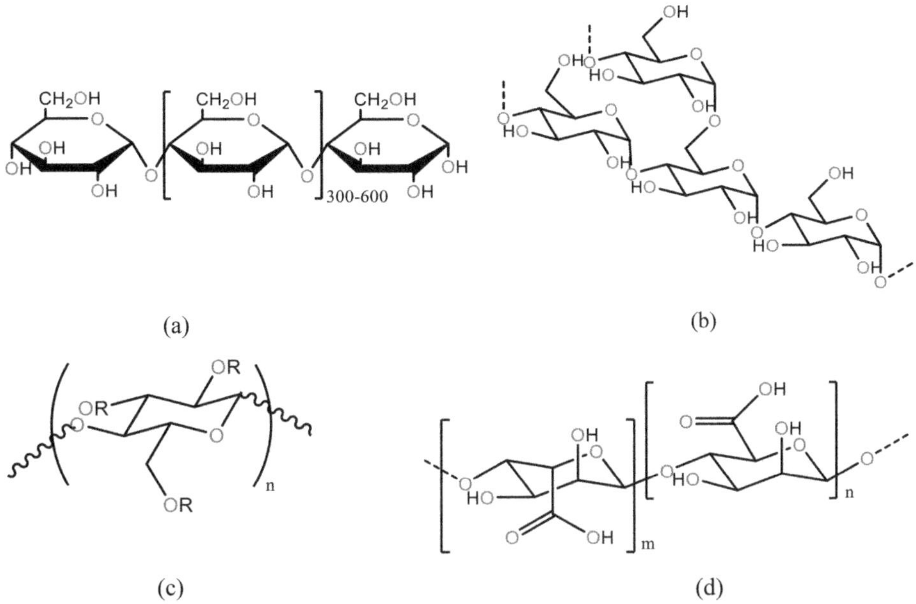

Figure 6.6 Chemical structures of (a) amylose and (b) amylopectin of starch. (c) Molecular structure of carboxymethyl cellulose (CMC) (R = H or CH_2COOH) or hydroxyethyl cellulose (HEC) (R = H or CH_2CH_2OH). (d) Structure of Alginate.[7] Reproduced from ref. 7 with permission from Elsevier, Copyright 2016.

starch was evaluated as a corrosion inhibitor.[43] A significant protective performance of 96% was obtained at 1000 mg L^{-1}, with the inhibitor obeying the Langmuir isotherm.

6.3.4 Cellulose

Cellulose is a polysaccharide having the general formula $(C_6H_{10}O_5)_n$. It consists of a linear chain having several hundred to many thousands β(1→4) linked D-glucose units. It is an important structural constituent of the primary cell wall of green plants, several algae, and the oomycetes. Some bacterial species secrete cellulose, and it is the most abundant organic polymer on earth. Carboxymethyl cellulose (CMC) (Figure 6.6c), or cellulose gum, is a derivative of cellulose-containing carboxymethyl groups bound to some of the –OH groups of the glucopyranose units making the cellulose framework. It was evaluated along with CS for API 5L X60 steel surface in a sweet medium using electrochemical and surface analysis.[49] CS and CMC were studied for API 5L X60 steel surface in a sweet environment *via* EIS, PDP, and surface analysis.[49] A mixed mode of inhibition was noted, and the adsorption of the inhibitor on the metal surface followed the Langmuir isotherm. CMC was studied as an inhibitor for mild steel in 2 M H_2SO_4.[50] Moderate efficiencies were observed in the temperature range of 30 °C to 60 °C, at a concentration of 0.5 g L^{-1}. Inhibitor adsorption followed the Langmuir isotherm.

Hydroxyethylcellulose (HEC) (Figure 6.6c) is a gelling and thickening agent derived from cellulose. It is widely used in cosmetics, cleaning solutions, and other household products. HEC was studied as an inhibitor for zinc-carbon battery, wherein the results explained its mixed-type behaviour.[51] Both physical and chemical modes of adsorption were observed. HEC was also studied as an inhibitor for mild steel and aluminium surface in 0.5 M H_2SO_4.[52] At 2 g L^{-1}, 93.61% protection was afforded for mild steel, and 64.18% was achieved for aluminium. EIS revealed charge transfer controlled behaviour, whereas the PDP measurements evidenced a cathodic-type inhibition. Hydroxypropyl cellulose (HPC) was studied as an inhibitor for aluminium in 0.5 M HCl and 2 M H_2SO_4.[54] HPC provided a mixed-type behaviour and anodic predominance.

6.3.5 Alginate

Alginate (ALG) (Figure 6.6d) is an anionic unbranched copolymer of (1–4)-linked β-D-mannuronic acid (M) and α-L-guluronic acid (G) constituting M, G, and MG block structures.[66] It is a carbohydrate polymer extracted from seaweed. ALG has attributes of biodegradability, non-toxic nature, and good biocompatibility. It can reduce blood glucose and lipid levels, clean the intestines, remove heavy metals, and other health functions. For Mg alloy AZ31 in a NaCl environment, an inhibitor formulation of HEC and ALG were studied.[53] Effective inhibition performance with 80.56% and 77.43% efficiencies were afforded. The coexistence of $Mg(OH)_2$ and adsorbed inhibitor complexes provided protection to the alloy surface. SECM, AFM, SEM, and EDX analyses provided evidence for the inhibitor adsorption and protection performance. For Mg alloy AZ91DF in 3.5% NaCl, sodium alginate (SA) was studied as a corrosion inhibitor.[55] 98% inhibition performance was obtained at 0.03 mol L^{-1}. For copper surface in 1 M HCl solution, SA was evaluated as a corrosion inhibitor.[56] The inhibition efficiency rose with concentration, and 83% protection was achieved at 0.1 mg L^{-1}. MC simulations provided evidence for the spontaneity of adsorption of SA on a copper surface. A synergistic combination of SA and chondroitin sulfate was studied for mild steel in 1 M HCl solution.[57] High efficiency of 96.88% was determined using a 400 mg L^{-1} dose each of SA and chondroitin sulfate. Elevation in temperature up to 60 °C resulted in a slight lowering in the inhibition performance from 93.50% to 77.33%. Surface studies *via* SEM and SECM supported inhibitor adsorption and surface protection. SA was also studied as an inhibitor for carbon steel in 0.5 M HCl medium[58] where 80% inhibition was obtained at 300 mg L^{-1}, reducing the corrosion rate from 223.4 mmpy to 44.28 mmpy. EIS data were fitted to a modified Randles circuit, and a capacitive control of the protection performance was evidenced. Cathodic-type inhibition was noted in the PDP analyses. For API X60 high strength carbon steel in 3.5% NaCl solution, SA was studied as an inhibitor.[59] An effective physical mode of adsorption was afforded by the presence of carboxylate oxygen on the metal surface. The efficiency of inhibition rose with inhibitor concentration, although decreased with a rise in the temperature. For Mg alloy AZ31 in NaCl environment, SA was studied.[60] Surface analytical measurements *via* SEM and FTIR supported inhibitor adsorption on a metallic substrate.

6.3.6 Pectin

Pectin is an acidic heteropolysaccharide found in the middle and primary lamella and terrestrial plant species cell walls. Mainly, pectin consists of galacturonic acid, which is a sugar derived from galactose. Pectin contains numerous D-galacturonic acid residues (Figure 6.7), a cyclic monosaccharide with one –COOH and 4 –OH side groups.[45] Commercially, pectin is produced in a white to light brown powder, mainly extracted from citrus fruits. Pectin was evaluated as an inhibitor for X60 steel in 0.5 M HCl solution.[44] Electrochemical and surface studies indicated that inhibitor adsorption caused a rise in surface hydrophobicity. For mild steel surface in 1 M HCl solution, pectin exhibited a high efficiency of 94.2% at an elevated temperature of 318 K at 2 g L^{-1}.[45] A complex between inhibitor and Fe^{2+} ions released during corrosion was attributed to providing the protection. For aluminium metal surface, a 92.1% efficiency was obtained at a dose of 2 g L^{-1}.[46] Thermodynamic parameters supported the spontaneous adsorption of inhibitors with a mixed mode of inhibition behaviour. Pectin extracted from tomato peel waste was applied as an inhibitor against tin corrosion in a solution of 2% NaCl, 1% acetic acid, and 0.5% citric acid.[47] At a concentration of 4 g L^{-1}, an efficiency of 72.98% was afforded, which was greater than that observed for commercially

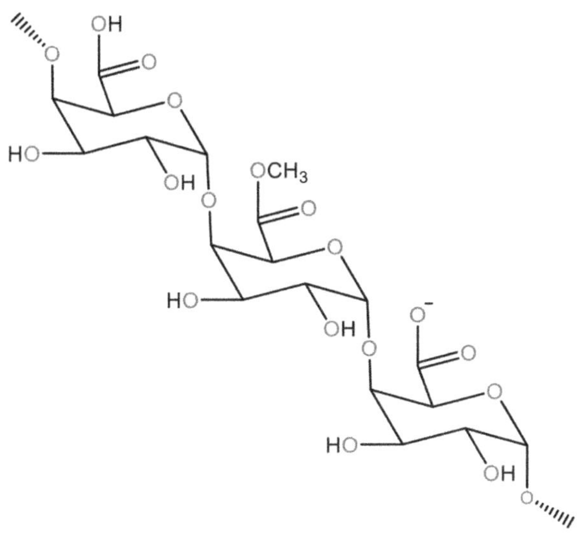

Figure 6.7 Molecular structure of pectin.[45] Reproduced from ref. 45 with permission from Elsevier, Copyright 2015.

obtained pectin from apple (60.02%). The influence of the addition of ceria (CeO_2) nanoparticles on corrosion inhibition efficiency of pectin was studied for X60 steel in 0.5 M HCl solution.[48] Both synergistic and antagonistic behaviours with the ceria nanoparticles were observed depending upon immersion time, temperature, and inhibitor dosage. Maximum efficiency of 91.3% at 500 mg L^{-1} pectin and 5 mM CeO_2 was achieved at 60 °C.

6.3.7 Dextran

Dextran (DEX) (Figure 6.8) is a complex branched polysaccharide extracted from wine. DEX has a framework of poly-α-D-glucosides having microbial origin and containing glycosidic bonds predominantly C-1 → C-6. For corrosion inhibition of Al alloy 6061 in 1 M HCl solution, DEX was explored as a green inhibitor.[61] Mixed-type inhibition behaviour was observed along with physisorption, which obeyed the Langmuir isotherm and 74.6% inhibition efficiency was observed at 0.4 g L^{-1}. In 0.25 M HCl solution, DEX was evaluated as a corrosion inhibitor on the surface of a 6061 Al—15%$_{(v)}$ $SiC_{(P)}$ composite. The electrochemical measurements of EIS and PDP analyses revealed a maximum efficiency of 80% at 303 K upon introduction of 0.4 g L^{-1} of DEX.[62] The inhibitor showed a mixed-type behaviour with a predominance of physical adsorption that was following Langmuir isotherm.

6.3.8 Proteins

Casein comes under the family of phosphoproteins. It is a constituent of about 60% of the proteins in human milk. Casein is composed of a high number of amino acid proline units. For the surface of mild steel in 0.1 M HCl solution, casein was studied as a corrosion inhibitor in the range of concentrations of 50 to 400 mg L^{-1}.[67] The incorporation of an inhibitor into the corrosive electrolyte brought a reduction in the corrosion rate from 98.4 mpy to 3.9 mpy. EIS indicated a capacitive performance, with a mixed mode of inhibition as shown by PDP. Casein with Na_2SnO_3 was studied as an organic/inorganic hybrid corrosion inhibitor for alkaline aluminium-air fuel cells.[68]

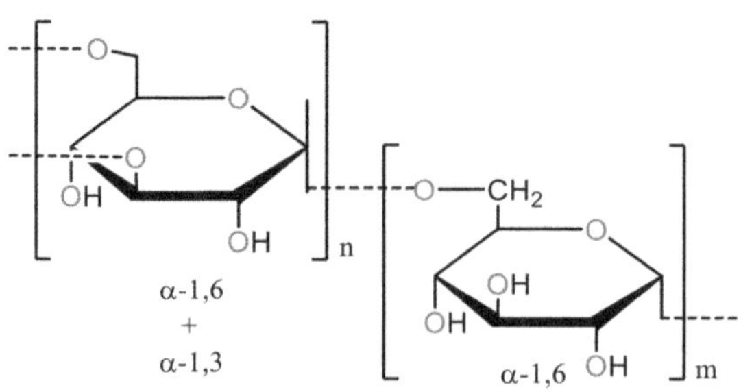

Figure 6.8 Structure of Dextran (DEX).[53] Reproduced from ref. 53 with permission from Elsevier, Copyright 2020.

The strongest inhibition performance was afforded in the presence of 0.6 g L^{-1} casein with 0.05 M Na_2SnO_3. In addition, casein promoted the deposition of a uniform and stable layer of tin on the aluminium surface. Casein was also evaluated for reinforcing steel in 0.1 mol L^{-1} NaOH solution in the presence of varying Cl^- ions and exhibited 82% protection.[69] Casein was also studied as a green inhibitor for 316L stainless steel in simulated body fluid.[70] The inhibitor showed a mixed mode of inhibition at 1000 mg L^{-1} after 120 min of immersion time with a moderate efficiency of 49.24%.

Gelatin is a colourless, translucent, and flavourless food ingredient extracted from collagen taken from animal body parts. It finds extensive application as a gelling agent in food, beverages, medication, drug and vitamin capsules, cosmetics, *etc.* For X60 steel, gelatin was studied as a corrosion inhibitor employing gravimetric and electrochemical studies in 15% HCl.[71] The addition of KI as a synergistic agent improved the inhibition efficiency from 70.42% to 84.51%. For aluminium and aluminium silicon alloys in 0.1 M NaOH solution, gelatin was also studied as a corrosion inhibitor using electrochemical studies.[72] Inhibitor adsorption followed the Flory-Huggins isotherm. Gelatin was studied for mild steel in 1 M HCl environment in the temperature range of 20–50 °C. It acted as a mixed-type of inhibitor and blocked both anodic and cathodic reactions, with the adsorption obeying the Langmuir isotherm.

Gluten is a seed storage protein found in some cereal grains (especially wheat). It is referred to as a combination of prolamin and glutelin proteins. Gluten hydrolysate was evaluated for mild steel corrosion inhibition in 1 M HCl employing gravimetric and electroanalytical studies.[73] Mixed-type inhibition behaviour was recorded *via* the formation of a protective inhibitor film on the steel substrate. Mixed physical and chemical modes of adsorption were observed, with 93% efficiency at 323 K. Maize gluten meal extract (MGME) and its major amino acid components glutamic acid, proline, and leucine were comparatively studied for steel in simulated concrete pore solution (SCPS) using experimental and theoretical studies.[74] Protection of 86.02% was afforded at a concentration of 1 g L^{-1}. In another study on SCPS using MGME as an inhibitor, high efficiency of 88.10% was afforded, and the inhibitor adsorption according to Langmuir isotherm.[75] SEM and AFM studies supported the inhibitor adsorption and the formation of protective film.

Soy or soyabean is a species of edible legume and has wide usages. Soybean (glycine max) plant extract was studied as a corrosion inhibitor for aluminium in 1 M HCl solution using weight loss, electrochemical and surface studies.[76] Mixed-type inhibition was observed with a maximum inhibition efficiency of 77.39% at 300 mg L^{-1}. Adsorption of the inhibitor on the metal surface was in agreement with the Temkin isotherm. Along with polyvinyl pyrrolidinone (PVP), soy polymer was studied for carbon steel in 1 M H_2SO_4 at a temperature range of 30–60 °C, wherein an efficiency of 89.5% was achieved.[77] A mixed mode of inhibition behaviour with alignment to Langmuir isotherm was noticed. Table 6.2 lists some of the different protein-based biopolymers that have been used as corrosion inhibitors for metals and alloys.

Table 6.2 Performance of some protein-based polymers as corrosion inhibitors.

Name	Metal surface/medium	Inhibition efficiency/ concentration	Reference
Casein	Mild steel/0.1 M HCl	96.41%/400 mg L^{-1}	67
Casein + Na_2SnO_3 (0.05 M)	Aluminium/4 M NaOH	83.10%/0.6 g L^{-1}	68
Casein	Steel/0.1 mol L^{-1} NaOH	82%/9.7 × 10^{-4} M	69
Casein	316L stainless steel/ Simulated body fluid	49.24%/1 g L^{-1}	70
Gelatin	X60 steel/15% HCl	70.42%/2.5% w/v	71
Gelatin	Al alloy/0.1 M NaOH	65.71%/2 g L^{-1}	72
Gluten hydrolysate	Mild steel/1 M HCl	92.8%/1 g L^{-1}	73
Maize gluten meal extract (MGME)	Steel/Simulated concrete pore solution (SCPS)	88%/2 g L^{-1}	74
MGME	Steel/SCPS	88.10%/2 g L^{-1}	75
Soybean (Glycine Max) plant extract	Aluminium in 1 M HCl	77.39%/300 mg L^{-1}	76
Soy polymer	Carbon steel/1 M H_2SO_4	89.51%/5 g L^{-1}	77
Soy polymer + PVP (2 g L^{-1})		96.09%/3 g L^{-1}	

6.4 Issues in Using Natural Polymers as Corrosion Inhibitors

In comparison to the monomer forms of carbohydrates, such as glucose, sucrose, fructose, *etc.*, due to their considerable molecular weight, carbohydrate polymers provide coverage to a greater surface area of the metal. However, due to the large molecular weights, these polymers pose solubility issues in the aqueous corrosive media. Also, most of these polymers contain N and O functionalities and limited functional groups, which restrict the adsorption and inhibition performance. In fact, despite the cheap and abundant availability, the abovementioned issues pose a huge concern in the commercialization of these inhibitors. To address to above issues, in recent years, researchers have considered the prospect of performing simple chemical modification of the carbohydrate polymers to improve the solubility and incorporate additional adsorption sites, as discussed in the below sections.

6.5 Chemically Modified Biopolymers as Corrosion Inhibitors

The synthesis and chemical functionalization of biopolymers can be accomplished using simple single-step chemical methods. The structural properties, such as the existence of functional groups, molecular weight, and thermal stability, can be adequately analysed using the techniques of FTIR, NMR (^{1}H NMR, ^{13}C NMR), dynamic light scattering (DLS), and thermogravimetric analysis (TGA). The quantum chemical methods such as DFT-based studies and molecular simulation methods (*e.g.*, MD, MC studies) can be utilized to determine the important parameters crucial for energy, structural stability, major adsorption sites, *etc.* biopolymer-based corrosion inhibitors are designed using DFT-based computational studies.[78] The structures and corrosion inhibition performance of different chemically modified natural polymers are shown in Figure 6.9.

Glycerin-grafted starch was studied as a corrosion inhibitor for C–Mn steel in 1 M HCl solution. At 300 mg L^{-1} of concentration, high efficiency of 94% was achieved.[79] The inhibition efficiency remained largely unchanged in the temperature range of 298 K to 323 K. The inhibitor adsorption was correlated to the Freundlich adsorption isotherm. Chemically modified cassava starches *viz.* activated starch (AS) and a carboxymethylated starch (CMS) having varying degrees of substitution (DS) were prepared.[80] The inhibitors were analysed for carbon steel in 200 mg L^{-1} NaCl under alkaline conditions. It was found that the inhibition performance of CMS increased with an increase in DS. Cyanoethyl cellulose (CEC) was prepared, and in a subsequent reaction step, a tetrazole derivative of cellulose (CTZ) was synthesized.[81] At 100 mg L^{-1} concentration of CTZ, high efficiency of 94.2% was achieved. EIS indicated a capacitive performance, and the adsorption was correlated to the Langmuir isotherm. A sodium alginate-derived cationic surfactant (AS) was prepared, and its metal complexes with Cu, Co, and Zn were synthesized.[82] The metal complexes AS-Zn, AS-Co, and AS-Cu were studied as inhibitors for carbon steel in 1 M HCl. A maximum efficiency of 96.27% was obtained for AS-Cu at a concentration of 5 mM. All the studied inhibitors were of a mixed-type nature, having cathodic control. Amino acid functionalized DEX derivatives (LDT and S-LDT) were synthesized, and their inhibition behaviour was studied for CO_2 saturated oilfield produce water using electrochemical and surface studies.[83] High stability of corrosion inhibition (99.7% after 72 h) was observed. The inhibitors LDT and S-LDT showed efficiencies of 86.6% and 99.5% at a concentration of 250 mg L^{-1}.

It is noteworthy to mention that most of the studies carried out on chemically modified natural polymers in recent years have been reported on CS polymers. Using a simple method of Schiff base formation,[84,92–95] a few derivatives of CS have been produced and analysed in a wide variety of corrosive media. The structures and corrosion inhibition performance of different chemically modified CS derivatives are shown in Figure 6.10. The introduction of an imine linkage (–HC=N–) during Schiff base formation provides the active site for adsorption on a metallic substrate. For this purpose, microwave irradiation has been chosen as an effective method. Schiff bases of CS have been prepared using benzaldehyde, vanillin, *N*, *N*-dimethylaminobenzaldehyde, and cinnamaldehyde and analysed during the acid pickling of mild steel and copper in a HCl environment.[84,96] Further, Schiff bases prepared using piperonal (Pip-Cht), vanillin (Van-Cht), and cinnamaldehyde (Cin-Cht) were analysed on carbon steel in 15% HCl.[85–87] Schiff base prepared from CS and salicylaldehyde was evaluated as an inhibitor in a sweet corrosion environment for J55 steel.[88] Other chemically modified CS has also been studied

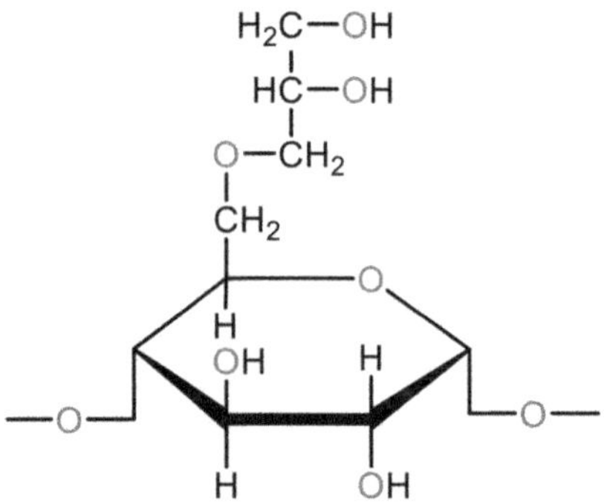

Glycerin-grafted starch
C-Mn steel/ 1M HCl
IE 94% (300 mgL^{-1}) [79]

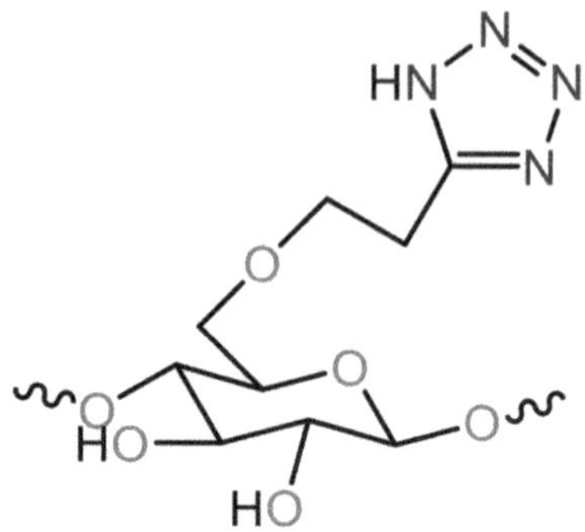

Cellulose tetrazole
Carbon steel/ 1M HCl
IE 94.2% (100 mgL^{-1}) [81]

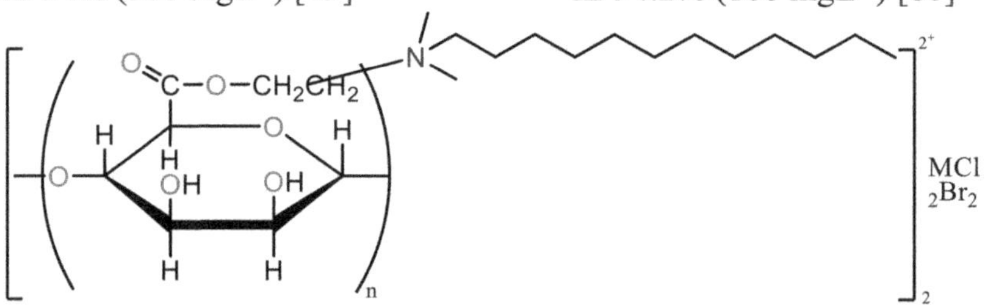

M = Co (AS-Co), Cu (AS-Cu), and Zn (AS-Zn)

Co, Cu, and Zn complexes of ALG derived cationic surfactant (AS)
Carbon steel/ 1M HCl
IE AS 88.03%; AS-Co 93.47%; AS-Cu 95.51%; AS-Zn 92.38% (5 mM) [82]

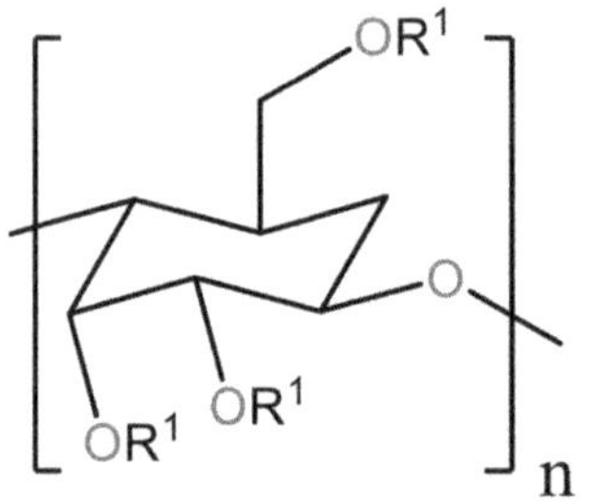

LDT

(R^1 = -$COCHNH_2CH_2SH$)

Amino acid functionalized DEX
Carbon steel/ CO_2-saturated oilfield produced water
IE 86.6% (250 mgL^{-1}) [83]

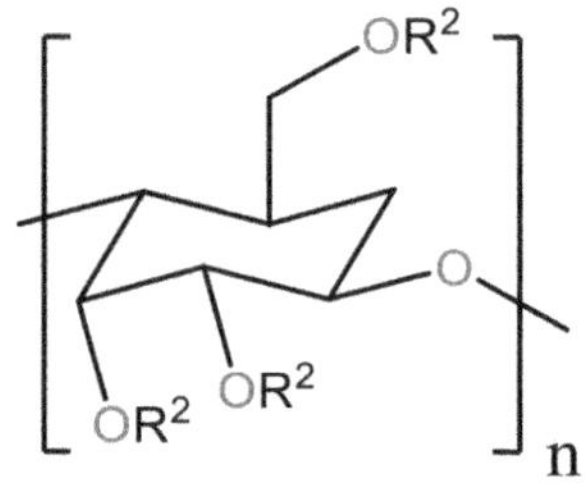

S-LDT

(R^2 = -$COCHNH_2CH_2SBn$)

Amino acid functionalized DEX
Carbon steel/ CO_2-saturated oilfield produced water
IE 99.5% (250 mgL^{-1}) [83]

Figure 6.9 Structures and corrosion inhibition efficiencies of different chemically modified polymers.

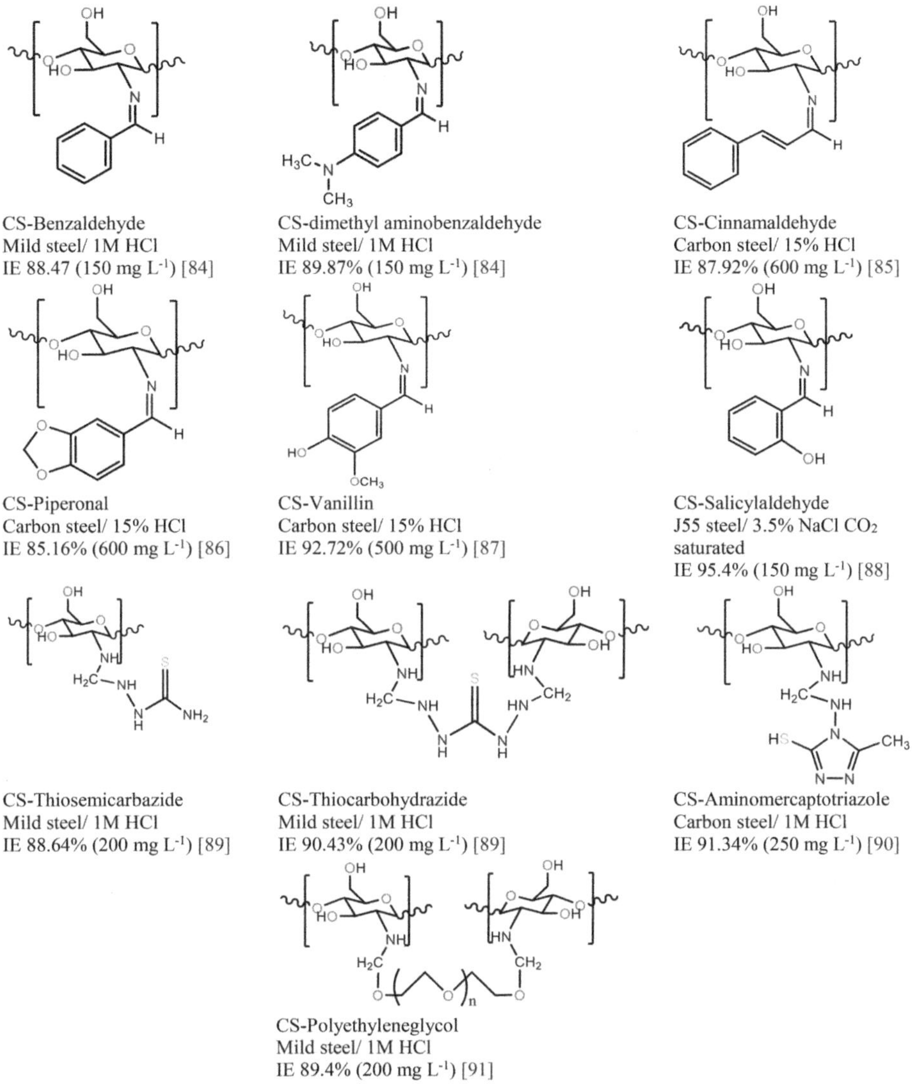

Figure 6.10 Structures and corrosion inhibition efficiencies of different chemically modified CS-based corrosion inhibitors.

on stainless steel surface in 3.5% NaCl,[97,98] on mild steel in 1 M HCl,[89–91] and in 1 M sulphamic acid[99] environment.

6.6 Conclusions and Prospects

This chapter presents the significance and application of some naturally occurring polymers in corrosion inhibition. Polysaccharides and proteins are considered green molecules, and their application for various purposes is in agreement with environmental sustainability. Herein we have provided an overview of the significance and application of some polysaccharides and protein molecules with structures and literary examples of each in corrosion inhibition of different metals and alloys. It is noteworthy to mention that extensive research is available on the application of natural polymers as corrosion inhibitors. It is important to note that these polymers show better protection and greater thermal stability compared to the corresponding monomeric carbohydrates, *e.g.* sugars, amino acids, *etc.* Some obvious drawbacks are the poor solubility that restricts the application in corrosive media. Furthermore, in order to improve the adsorption and protection behaviour, the researchers have also considered the chemical modification of these polymers. Accordingly, we have covered the chemically functionalized polymers in a separate section.

Further studies should focus on the influence of different functional groups, and detailed studies on the synergism of the polymer-based corrosion inhibitors. Influence of elevated temperatures, flow conditions, and other additives such as surfactants, oils, *etc.* is required to be explored in detail. Comparison of the existing natural polymer-based corrosion inhibitors with that commercially available and in practice should be carried out.

References

1. M. A. Quraishi, D. S. Chauhan and V. S. Saji, Heterocyclic biomolecules as green corrosion inhibitors, *J. Mol. Liq.*, 2021, **341**, 117265.
2. A. Singh, K. R. Ansari, D. S. Chauhan, M. A. Quraishi, H. Lgaz and I.-M. Chung, Comprehensive investigation of steel corrosion inhibition at macro/micro level by ecofriendly green corrosion inhibitor in 15% HCl medium, *J. Colloid Interface Sci.*, 2020, **560**, 225–236.
3. K. Xhanari, M. Finšgar, M. K. Hrnčič, U. Maver, Ž. Knez and B. Seiti, Green corrosion inhibitors for aluminium and its alloys: a review, *RSC Adv.*, 2017, 7, 27299–27330.
4. C. Verma, E. E. Ebenso, I. Bahadur and M. A. Quraishi, An overview on plant extracts as environmental sustainable and green corrosion inhibitors for metals and alloys in aggressive corrosive media, *J. Mol. Liq.*, 2018, **266**, 577–590.
5. C. Verma, E. E. Ebenso and M. A. Quraishi, Ionic liquids as green and sustainable corrosion inhibitors for metals and alloys: an overview, *J. Mol. Liq.*, 2017, **233**, 403–414.
6. C. Verma, D. S. Chauhan and M. A. Quraishi, Drugs as environmentally benign corrosion inhibitors for ferrous and nonferrous materials in acid environment: An overview, *J. Mater. Environ. Sci.*, 2017, **8**, 4040–4051.
7. S. A. Umoren and U. M. Eduok, Application of carbohydrate polymers as corrosion inhibitors for metal substrates in different media: a review, *Carbohydr. Polym.*, 2016, **140**, 314–341.
8. D. S. Chauhan, M. Quraishi, V. Srivastava, J. Haque and B. El Ebrahimi, Virgin and chemically functionalized amino acids as green corrosion

inhibitors: Influence of molecular structure through experimental and *in silico* studies, *J. Mol. Struct.*, 2020, **1226**, 129259.
9. A. Jha and A. Kumar, Biobased technologies for the efficient extraction of biopolymers from waste biomass, *Bioprocess Biosyst. Eng.*, 2019, **42**, 1893–1901.
10. G. A. Martău, M. Mihai and D. C. Vodnar, The use of chitosan, alginate, and pectin in the biomedical and food sector—biocompatibility, bioadhesiveness, and biodegradability, *Polymers*, 2019, **11**, 1837.
11. H. Sashiwa and S.-i. Aiba, Chemically modified chitin and chitosan as biomaterials, *Prog. Polym. Sci.*, 2004, **29**, 887–908.
12. L. Gritsch, C. Lovell, W. H. Goldmann and A. R. Boccaccini, Fabrication and characterization of copper (II)-chitosan complexes as antibiotic-free antibacterial biomaterial, *Carbohydr. Polym.*, 2017, **179**, 370–379.
13. J. Karbowniczek, L. Cordero-Arias, S. Virtanen, S. K. Misra, E. Valsami-Jones, L. Tuchscherr, B. Rutkowski, K. Górecki, P. Bała and A. Czyrska-Filemonowicz, Electrophoretic deposition of organic/inorganic composite coatings containing ZnO nanoparticles exhibiting antibacterial properties, *Mater. Sci. Eng. C*, 2017, **77**, 780–789.
14. J. E. dos Santos, E. R. Dockal and E. T. Cavalheiro, Synthesis and characterization of Schiff bases from chitosan and salicylaldehyde derivatives, *Carbohydr. Polym.*, 2005, **60**, 277–282.
15. V. Wagener, A. Boccaccini and S. Virtanen, Protein-adsorption and Ca-phosphate formation on chitosan-bioactive glass composite coatings, *Appl. Surf. Sci.*, 2017, **416**, 454–460.
16. A. Simchi, F. Pishbin and A. Boccaccini, Electrophoretic deposition of chitosan, *Mater. Lett.*, 2009, **63**, 2253–2256.
17. A. Boccaccini, S. Keim, R. Ma, Y. Li and I. Zhitomirsky, Electrophoretic deposition of biomaterials, *J. R. Soc., Interface*, 2010, **7**, S581–S613.
18. Q. Yao, P. Nooeaid, J. A. Roether, Y. Dong, Q. Zhang and A. R. Boccaccini, Bioglass®-based scaffolds incorporating polycaprolactone and chitosan coatings for controlled vancomycin delivery, *Ceram. Int.*, 2013, **39**, 7517–7522.
19. S. A. Umoren, M. J. Banera, T. Alonso-Garcia, C. A. Gervasi and M. V. Mirífico, Inhibition of mild steel corrosion in HCl solution using chitosan, *Cellulose*, 2013, **20**, 2529–2545.
20. M. Rhazi, J. Desbrières, A. Tolaimate, M. Rinaudo, P. Vottero and A. Alagui, Contribution to the study of the complexation of copper by chitosan and oligomers, *Polymer*, 2002, **43**, 1267–1276.
21. M. Rhazi, J. Desbrières, A. Tolaimate, M. Rinaudo, P. Vottero, A. Alagui and M. El Meray, Influence of the nature of the metal ions on the complexation with chitosan. Application to the treatment of liquid waste, *Eur. Polym. J.*, 2002, **38**, 1523–1530.
22. L. S. Lima and C. Airoldi, Interaction of copper with chitosan and succinic anhydride derivative—a factorial design evaluation of the chemisorption process, *Colloids Surf., A*, 2003, **229**, 129–136.
23. L. S. Lima and C. Airoldi, A thermodynamic investigation on chitosan–divalent cation interactions, *Thermochim. Acta*, 2004, **421**, 133–139.

24. M. N. El-Haddad, Chitosan as a green inhibitor for copper corrosion in acidic medium, *Int. J. Biol. Macromol.*, 2013, **55**, 142–149.
25. K. El Mouaden, B. El Ibrahimi, R. Oukhrib, L. Bazzi, B. Hammouti, O. Jbara, A. Tara, D. S. Chauhan and M. A. Quraishi, Chitosan polymer as a green corrosion inhibitor for copper in sulfide-containing synthetic seawater, *Int. J. Biol. Macromol.*, 2018, **119**, 1311–1323.
26. N. K. Gupta, P. G. Joshi, V. Srivastava and M. A. Quraishi, Chitosan: A macromolecule as green corrosion inhibitor for mild steel in sulfamic acid useful for sugar industry, *Int. J. Biol. Macromol.*, 2018, **106**, 704–711.
27. R. F. Elshaarawy, F. H. Mustafa, A. Herbst, A. E. Farag and C. Janiak, Surface functionalization of chitosan isolated from shrimp shells, using salicylaldehyde ionic liquids in exploration for novel economic and ecofriendly antibiofoulants, *RSC Adv.*, 2016, **6**, 20901–20915.
28. M. Abdallah, Guar gum as corrosion inhibitor for carbon steel in sulfuric acid solutions, *Port. Electrochim. Acta*, 2004, **22**, 161–175.
29. G. Palumbo, K. Berent, E. Proniewicz and J. Banaś, Guar gum as an eco-friendly corrosion inhibitor for pure aluminium in 1-M HCl solution, *Materials*, 2019, **12**, 2620.
30. M. Messali, H. Lgaz, R. Dassanayake, R. Salghi, S. Jodeh, N. Abidi and O. Hamed, Guar gum as efficient non-toxic inhibitor of carbon steel corrosion in phosphoric acid medium: Electrochemical, surface, DFT and MD simulations studies, *J. Mol. Struct.*, 2017, **1145**, 43–54.
31. N. Thombare, U. Jha, S. Mishra and M. Siddiqui, Guar gum as a promising starting material for diverse applications: A review, *Int. J. Biol. Macromol.*, 2016, **88**, 361–372.
32. H. A. Swenson, H. M. Kaustinen, O. A. Kaustinen and N. S. Thompson, Structure of gum arabic and its configuration in solution, *J. Polym. Sci., Part A-2*, 1968, **6**, 1593–1606.
33. K. Azzaoui, E. Mejdoubi, S. Jodeh, A. Lamhamdi, E. Rodriguez-Castellón, M. Algarra, A. Zarrouk, A. Errich, R. Salghi and H. Lgaz, Eco friendly green inhibitor Gum Arabic (GA) for the corrosion control of mild steel in hydrochloric acid medium, *Corros. Sci.*, 2017, **129**, 70–81.
34. C. Shen, V. Alvarez, J. D. Koenig and J.-L. Luo, Gum Arabic as corrosion inhibitor in the oil industry: experimental and theoretical studies, *Corros. Eng., Sci. Technol.*, 2019, **54**, 444–454.
35. G. Palumbo, M. Górny and J. Banaś, Corrosion inhibition of pipeline carbon steel (N80) in CO_2-saturated chloride (0.5 M of KCl) solution using Gum arabic as a possible environmentally friendly corrosion inhibitor for shale gas industry, *J. Mater. Eng. Perform.*, 2019, **28**, 6458–6470.
36. S. Umoren, I. Obot, E. Ebenso, P. Okafor, O. Ogbobe and E. Oguzie, Gum arabic as a potential corrosion inhibitor for aluminium in alkaline medium and its adsorption characteristics, *Anti-Corros. Methods Mater.*, 2006, **53**, 277–282.
37. M. Abu-Dalo, A. Othman and N. Al-Rawashdeh, Exudate gum from acacia trees as green corrosion inhibitor for mild steel in acidic media, *Int. J. Electrochem. Sci.*, 2012, **7**, 9303–9324.

38. M. Mobin and M. Rizvi, Inhibitory effect of xanthan gum and synergistic surfactant additives for mild steel corrosion in 1 M HCl, *Carbohydr. Polym.*, 2016, **136**, 384–393.
39. V. Pokhmurs'kyi, I. Zin, M. Tymus, S. Kornii, O. Karpenko, O. Khlopyk and N. Korets'ka, Inhibition of the corrosion of carbon steel by Xanthan biopolymer, *Mater. Sci.*, 2020, **55**, 522–528.
40. I. Arukalam, C. Alaohuru, C. Ugbo, K. Jideofor, P. Ehirim and I. Madufor, Effect of xanthan gum on the corrosion protection of aluminium in HCl medium, *Int. J. Sci. Eng. Res.*, 2014, **3**, 5–16.
41. M. Mobin, M. Khan and M. Parveen, Inhibition of mild steel corrosion in acidic medium using starch and surfactants additives, *J. Appl. Polym. Sci.*, 2011, **121**, 1558–1565.
42. S. C. Nwanonenyi, I. C. Madufor, P. C. Uzoma and I. C. Chukwujike, a. chemistry, Corrosion inhibition of mild steel in sulphuric acid environment using millet starch and potassium iodide, *Int. Res. J. Pure Appl. Chem.*, 2016, **12**, 1–15.
43. R. Rosliza and W. W. Nik, Improvement of corrosion resistance of AA6061 alloy by tapioca starch in seawater, *Curr. Appl. Phys.*, 2010, **10**, 221–229.
44. S. A. Umoren, I. B. Obot, A. Madhankumar and Z. M. Gasem, Performance evaluation of pectin as ecofriendly corrosion inhibitor for X60 pipeline steel in acid medium: experimental and theoretical approaches, *Carbohydr. Polym.*, 2015, **124**, 280–291.
45. M. V. Fiori-Bimbi, P. E. Alvarez, H. Vaca and C. A. Gervasi, Corrosion inhibition of mild steel in HCl solution by pectin, *Corros. Sci.*, 2015, **92**, 192–199.
46. M. M. Fares, A. Maayta and M. M. Al-Qudah, Pectin as promising green corrosion inhibitor of aluminum in hydrochloric acid solution, *Corros. Sci.*, 2012, **60**, 112–117.
47. A. N. Grassino, J. Halambek, S. Djaković, S. R. Brnčić, M. Dent and Z. Grabarić, Utilization of tomato peel waste from canning factory as a potential source for pectin production and application as tin corrosion inhibitor, *Food Hydrocolloids*, 2016, **52**, 265–274.
48. S. A. Umoren and A. Madhankumar, Effect of addition of CeO_2 nanoparticles to pectin as inhibitor of X60 steel corrosion in HCl medium, *J. Mol. Liq.*, 2016, **224**, 72–82.
49. S. A. Umoren, A. A. AlAhmary, Z. M. Gasem and M. M. Solomon, Evaluation of chitosan and carboxymethyl cellulose as ecofriendly corrosion inhibitors for steel, *Int. J. Biol. Macromol.*, 2018, **117**, 1017–1028.
50. M. Solomon, S. Umoren, I. Udosoro and A. Udoh, Inhibitive and adsorption behaviour of carboxymethyl cellulose on mild steel corrosion in sulphuric acid solution, *Corros. Sci.*, 2010, **52**, 1317–1325.
51. M. Deyab, Hydroxyethyl cellulose as efficient organic inhibitor of zinc–carbon battery corrosion in ammonium chloride solution: Electrochemical and surface morphology studies, *J. Power Sources*, 2015, **280**, 190–194.
52. I. Arukalam, I. Madufor, O. Ogbobe and E. Oguzie, Experimental and theoretical studies of hydroxyethyl cellulose as inhibitor for acid corrosion inhibition of mild steel and aluminium, *Open Corros. J.*, 2014, **6**, 1–10.

53. S. A. Umoren, M. M. Solomon, A. Madhankumar and I. B. Obot, Exploration of natural polymers for use as green corrosion inhibitors for AZ31 magnesium alloy in saline environment, *Carbohydr. Polym.*, 2020, **230**, 115466.
54. S. Nwanonenyi, H. Obasi and I. Eze, Hydroxypropyl cellulose as an efficient corrosion inhibitor for aluminium in acidic environments: Experimental and theoretical approach, *Chem. Afr.*, 2019, **2**, 471–482.
55. Y. Li, Z. X. Ba, Y. L. Li, Y. Ge and X. C. Zhu, Influence of sodium alginate inhibitor addition on the corrosion protection performance of AZ91D magnesium alloy in NaCl solution, *Anti-Corros. Methods Mater.*, 2017, **64**, 486–491.
56. A. Jmiai, B. El Ibrahimi, A. Tara, S. El Issami, O. Jbara and L. Bazzi, Alginate biopolymer as green corrosion inhibitor for copper in 1 M hydrochloric acid: experimental and theoretical approaches, *J. Mol. Struct.*, 2018, **1157**, 408–417.
57. W. Zhang, B. Nie, H.-J. Li, Q. Li, C. Li and Y.-C. Wu, Inhibition of mild steel corrosion in 1 M HCl by chondroitin sulfate and its synergistic effect with sodium alginate, *Carbohydr. Polym.*, 2021, **260**, 117842.
58. A. M. Al-Bonayan, Sodium aliginate as corrosion inhibitor for carbon steel in 0.5 M HCl solutions, *Int. J. Sci. Eng. Res.*, 2014, **5**, 611–618.
59. I. Obot, I. B. Onyeachu and A. M. Kumar, Sodium alginate: A promising biopolymer for corrosion protection of API X60 high strength carbon steel in saline medium, *Carbohydr. Polym.*, 2017, **178**, 200–208.
60. N. Dang, Y. Wei, L. Hou, Y. Li and C. Guo, Investigation of the inhibition effect of the environmentally friendly inhibitor sodium alginate on magnesium alloy in sodium chloride solution, *Mater. Corros.*, 2015, **66**, 1354–1362.
61. B. P. Charitha and P. Rao, Electrochemical and adsorption studies for the corrosion control of 6061 Al alloy using eco-friendly inhibitor, *Surf. Eng. Appl. Electrochem.*, 2017, **53**, 551–559.
62. B. Charitha and P. Rao, Inhibitory action of dextran on the corrosion of 6061 Al–15%$_{(v)}$ SiC$_{(P)}$ composite in HCl medium, *Perspect. Sci.*, 2016, **8**, 176–178.
63. M. Hamcerencu, J. Desbrieres, M. Popa, A. Khoukh and G. Riess, New unsaturated derivatives of Xanthan gum: Synthesis and characterization, *Polymer*, 2007, **48**, 1921–1929.
64. P.-E. Jansson, L. Kenne and B. Lindberg, Structure of the extracellular polysaccharide from *Xanthomonas campestris*, *Carbohydr. Res.*, 1975, **45**, 275–282.
65. L. D. Melton, L. Mindt and D. A. Rees, Covalent structure of the extracellular polysaccharide from *Xanthomonas campestris*: evidence from partial hydrolysis studies, *Carbohydr. Res.*, 1976, **46**, 245–257.
66. W. R. Gombotz and S. Wee, Protein release from alginate matrices, *Adv. Drug Delivery Rev.*, 1998, **31**, 267–285.
67. T. Rabizadeh and S. K. Asl, Casein as a natural protein to inhibit the corrosion of mild steel in HCl solution, *J. Mol. Liq.*, 2019, **276**, 694–704.

68. Y. Nie, J. Gao, E. Wang, L. Jiang, L. An and X. Wang, An effective hybrid organic/inorganic inhibitor for alkaline aluminum-air fuel cells, *Electrochim. Acta*, 2017, **248**, 478–485.
69. A. Zomorodian, R. Bagonyi and A. Al-Tabbaa, The efficiency of eco-friendly corrosion inhibitors in protecting steel reinforcement, *J. Build. Eng.*, 2021, **38**, 102171.
70. A. Hendy, J. M. Elnady, N. Hassan, T. Tamer, A. S. Ahmed and R. M. Abou Shahba, Corrosion inhibition of Casein as a natural protein for 316L stainless steel in simulated body fluid, *Al-Azhar J. Dent. Sci.*, 2020, **31**, 11–18.
71. K. Haruna, I. Obot, N. Ankah, A. Sorour and T. Saleh, Gelatin: A green corrosion inhibitor for carbon steel in oil well acidizing environment, *J. Mol. Liq.*, 2018, **264**, 515–525.
72. M. Abdallah, E. Kamar, A. El-Etre and S. Eid, Gelatin as corrosion inhibitor for aluminum and aluminum silicon alloys in sodium hydroxide solutions, *Prot. Met. Phys. Chem. Surf.*, 2016, **52**, 140–148.
73. P. Roy, T. Maji, S. Dey and D. Sukul, Adsorption behaviour of gluten hydrolysate on mild steel in 1 M HCl and its role as a green corrosion inhibitor, *RSC Adv.*, 2015, **5**, 61170–61178.
74. Z. Zhang, H. Ba, Z. Wu and Y. Zhu, The inhibition mechanism of maize gluten meal extract as green inhibitor for steel in concrete *via* experimental and theoretical elucidation, *Constr. Build. Mater.*, 2019, **198**, 288–298.
75. Z. Zhang, H. Ba and Z. Wu, Sustainable corrosion inhibitor for steel in simulated concrete pore solution by maize gluten meal extract: electrochemical and adsorption behavior studies, *Constr. Build. Mater.*, 2019, **227**, 117080.
76. A. M. El-Azaly, Influence of Soybean (Glycine Max) plant extract on corrosion of aluminum in 1 M HCl, *Int. J. Electrochem. Sci.*, 2019, **14**, 2714–2731.
77. S. Nwanonenyi, H. Obasi, I. Chukwujike, M. Chidiebere and E. Oguzie, Inhibition of carbon steel corrosion in 1 M H_2SO_4 using soy polymer and polyvinylpyrrolidone, *Chem. Afr.*, 2019, **2**, 277–289.
78. I. Obot, D. Macdonald and Z. Gasem, Density functional theory (DFT) as a powerful tool for designing new organic corrosion inhibitors. Part 1: an overview, *Corros. Sci.*, 2015, **99**, 1–30.
79. L. Sihem, B. Abderrahim, B. Brahim, M. Asma, T. Lahcene and S. Marzorati, Glycerin-grafted starch as corrosion inhibitor of C-Mn steel in 1 M HCl solution, *Appl. Sci.*, 2019, **9**, 4684.
80. M. Bello, N. Ochoa, V. Balsamo, F. Lopez-Carrasquero, S. Coll, A. Monsalve and G. González, Modified cassava starches as corrosion inhibitors of carbon steel: An electrochemical and morphological approach, *Carbohydr. Polym.*, 2010, **82**, 561–568.
81. S. A. Al Kiey, M. S. Hasanin and S. Dacrory, Potential anticorrosive performance of green and sustainable inhibitor based on cellulose derivatives for carbon steel, *J. Mol. Liq.*, 2021, **338**, 116604.
82. S. M. Tawfik, Alginate surfactant derivatives as an ecofriendly corrosion inhibitor for carbon steel in acidic environments, *RSC Adv.*, 2015, **5**, 104535–104550.

83. Q. Zhang, B. Hou, Y. Li, G. Zhu, Y. Lei, X. Wang, H. Liu and G. Zhang, Dextran derivatives as highly efficient green corrosion inhibitors for carbon steel in CO_2-saturated oilfield produced water: Experimental and theoretical approaches, *Chem. Eng. J.*, 2021, **424**, 130519.
84. J. Haque, V. Srivastava, D. S. Chauhan, H. Lgaz and M. A. Quraishi, Microwave-induced synthesis of chitosan Schiff bases and their application as novel and green corrosion inhibitors: Experimental and theoretical approach, *ACS Omega*, 2018, **3**, 5654–5668.
85. D. S. Chauhan, M. A. J. Mazumder, M. A. Quraishi and K. R. Ansari, Chitosan-cinnamaldehyde Schiff base: A bioinspired macromolecule as corrosion inhibitor for oil and gas industry, *Int. J. Biol. Macromol.*, 2020, **158**, 127–138.
86. D. S. Chauhan, M. A. J. Mazumder, M. A. Quraishi, K. R. Ansari and R. Suleiman, Microwave-assisted synthesis of a new Piperonal-Chitosan Schiff base as a bio-inspired corrosion inhibitor for oil-well acidizing, *Int. J. Biol. Macromol.*, 2020, **158**, 231–243.
87. M. A. Quraishi, K. R. Ansari, D. S. Chauhan, S. A. Umoren and M. A. J. Mazumder, Vanillin modified chitosan as a new bio-inspired corrosion inhibitor for carbon steel in oil-well acidizing relevant to petroleum industry, *Cellulose*, 2020, **27**, 6425–6443.
88. K. R. Ansari, D. S. Chauhan, M. A. Quraishi, M. A. J. Mazumder and A. Singh, Chitosan Schiff base: an environmentally benign biological macromolecule as a new corrosion inhibitor for oil & gas industries, *Int. J. Biol. Macromol.*, 2020, **144**, 305–315.
89. D. S. Chauhan, K. R. Ansari, A. A. Sorour, M. A. Quraishi, H. Lgaz and R. Salghi, Thiosemicarbazide and thiocarbohydrazide functionalized chitosan as ecofriendly corrosion inhibitors for carbon steel in hydrochloric acid solution, *Int. J. Biol. Macromol.*, 2018, **107**, 1747–1757.
90. D. S. Chauhan, M. A. Quraishi, A. A. Sorour, S. K. Saha and P. Banerjee, Triazole-modified chitosan: a biomacromolecule as a new environmentally benign corrosion inhibitor for carbon steel in a hydrochloric acid solution, *RSC Adv.*, 2019, **9**, 14990–15003.
91. V. Srivastava, D. S. Chauhan, P. G. Joshi, V. Maruthapandian, A. A. Sorour and M. A. Quraishi, PEG-functionalized chitosan: A biological macromolecule as a novel corrosion inhibitor, *ChemistrySelect*, 2018, **3**, 1990–1998.
92. K. R. Ansari and M. A. Quraishi, Bis-Schiff bases of isatin as new and environmentally benign corrosion inhibitor for mild steel, *J. Ind. Eng. Chem.*, 2014, **20**, 2819–2829.
93. K. R. Ansari, M. A. Quraishi and A. Singh, Schiff's base of pyridyl substituted triazoles as new and effective corrosion inhibitors for mild steel in hydrochloric acid solution, *Corros. Sci.*, 2014, **79**, 5–15.
94. N. K. Gupta, C. Verma, M. A. Quraishi and A. K. Mukherjee, Schiff's bases derived from L-lysine and aromatic aldehydes as green corrosion inhibitors for mild steel: Experimental and theoretical studies, *J. Mol. Liq.*, 2016, **215**, 47–57.

95. P. Singh and M. A. Quraishi, Corrosion inhibition of mild steel using Novel Bis Schiff's Bases as corrosion inhibitors: electrochemical and surface measurement, *Measurement*, 2016, **86**, 114–124.
96. K. E. Mouaden, D. S. Chauhan, M. Quraishi, L. Bazzi and M. Hilali, Cinnamaldehyde-modified chitosan as a bio-derived corrosion inhibitor for acid pickling of copper: Microwave synthesis, experimental and computational study, *Int. J. Biol. Macromol.*, 2020, **164**, 3709–3717.
97. D. S. Chauhan, K. E. Mouaden, M. A. Quraishi and L. Bazzi, Aminotriazolethiol-functionalized chitosan as a macromolecule-based bioinspired corrosion inhibitor for surface protection of stainless steel in 3.5% NaCl, *Int. J. Biol. Macromol.*, 2020, **152**, 234–241.
98. K. EL Mouaden, D. S. Chauhan, M. A. Quraishi and L. Bazzi, Thiocarbohydrazide-crosslinked chitosan as a bioinspired corrosion inhibitor for protection of stainless steel in 3.5% NaCl, *Sustain. Chem. Pharm.*, 2020, **15**, 100213.
99. D. S. Chauhan, V. Srivastava, P. G. Joshi and M. A. Quraishi, PEG cross-linked Chitosan: a biomacromolecule as corrosion inhibitor for sugar industry, *Int. J. Ind. Chem.*, 2018, **9**, 363–377.

CHAPTER 7

Amino Acids and Their Derivatives as Green Corrosion Inhibitors

7.1 Introduction

Amino acids are organic molecules consisting mainly of nitrogen, oxygen, and hydrogen. These molecules are the building blocks of proteins.[1] Inside the human body, amino acids are involved in biological reactions, transport of nutrients, illness prevention, and other functions. The deficiency of amino acids inside the human body can cause a lowering in immunity, digestive troubles, fertility issues, depression, mental problems, stunted growth in kids, *etc.* The amino acid molecules contain an $-NH_2$ group, a $-COOH$ group, and a side chain (aliphatic/aromatic) specific to each amino acid molecule. These functional groups can adsorb on a desired metallic substrate, thereby promoting the corrosion protection of amino acids. The amino acids exist in the aqueous solutions as zwitterions, *i.e.*, containing separate positively ($-NH_3^+$) and negatively charged ($-COO^-$) groups, which imparts a high solubility. Moreover, amino acids are naturally occurring, biodegradable, cost-effective, and present as effective alternatives *versus* commercially used toxic corrosion inhibitors. Several reports have come up in the literature on the application of amino acids as corrosion inhibitors.[2,3] The chemical modification of the amino acids *via* functionalization strategies to develop amino acid derivatives-based corrosion inhibitors has also been reported.

The aim of the present chapter is to overview the contribution of amino acids and amino acid-based green and sustainable molecules as corrosion inhibitors. A brief account of the significance of amino acids in corrosion

Green Chemistry Series No. 73
Green Corrosion Inhibition: Fundamentals, Design, Synthesis and Applications
By Chandrabhan Verma, Mumtaz A. Quraishi and Dheeraj Singh Chauhan

Published by the Royal Society of Chemistry, www.rsc.org

inhibition is described, followed by some examples of amino acid derivatives as corrosion inhibitors. The impact of modern computational techniques in the estimation of amino acid-based inhibitors[4] is also explored.

7.2 Significance of Amino Acids as Corrosion Inhibitors

The foremost requirements of an industrially applicable corrosion inhibitor are the ease of availability, cost-effectiveness, ease of scalability of synthesis, *etc.* In this context, amino acids prove to be considerably cheaper in comparison to conventional commercial corrosion inhibitors used in the industry. Amino acids are produced at the commercial level at the million-ton scale, and still, the demand is on the rise at an annual growth rate of 5–7%. Furthermore, after production, owing to their excellent solubility, the amino acids can be as such applied in practical applications. This can check the cost of any requirement of co-solvent[5] for promoting the solubility in the corrosive media. Having a biological nature, amino acids do not pose any health hazards when discharged into the environment after usage. Further, there are no issues of prolonged shelf life associated with amino acids, and no requirement of tedious extraction or purification steps for their production.

Considering these performance aspects, as discussed above, the high solubility is a plus point for ease of mixing in any corrosion inhibitor formulation. The presence of N, O, and S atoms, heterocycles, functional groups, additional alkyl chains, *etc.*, provide these molecules with excellent sites for adsorption on the target metallic substrate.[6] It is well known that at pH < isoelectric point, an amino acid would undergo protonation, resulting in protonation of that amino acid and direct adsorption on the metallic substrate *via* electrostatic interaction. The functional groups can also provide the possibility of reaction/chemical modification in the molecular structure for improving the corrosion protection behaviour. Amino acids, therefore, fill the gap required for the green and sustainable corrosion inhibitors required in the industry. Table 7.1 presents the list of abbreviated names for some of the amino acids and their derivatives.

7.3 Application of Amino Acids in Corrosion Inhibition

The amino acids are, in general, categorized based on their molecular structure. Depending upon the location of the functional groups, they can be named alpha (α-), beta (β-), gamma (γ-), and delta (δ-) amino acids. Other categories are based on the polarity, ionization, and the nature of the side chains (aliphatic, acyclic, aromatic, containing –OH or –S groups, *etc.*). Figure 7.1 shows the names and molecular structures of the twenty physiological amino acids. In the below section, we have provided a brief literature overview of the application of amino acids as corrosion inhibitors.

Table 7.1 List of abbreviations for amino acid names.

Abbreviated name of amino acid	Full form	Abbreviated name of amino acid	Full form
Ala	Alanine	Met	Methionine
Abu	Aminobutyric acid	Orn	Ornithine
Asn	Asparagine	PAsp	Polyaspartic acid
Asp	Aspartic acid	Phe	Phenylalanine
Cys	Cysteine	Pro	Proline
Cyt	Cystine	Ser	Serine
Glu	Glutamic acid	*S*-MCys	*S*-Methyl cysteine
Gln	Glutamine	Try	Tryptophan
Gly	Glycine	Tyr	Tyrosine
His	Histidine	Val	Valine
Ile	Isoleucine	Met	Methionine
Leu	Leucine		

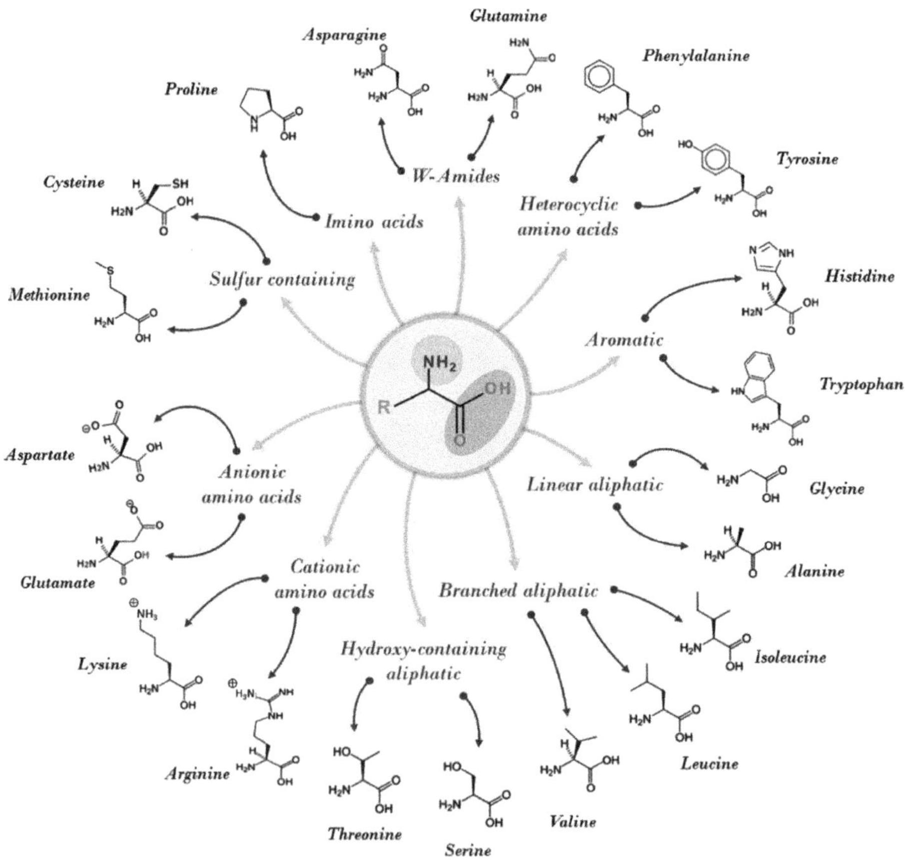

Figure 7.1 Structures of the twenty physiological amino acids and their chemical classification.[2] Reproduced from ref. 2 with permission from Elsevier, Copyright 2020.

Gly, Ala, and Leu were studied *via* gravimetric, gasometric methods, and quantum chemical investigations and high efficiencies of 65.98%, 76.03%, and 88.92%, respectively, were obtained.[7] Adsorption on the metallic surface obeyed the Langmuir isotherm, with a physical mode of adsorption. For $Cu_{10}Ni$ alloy surface, in sulfide-polluted seawater, Val, Ala, and Gly were evaluated, and the inhibition efficiencies followed the order Val (78.6%) > Ala (63.3%) > Gly (52.2%).[8] The introduction of I^- ions in the form of KI produced a synergistic improvement in the protection performance. The EDX analysis supported the inhibitor adsorption on the metal surface *via* C and N atoms of the amino acid molecules. The predominance of the physical mode of adsorption was observed. Gln was studied for mild steel surface in 1 M HCl environment *via* gravimetric and computational studies.[9] At a concentration of 100 mg L^{-1}, 96% efficiency of inhibition was observed. Theoretical studies revealed the inhibitor adsorption *via* N atoms on the metal surface. A list of amino acids and their corrosion inhibition performance is described in Table 7.2.

Table 7.2 Performance of some amino acids as corrosion inhibitors.

Name	Metal surface/medium	Inhibition efficiency/concentration	Ref.
Gly	Mild steel/0.5 M H_2SO_4	65.98%/500 mM	7
Ala		76.03%/500 mM	
Leu		88.92%/500 mM	
Gln	Mild steel/1 M HCl	96%/100 mg L^{-1}	9
Gln	Computational study	Order of IE: Gln > Asn > Asp > Glu	10
Asn			
Asp			
Glu			
Arg	Mild steel/3.5% NaCl	78.54%/900 mg L^{-1}	11
Ser	Cold-rolled steel/1 M HCl	Order of IE: Glu > Ser > Thr	12
Thr	(computational study)		
Glu			
Cys	Mild steel/0.5 M H_2SO_4	96%/10 mM	13
Cyt	Mild steel/0.5 M H_2SO_4	86%/1 mM	14
Met	Iron/0.1 M HCl	97.7%/100 mg L^{-1}	15
Tyr		89.1%/100 mg L^{-1}	
Cys	Mild steel/0.1 M HCl	82.21%/500 mg L^{-1}	16
Gly		56.50%/500 mg L^{-1}	
Leu		64.42%/500 mg L^{-1}	
Ala		56.56%/500 mg L^{-1}	
Ala	Iron/1 M HCl	76.89%/1 mM	17
L-Cys	Mild steel/1 M HCl	90.40%/100 mM	18
	Mild steel/0.5 M H_2SO_4	85.01%/100 mM	
Met	Carbon steel/0.5 M H_2SO_4	85.92%; 60.03%/10 mM	19
Cys	Copper/0.5 M H_2SO_4	81.47%; 58.98%/10 mM	
Ser		72.57%; 42.82%/10 mM	
Arg		61.24%; 93.70%/10 mM	
Gln		53.08%; 88.50%/10 mM	
Asn		43.18%; 74.91%/10 mM	

(continued)

Table 7.2 *(continued)*

Name	Metal surface/medium	Inhibition efficiency/concentration	Ref.
Val	Cu/1 M HNO_3	−15%/1 mM	20
Gly		−4%/1 mM	
Arg		38%/1 mM	
Lys		54%/1 mM	
Cys		61%/1 mM	
Asp	Cu/1 M HNO_3	24.51%/1 mM	21
Glu		45.01%/1 mM	
Ala		24.51%/1 mM	
Asn		45.01%/1 mM	
Gln		24.51%/1 mM	
Leu		45.01%/1 mM	
Met		80.38%/1 mM	
Thr		50.41%/1 mM	
Pro	Cu/1 M HNO_3	29.7%/1 mM	22
Phe		36.0%/1 mM	
Tyr		67.2%/1 mM	
Trp		81.9%/1 mM	
Ser	Cu/0.5 M HCl	83.1%/1 mM	23
Thr		87.7%/1 mM	
Glu		94.5%/1 mM	
Cys	$Cu_{10}Ni$ alloy/sulfide-polluted seawater	83.8%/80 mg L^{-1}	24
Pro	Cu/8 M H_3PO_4	39.60%/1 M	25
Cys		52.17%/1 M	
Phe		47.82%/1 M	
Ala		36.00%/1 M	
His		54.34%/1 M	
Gly		32.90%/1 M	
Cys	Cu/neutral Na_2SO_4	75.68%/60 mM	26
	Cu/alkaline Na_2SO_4	88.16%/60 mM	
Cys	AA5052 alloy/4 M NaOH	Computational study	27
Cys	Mild steel/0.1 M H_2SO_4	88.17%/40 mM	28
Ser		76.04%/40 mM	
Abu		70.03%/40 mM	
Thr		67.24%/40 mM	
Ala		58.33%/40 mM	
Val		52.12%/40 mM	
Trp		91.32%/40 mM	
Phe		87.21%/40 mM	
Tyr		82.44%/40 mM	
L-Try	Carbon steel/1 M HCl	92.7%/10 mM	29
Trp	AA2024 alloy/1 M HCl	88.7%/8 mM	30
	20% $CaCl_2$	83.32%/8 mM	
	3.5% NaCl	80.55%/8 mM	

In a theoretical analysis of Asn, Asp, Gln, and Glu, it was noted that there was a considerable affinity for bioaccumulation[31] in amino acids (from the negative values of log P).[10] The inhibition efficiency was in accordance with the order Gln > Asn > Asp > Glu. Arg was evaluated for mild steel surface in 3.5% NaCl and produced an efficiency of 78.54% at a high concentration of 900 mg L^{-1}.[11] Corrosion rates observed using gravimetric and electrochemical methods were in agreement, and computational study *via* DFT and MC simulations supported the likelihood of inhibitor adsorption on Fe (111) plane. Ser, Thr, and Glu were studied for cold-rolled steel in 1 M HCl medium.[12] The experimentally determined inhibition efficiencies were in close agreement with the theoretically determined molecular volume and were further supported by the reactivity descriptors obtained by computational analysis.

For the surface of cobalt, a series of amino acids *viz.* Ala, Gly, Leu, Val, Ile, Ser, Thr, Met, Phe, Tyr, Trp, Asp, Asn, Lys, and Glu were analysed *via* DFT and semiempirical calculations.[32] The involvement of mainly oxygen atoms was revealed in the adsorption process. The amino acid adsorption and inhibition actions were mainly in the form of the blocking type. In another DFT-based study on a Ni surface, twelve amino acids were evaluated, among which His, in the neutral form, showed the best performance on the basis of molecular orbital energy gap (ΔE) in both gaseous and aqueous media.[33] Lys, Gly, and Ala, in the protonated forms, showed the lowest ΔE in the aqueous medium.

Nitrogen-containing corrosion inhibitors are in general considered the most commonly explored. This is attributed to the presence of lone pair of electrons on the N atom that allows the adsorption over a metal surface. However, methionine and cysteine are two amino acids that contain an S atom and also possess the above attributes, and hence these are extensively explored as corrosion inhibitors. The –SH group contained in cysteine can form a self-assembled monolayer (SAM) over a metal surface, which can lead to strong chemical adsorption. The SAMs can be explained as a thin organic film having a thickness in the range of a single molecular layer. The SAMs are homogeneous, have a dense packing, and can provide an effective barrier against corrosion damage. The development of SAMs-forming inhibitors is considered an excellent strategy against the corrosion of copper.[34] Although several studies are available for steel surfaces also reporting amino acids as corrosion inhibitors. The influence of oxygen and oxidant was studied on SAMs of Cys on a Cu surface *via* computational technique.[35] In comparison to a Cu surface, Cys preferentially underwent adsorption on Cu_2O substrate to afford corrosion inhibition. The introduction of $KMnO_4$ in traces provided considerable improvement in the protection performance of Cys.

On mild steel in 0.5 M H_2SO_4, Cys provided a high efficiency of 96%.[13] EIS revealed that the adsorption of inhibitor on the metallic substrate completely suppressed the inductance loop occurring in the low-frequency region. Cystine (Cyt), which is formed by the condensation of two Cys units, produced a high efficiency of 86% for mild steel in 0.5 M H_2SO_4 at 5 mM concentration.[14] Single capacitive loops were observed for the EIS data. Met and Tyr, when evaluated for 0.1 M HCl on the surface of iron, provided 97.7% and 89.1% efficiencies.[15] L-Cys (IE 85.1%), L-Try (IE 96.3%), L-His (IE 93.4%), and L-Ser (IE 41.5%) were studied for mild steel in 1 M HCl *via* experimental and theoretical techniques.[36] The adsorption of L-Cys followed the thiol SAMs formation on the metallic substrate as discussed above. Cys (IE 82.21%), Gly (IE 56.50%), Leu (IE 64.42%), and Ala (IE 56.56%), when studied for mild steel in 1 M HCl environment, produced high efficiencies, supported by the trends of QSAR studies.[16] The N atoms of the $-NH_2$ groups were attributed as the major sites of adsorption on the metallic surface.

The adsorption of Ala was evaluated on the Fe (111) plane *via* experimental studies, and high efficiencies of 76.89% were obtained, supported by the MC simulations.[17] The adsorption of the amino acid on the oxidized surface of the metal was *via* H-bond formation. L-Cys was evaluated as an inhibitor for mild steel in 1 M HCl medium and 0.5 M H_2SO_4.[18] 90.40% protection in HCl and 85.01% was achieved. Theoretical studies revealed that in both environments, the adsorption of Cys was *via* the formation of an adsorbed film of the electrolyte anion. A series of sixteen amino acids were evaluated for Armco iron surface in 1 M HCl medium using experimental and computational methods.[37] The studied amino acids behaved as cathodic-type inhibitors, and Met, Cys, and Cyt containing S atoms in the molecular structure provided the best overall performance. The S atom was ascribed as the major site for adsorption. Met provided the best adsorption/inhibition behaviour and obeyed the Frumkin adsorption isotherm. The inhibition efficiency rose with an increase in temperature, which was a notable observation. Carbon steel and copper corrosion inhibition was studied in the presence of six amino acids employing electrochemical measurements and supported with computational techniques.[19] For steel surface, Met, and for copper, Arg provided the best overall performance. Adsorption of the inhibitors took place *via* electron donation and back-donation processes. A copper surface was analysed in 1 M HNO_3 in the presence of Val, Gly, Arg, Lys, and Cys,[20] wherein it was observed that while Val and Gly acted as corrosion promoters, the other amino acids provided inhibition behaviour, and Cys acted as the best inhibitor. In a subsequent computational study employing the DFT and MC investigations, it was revealed that Arg provided the best inhibition performance.[38]

On Cu surface in 1 M HNO_3,[21] Asp, Glu, Ala, Asn, Gln, Met, and Thr were evaluated as inhibitors wherein the best performance was revealed for Met, Thr, and Asn. The corrosion inhibition performance was afforded by the interaction of a copper surface with N, S, and O atoms, and the inhibitors acted by blocking the metal surface from the aggressive electrolyte. In a subsequent study, Pro, Phe, Tyr, and Trp were studied, wherein Tyr provided the best performance.[22] Computational studies evidenced that the inhibitor adsorption on the metallic surface was *via* the heteroatoms and the π-electrons.

Ser, Thr, and Glu were studied for Cu surface *via* electrochemical techniques in a 0.5 M HCl environment.[23] The studied amino acids behaved as a mixed type of corrosion inhibitor, and the EIS data revealed a diffusion-controlled performance. Glu carrying a smaller net positive charge on N atom and the greater net positive charge on O atom was observed as the best inhibitor.

On the surface of $Cu_{10}Ni$ alloy in sulfide-polluted seawater, Cys provided 83.8% efficiency, with mixed-type behaviour and cathodic predominance.[24] The inhibitor adsorption was in accordance with the Langmuir adsorption isotherm, and a synergistic improvement in the performance was noted with the addition of KI. Six amino acids, namely Pro, Cys, Phe, Ala, His, and Gly were studied on copper surface in 8 M H_3PO_4.[25] All the amino acids showed low to moderate efficiencies, wherein Cys provided the best performance with 52.1% protection. Thermodynamic parameters indicated a physical mode of adsorption. In neutral and alkaline Na_2SO_4, on the surface of copper, Cys was also studied, and high efficiencies of 75.68% and 88.16% were observed.[26] The inhibition performance was ascribed to the formation of Cu(I)-Cys surface film at the lower anodic potentials, which disturbed at greater anodic potentials due to the formation of $Cu(II)_{ads}$ complex.

On the surface of AA5052 aluminium alloy in 4 M NaOH solution, Cys adsorption was studied.[27] The polarization curves indicated that Cys acted as a cathodic corrosion inhibitor and provided a geometric covering effect that retarded the corrosion reaction. The inhibitor adsorption was attributed to the coordinate bond formation between N, O, S atoms and the atoms of the Al surface. Nine amino acids, namely Cys, Ser, Abu, Thr, Val, Ala, Trp, Phe, and Tyr were analysed for mild steel surface.[28] It was observed that the presence of –SH and –OH functional groups aided in the adsorption on the metallic substrate. In the phenyl ring containing amino acids, the ring was the major site for adsorption on the metal surface *via* electrophilic attack. Some amino acids were studied for pitting corrosion of Al alloy Al7075 in NaCl solution *via* computational analysis.[39] The ΔE values calculated *via* the calculations in the aqueous phase using AM1 and PM3 methods suggested a strong tendency to adsorb at the metallic substrate. L-Try was studied for low carbon steel in 1 M HCl solution, and a moderate efficiency of 67.3% was derived at 328 K.[29] The presence of heterocyclic rings in the form of indole and imidazole groups in L-Try, and L-His in their structural framework were responsible for the adsorption of the amino acids on the metallic substrate.

7.4 Issues in the Use of Amino Acids as Corrosion Inhibitors

Amino acids are naturally green and sustainable, are available in abundance, and are biodegradable. These molecules show excellent solubility in the aqueous corrosive media, which makes them ideal candidates for the development of corrosion inhibitors. In addition, since all of them have proven structures and properties, there is no additional requirement for research in structural analysis. However, it has been observed that, despite being effective inhibitors, the simple chemical structures lead to poor surface coverage that results in a restriction in their potency to adsorb on the metal surface. A prospective strategy could be to perform simple single-step chemical modifications of the amino acid structure, as discussed below.

7.5 Chemically Modified Amino Acids as Corrosion Inhibitors

Chemical modifications of amino acids can lead to incorporating alkyl chains, increasing heteroatoms, heterocycles, phenyl rings, functional groups, *etc.* This method can also result in the formation of amino acid derivatives, such as Schiff bases, ionic liquids, surfactants, *etc.* Another strategy is *via* the combination of two or more amino acids, which has been fruitful in developing efficient corrosion inhibitors.[40,41] The amino acid derivatives so formed can have improved solubility, better adsorption, and efficient inhibition performance.[42] A list of some chemically modified amino acids and their corrosion inhibition performance is given in Table 7.3. The below section shows an overview of some simple methodologies to develop amino acid-based inhibitors.

7.5.1 Schiff Base Formation

The major benefits associated with Schiff base corrosion inhibitors[43–45] are the single-step preparation in most cases, and the π-bond in the imine linkage (–HC=NH–) that can promote adsorption on target metallic substrates. Accordingly, several studies have explored the synthesis and application of amino acid Schiff bases as corrosion inhibitors. A Schiff base of Lys was prepared using salicylaldehyde, anisaldehyde, and dimethylamino benzaldehyde following a single-step room temperature reaction protocol.[46] This resulted in the incorporation of an aliphatic carbon chain, phenyl ring, imine bond, and the polar functional groups –OH (SB-1), $-OCH_3$ (SB-2), and $-N(CH_3)_2$ (SB-3) in the prepared molecules. The trends observed in the inhibition efficiencies were in direct agreement with the series of Hammett substitution constants for $-OCH_3$ (−0.22) and $-N(CH_3)_2$ (−0.83). Schiff bases of Trp synthesized using salicylaldehyde were explored as inhibitors for stainless steel in 1 M HCl.[47] The inhibitor adsorption on the metallic substrate was *via* π-electrons and the lone pair electrons on the heteroatoms. Schiff bases of amino acids Gly (ASB-1), Ala (ASB-2), and Leu (ASB-3) were evaluated as inhibitors for sweet corrosion of carbon steel (3.5% NaCl saturated with CO_2).[48] At a concentration of 1 mM, high efficiencies of >90% were obtained for all the studied inhibitors. The inhibitors behaved as mixed type, and their adsorption followed the Langmuir isotherm. Cinnamaldehyde Schiff bases prepared using three amino acids, namely Gly (CGSB), His (CHSB), and Trp (CTSB), were studied as inhibitors for mild steel in 1 M HCl solution.[49] Effective adsorption and corrosion inhibition performance was observed at 60 °C after an immersion period of 96 h. Thermodynamic parameters indicated the chemical adsorption of the inhibitors on the metal surface. Three Green Schiff bases of Cys, namely CSB-1, CSB-2, and CSB-3, were prepared and evaluated for mild steel in 1 M HCl solution.[50] High efficiency of 97% was observed at 200 mg L^{-1} concentration. Electrochemical measurements indicated a charge transfer controlled mechanism with mixed-type behaviour.

Table 7.3 Performance of some chemically modified amino acids as corrosion inhibitors.

Name	Metal surface/medium	Inhibition efficiency/concentration	Ref.
Ala	Iron/1 M HCl	77.3%/5 mM	51
Cys		94.2%/5 mM	
S-MCys		86.0%/5 mM	
Glu	Cu/0.5 M HCl	61.6%/15 mM	41
Cys		92.9%/15 mM	
Gly		60%/15 mM	
Glutathione		95.3%/5 mM	
NACYS	Mild steel/1 M HCl	92.5%/10 mM	52
NASBCYS		95.7%/10 mM	
NASHCYS		92.4%/10 mM	
Cys		90.4%/10 mM	
Met	Mild steel/0.5 M H_2SO_4	93.21%/50 mM	53
Met sulfoxide		94.00%/50 mM	
Met sulfone		98.20%/50 mM	
Met	Cu/1 M HNO_3	87.6%/5 mM	54
Met sulfoxide		91.2%/5 mM	
Met sulfone		94.1%/5 mM	
OPEM	Mild steel/15% HCl	97.5%/200 mg L^{-1}	55
OPEA		95.8%/200 mg L^{-1}	
AIPA	Mild steel/1 M HCl	96.52%/0.456 mM	40
Phe-triazole	Mild steel/1 M HCl	89.3%/1 mM	56 and 57
Thr-triazole		86.3%/1 mM	
Tetra-*n*-butyl ammonium methioninate	Mild steel/1 M HCl	95.1%/1.59 mM	58
AIZ-1	Mild steel/1 M HCl	90%/0.55 mM	42 and 59
AIZ-2		93.47%/0.55 mM	
AIZ-3		96.08%/0.55 mM	
Z-1	Mild steel/1 M HCl	94.24%/0.006 mM	59
Z-2		97.10%/0.150 mM	
GZC-8	Steel/3.5% NaCl CO_2 saturated	61.47%/25 mg L^{-1}	60
GZC-12		87.28%/25 mg L^{-1}	
GZC-18		90.14%/25 mg L^{-1}	
S_1	Carbon steel/1 M HCl	83.59%/1 mM	61
S_2		94.19%/1 mM	
S_3		89.55%/1 mM	
SB-1	Mild steel/1 M HCl	93.9%/1 mM	46
SB-2		94.7%/1 mM	
SB-3		95.6%/1 mM	
S1	Stainless steel/1 M HCl	71.35%/0.5 mM	47
S2		81.03%/0.5 mM	

7.5.2 Ionic Liquids

Room temperature ionic liquids (RTILs or ILs) are considered environmentally benign chemicals and have found wide applicability in diverse areas.[62–64] ILs are the salts of organic molecules and have melting points <100 °C. These molecules consist of an organic cation along with different organic and inorganic anions.[65] ILs have some unique features, *viz.* high polarity, low toxicity, thermal stability, low vapour pressure, and non-flammability, and exhibit excellent solubility.[63,66] Considerable research has been devoted to the application of ILs as corrosion inhibitors in the past few years.[67,68] A few reports have appeared in the literature on the chemical functionalization of amino acids to develop ILs that can be utilized as effective corrosion inhibitors, as discussed below. Tetra-*n*-butyl ammonium methioninate, which is derived from Met was studied for mild steel in 1 M HCl medium, and 95.1% efficiency was obtained at 1.59 mM.[58] Mixed-type inhibition behaviour with anodic predominance was observed. The formation of the surface complex $(Fe[(H_2O)_xCl]_{ads}$ (TBA)) was assigned to the protection of metallic surface anodic sites. In a single-step chemical transformation, Gly, Ala, and Phe reacted with glyoxal, HCHO, and CH_3COOH to afford corresponding imidazolium zwitterions AIZ-1 AIZ-2, and AIZ-3.[42] The zwitterions provided high solubility with effective corrosion protection performance. Mixed-type behaviour with a charge transfer control of the inhibitor adsorption was afforded at a low dose of 0.55 mM. The incorporation of methyl groups and phenyl rings was attributed to high protection performance. In a subsequent study, Ser, and Met were modified to produced zwitterions Z-1, and Z-2, which produced high efficiencies at 0.006 mM, and 0.150 mM respectively.[59] On the other hand, the incorporation of S atoms did not produce any notable improvement in the adsorption behaviour. Three amino acid methyl ester sacchrinate-based ILs namely Phe, Leu, and Ala were analysed as corrosion inhibitors for mild steel in 1 M HCl solution.[69] The studied amino acids behaved as mixed-type inhibitors and their adsorption was in accordance with Langmuir isotherm. 1-Alkyl-3-methylimidazolium proline ([Omi[Pro]) was studied for corrosion of copper in 3.5% NaCl.[70] 90% efficiency was obtained at a concentration of 0.001 mol L^{-1}. A mixed-type behaviour with anodic predominance was observed.

7.5.3 Surfactants

Surfactants are amphiphilic molecules that promote the adsorption at electrochemical interfaces of metal/metal oxide/water *etc.*, in case of corrosion inhibition. Several surfactants have been reported to promote the adsorption of organic corrosion inhibitors on the target metallic surfaces.[71–73] In this way, the role of surfactants in corrosion inhibition is to aid in solubility and adsorption and therefore act as a constituent of a corrosion inhibitor formulation. However, a number of reports have come into the literature stating the application of surfactants as corrosion inhibitors.[74,75] Zwitterionic surfactants developed using amino acids having varying carbon chain lengths, namely GZC-8, GZC-12, and GZC-18, were studied as inhibitors for sweet corrosion of steel.[60] The studied amino acids were of a mixed-type nature, and at a low concentration of 25 mg L^{-1}, high efficiencies were observed. EIS revealed a major difference in the electrochemical mechanism from inductive (GZC-8), capacitive (GZC-12), and diffusion-control (GZC-18) for the three inhibitors. An increase in the carbon chain length produced an increment in the inhibition efficiency. Leu, Phe, and Met were reacted to oleic acid and PEG to produce three non-ionic surfactants S_1, S_2, and S_3.[61] Spontaneous adsorption of the surfactants on the metallic substrates was observed from the analysis of thermodynamic parameters. Surface tension and critical micellar concentration (CMC) were the lowest for S_2.

7.5.4 Other Chemical Modifications

For corrosion of iron in 1 M HCl, Ala, Cys, and *S*-methyl Cys were evaluated as inhibitors.[51] A schematic of the different modes of adsorption of the three amino acids is shown in Figure 7.2. Computational parameters suggested agreement with the order of inhibition efficiencies determined experimentally. Glutathione, a derivative of three amino acids Glu, Cys, and Gly, was analysed as an inhibitor for a copper surface in 0.5 M HCl.[41] The highest efficiency of glutathione was ascribed to the existence of intramolecular synergism of Gly, Cys, and Glu. Different derivatives of the amino acid Cys were studied for mild steel, wherein the introduction of $-OCH_3$ group, attached to –NH, *S*-benzyl group, and *S*-hexyl chain along with Cys were studied.[52] The presence of *S*-benzyl group (IE 95.7%) supported its better inhibition performance.

On the surface of copper, Met and its derivatives sulfide, sulfoxide, and sulfone were explored experimentally and *via* theoretical means.[53,54] A mixed mode of inhibition behaivor was afforded. Same derivatives were also evaluated for the copper surface in HNO_3 and high inhibition efficiency was observed.[54] Two amino acid derivatives, OPEM and OPEA, derived from Gly and Cys, were evaluated for mild steel in 15% HCl, simulating the oil-well acidizing condition, and high efficiency was afforded.[55] A condensation product of Cys and His, namely AIPA, was analysed at a quite low dose of 0.456 mM.[40] The presence of the imidazole ring and the –S– group were ascribed to the obtained high inhibition efficiency. It has been established that heterocyclic organic corrosion inhibitors have shown excellent performance in a wide variety of corrosive media. In this connection, triazole derivatives of amino acids Phe and Thr were prepared *via* microwave irradiation and studied for mild steel in 1 M HCl, which resulted in high inhibition efficiencies.[56,57] The structures and performances of some chemically modified amino acids are given in Figure 7.3.

7.6 Polymeric Amino Acids as Corrosion Inhibitors

As discussed in the previous chapter, polymeric corrosion inhibitors are established for their effective inhibition performance. Due to high molecular weight, the polymers can afford a great surface coverage on the metallic substrate. The presence of a higher number of functional groups on the metallic substrate can further aid in the adsorption on the metal surface. Therefore, the polymers developed from amino acids are considered effective corrosion

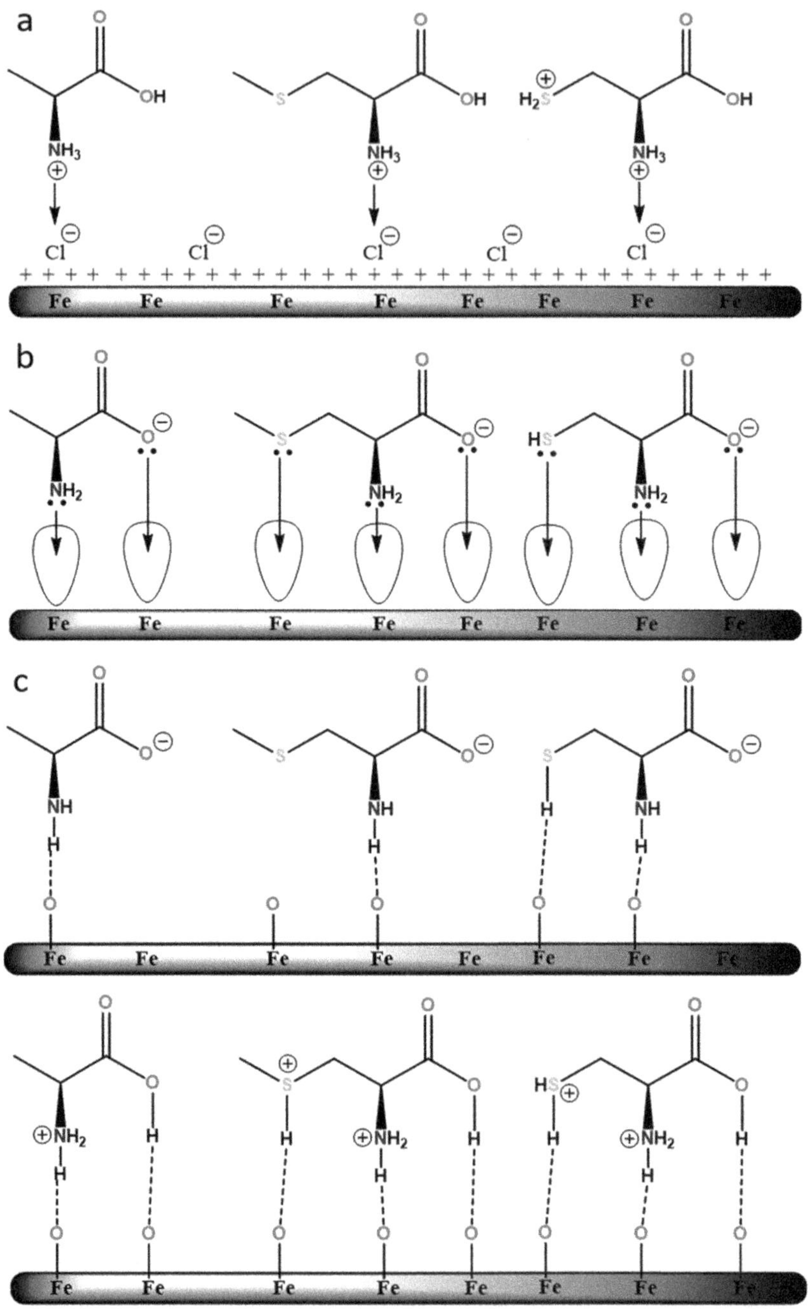

Figure 7.2 Different modes of adsorption for Ala, *S*-methyl Cys, and Cys on the surface of mild steel, (a) physical, (b) chemical and (c) H-bonding.[51] Reproduced from ref. 51 with permission from Elsevier Copyright 2010.

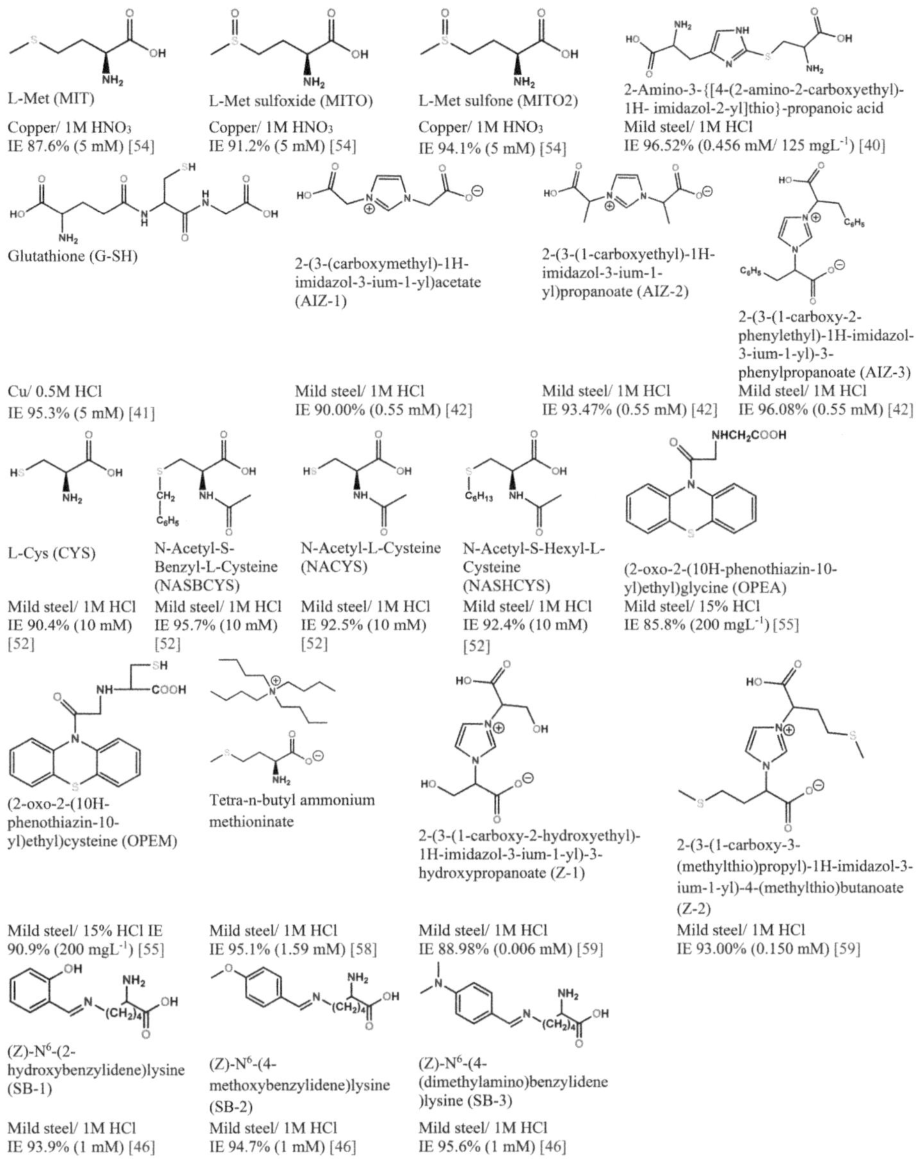

Figure 7.3 Structures and corrosion inhibition performance of some chemically modified amino acids.

inhibitors.[76–78] A Met ester-derived cycloterpolymer was prepared and evaluated as an inhibitor for elevated temperatures of up to 60 °C.[79] High inhibition performance of IE 95.7% was observed. A polysaccharide, namely Konjac glucomannan (KGM), was modified using His and Arg to afford polysaccharide esters KGMH, and KGMA, respectively.[80] The inhibitors were evaluated

Table 7.4 Performance of polymeric amino acids as corrosion inhibitors.

Name	Metal surface/medium	Inhibition efficiency/ concentration	Ref.
Polymeric sulfide	Mild steel/1 M HCl	93%/175 μM	79
Polymeric sulfoxide		96.8%/175 μM	
Polymeric sulfone		95.7%/175 μM	
KGM-His	Mild steel/0.5 M HCl	92.4%/2 g L^{-1}	80
KGM-Arg		89.9%/2 g L^{-1}	
PAsp + imidazole	Cu/3% amiosulfonic acid	89.8%/1 g L^{-1}	83
PAsp + 1 mM KI	Mild steel/0.5 M H_2SO_4	96.3%/2 g L^{-1}	84
PAsp	Mild steel/3% NaCl	61%/2 g L^{-1}	85
PAsp + 0.01 g L^{-1} Zn^{2+} ions		97%/0.5 g L^{-1}	

for mild steel in HCl environment. Tyr, His, Phe, and Leu were prepared with polyaspartic acid (PAsp) in detergent solutions without/with phosphate/silicate in an early report.[81] Several studies have subsequently shown the application of PAsp as a corrosion inhibitor. The influence of pH, temperature, and hydrodynamic condition was evaluated on the steel corrosion inhibition by PAsp.[82] PAsp accelerated corrosion at pH < 7, and promoted inhibition at pH > 10. For copper corrosion also, PAsp was reported as an inhibitor in aminosulfonic acid, and at 1 g L^{-1}, high efficiency was obtained.[83] The introduction of imidazole introduced synergism along with PAsp for mild steel surface in 0.5 M H_2SO_4 and resulted in an improvement in the protection performance.[84] In another study, the synergism of Zn^{2+} ions with PAsp significantly improved the inhibition performance.[85] Table 7.4 summarizes the performance of some polymeric amino acids as corrosion inhibitors.

7.7 Conclusions and Prospects

Amino acids are biomolecules that play various roles in biochemical processes. Due to high aqueous solubility and biocompatibility, amino acids are utilized as green corrosion inhibitors. Several research reports are available mentioning the application of amino acids as corrosion inhibitors for different metals and alloys. The presence of N and O heteroatoms in the amino acids allows effective adsorption. Further, the –SH groups have been reported to form self-assembled monolayers on the metal surfaces. The simple structures suggest that the computational studies can be performed on amino acid molecules in comparatively less time than on other complex organic molecules, drugs, polymers, *etc.* Furthermore, considering the remarkable adsorption and protection performance, some reports are available on amino acid derivatives as inhibitors. Effect of variation in alkyl chain length, incorporation of heteroatoms, heterocycles, phenyl rings, *etc.* have been studied on amino acids and derivatives. Researchers have studied Schiff bases, ionic liquids, and surfactants derived from amino acids, and some polymeric amino acids as inhibitors.

Prospective studies should focus on the development of corrosion inhibitor formulations using amino acids or derivatives as the major active constituents. These formulations should be evaluated at high temperatures and in boiling conditions, under flow regimes, and in interaction with corrosion products. Comparison with corrosion inhibitors commonly in practice in industrial environments should be conducted with amino acid-based inhibitors. The cost of the amino acid inhibitors and detailed toxicity profiles should be assessed to get a practical understanding of the prospects of commercialization.

References

1. D. L. Nelson, A. L. Lehninger and M. M. Cox, *Lehninger Principles of Biochemistry*, Macmillan, 2008.
2. B. El Ibrahimi, A. Jmiai, L. Bazzi and S. El Issami, Amino acids and their derivatives as corrosion inhibitors for metals and alloys, *Arabian J. Chem.*, 2020, **13**, 740–771.
3. D. S. Chauhan, M. Quraishi, V. Srivastava, J. Haque and B. El ibrahimi, Virgin and chemically functionalized amino acids as green corrosion inhibitors: Influence of molecular structure through experimental and in silico studies, *J. Mol. Struct.*, 2021, **1226**, 129259.
4. I. Obot, K. Haruna and T. Saleh, Atomistic simulation: A unique and powerful computational tool for corrosion inhibition research, *Arabian J. Sci. Eng.*, 2019, **44**, 1–32.
5. M. Finšgar and J. Jackson, Application of corrosion inhibitors for steels in acidic media for the oil and gas industry: a review, *Corros. Sci.*, 2014, **86**, 17–41.
6. M. A. Quraishi, D. S. Chauhan and V. S. Saji, Heterocyclic biomolecules as green corrosion inhibitors, *J. Mol. Liq.*, 2021, **341**, 117265.
7. N. O. Eddy, U. J. Ibok and B. I. Ita, QSAR and quantum chemical studies on the inhibition potentials of some amino acids for the corrosion of mild steel in H_2SO_4, *J. Comput. Methods Sci. Eng.*, 2011, **11**, 25–43.
8. A. Fouda, A. A. Nazeer and E. Ashour, Amino acids as environmentally-friendly corrosion inhibitors for $Cu_{10}Ni$ alloy in sulfide-polluted salt water: Experimental and theoretical study, *Zaštita Materijala*, 2011, **52**, 21–34.
9. A. Singh and E. E. Ebenso, Use of Glutamine as a new and effective corrosion inhibitor for mild steel in 1 M HCl solution, *Int. J. Electrochem. Sci.*, 2013, **8**, 12874–12883.
10. N. O. Eddy, Part 3. Theoretical study on some amino acids and their potential activity as corrosion inhibitors for mild steel in HCl, *Mol. Simul.*, 2010, **36**, 354–363.
11. K. Khaled and S. R. Al-Mhyawi, Electrochemical and density function theory investigations of L-Arginine as corrosion inhibitor for Steel in 3.5% NaCl, *Int. J. Electrochem. Sci.*, 2013, **8**, 4055–4072.
12. F. Kandemirli, M. Saracoglu, M. A. Amin, M. A. Basaran and C. D. Vurdu, The Quantum chemical calculations of serine, therionine and glutamine, *Int. J. Electrochem. Sci.*, 2014, **9**, 3819–3827.

13. M. Özcan, F. Karadağ and I. Dehri, Interfacial behavior of cysteine between mild steel and sulfuric acid as corrosion inhibitor, *Acta Phys.-Chim. Sin.*, 2008, **24**, 1387–1392.
14. M. Özcan, AC impedance measurement of cystine adsorption at mild steel/sulfuric acid interface as corrosion inhibitor, *J. Solid State Electrochem.*, 2008, **12**, 1653–1661.
15. S. Zor, F. Kandemirli and M. Bingul, Inhibition effects of methionine and tyrosine on corrosion of iron in HCl solution: electrochemical, FTIR, and quantum-chemical study, *Prot. Met. Phys. Chem. Surf.*, 2009, **45**, 46–53.
16. N. O. Eddy, F. E. Awe, C. E. Gimba, N. O. Ibisi and E. E. Ebenso, QSAR, Experimental and computational chemistry simulation studies on the inhibition potentials of some amino acids for the corrosion of mild steel in 0.1 M HCl, *Int. J. Electrochem. Sci.*, 2011, **6**, 931–957.
17. K. Khaled, N. Abdelshafi, A. El-Maghraby, A. Aouniti, N. Al-Mobarak and B. Hammouti, Alanine as corrosion inhibitor for iron in acid medium: a molecular level study, *Int. J. Electrochem. Sci.*, 2012, **7**, 12706–12719.
18. H. Cang, Z. Fei, W. Shi and Q. Xu, Experimental and theoretical study for corrosion inhibition of mild steel by L-Cysteine, *Int. J. Electrochem. Sci.*, 2012, **7**, 10121–10131.
19. G. L. Mendonça, S. N. Costa, V. N. Freire, P. N. Casciano, A. N. Correia and P. de Lima-Neto, Understanding the corrosion inhibition of carbon steel and copper in sulphuric acid medium by amino acids using electrochemical techniques allied to molecular modelling methods, *Corros. Sci.*, 2017, **115**, 41–55.
20. K. Barouni, L. Bazzi, R. Salghi, M. Mihit, B. Hammouti, A. Albourine and S. El Issami, Some amino acids as corrosion inhibitors for copper in nitric acid solution, *Mater. Lett.*, 2008, **62**, 3325–3327.
21. K. Barouni, A. Kassale, A. Albourine, O. Jbara, B. Hammouti and L. Bazzi, Amino acids as corrosion inhibitors for copper in nitric acid medium: Experimental and theoretical study, *J. Mater. Environ. Sci.*, 2014, **5**, 456–463.
22. K. Barouni, A. Kassale, L. Bazzi, R. Salghi, B. Hammouti, A. Albourine, S. El Issami, O. Jbara and M. Bouachrine, Inhibition of corrosion of copper in nitric acid solution by four amino acids, *Res. Chem. Intermed.*, 2014, **40**, 991–1002.
23. D.-Q. Zhang, Q.-R. Cai, L.-X. Gao and K. Y. Lee, Effect of serine, threonine and glutamic acid on the corrosion of copper in aerated hydrochloric acid solution, *Corros. Sci.*, 2008, **50**, 3615–3621.
24. A. A. Nazeer, N. K. Allam, A. Fouda and E. Ashour, Effect of cysteine on the electrochemical behavior of $Cu_{10}Ni$ alloy in sulfide polluted environments: experimental and theoretical aspects, *Mater. Chem. Phys.*, 2012, **136**, 1–9.
25. H. A. Rahman, A. Moustafa and M. Awad, Potentiodynamic and quantum studies of some amino acids as corrosion inhibitors for copper, *Int. J. Electrochem. Sci.*, 2012, **7**, 1266–1287.
26. M. B. Petrović, M. B. Radovanović, A. T. Simonović, S. M. Milić and M. M. Antonijević, The effect of cysteine on the behaviour of copper in neutral and alkaline sulphate solutions, *Int. J. Electrochem. Sci.*, 2012, **7**, 9043–9057.

27. D. Wang, L. Gao, D. Zhang, D. Yang, H. Wang and T. Lin, Experimental and theoretical investigation on corrosion inhibition of AA5052 aluminium alloy by l-cysteine in alkaline solution, *Mater. Chem. Phys.*, 2016, **169**, 142–151.
28. N. Eddy, Experimental and theoretical studies on some amino acids and their potential activity as inhibitors for the corrosion of mild steel, part 2, *J. Adv. Res.*, 2011, **2**, 35–47.
29. J.-J. Fu, S.-N. Li, L.-H. Cao, Y. Wang, L.-H. Yan and L.-D. Lu, L-Tryptophan as green corrosion inhibitor for low carbon steel in hydrochloric acid solution, *J. Mater. Sci.*, 2010, **45**, 979–986.
30. X. Li, B. Xiang, X.-l. Zuo, Q. Wang and Z.-d. Wei, performance, Inhibition of tryptophan on AA 2024 in chloride-containing solutions, *J. Mater. Eng. Perform.*, 2011, **20**, 265–270.
31. V. S. Sastri, *Green Corrosion Inhibitors: Theory and Practice*, John Wiley & Sons, 2012.
32. G. Gece, S. Bilgic and Ö. Türkşen, Quantum chemical studies of some amino acids on the corrosion of cobalt in sulfuric acid solution, *Mater. Corros.*, 2010, **61**, 141–146.
33. G. Gece and S. Bilgic, A theoretical study on the inhibition efficiencies of some amino acids as corrosion inhibitors of nickel, *Corros. Sci.*, 2010, **52**, 3435–3443.
34. C. M. Whelan, M. Kinsella, H. M. Ho and K. Maex, Corrosion inhibition by thiol-derived SAMs for enhanced wire bonding on Cu surfaces, *J. Electrochem. Soc.*, 2004, **151**, B33.
35. H. jing Zeng, L. Zhang, P. Liu and L. X. Gao, Influence of oxygen and oxidant on corrosion inhibition of cysteine self-assembled membranes for copper, *Colloids Surf., A*, 2014, **445**, 105–110.
36. J.-j. Fu, S.-n. Li, Y. Wang and L.-h. Cao, Computational and electrochemical studies of some amino acid compounds as corrosion inhibitors for mild steel in hydrochloric acid solution, *J. Mater. Sci.*, 2010, **45**, 6255–6265.
37. A. Aouniti, K. Khaled and B. Hammouti, Correlation between inhibition efficiency and chemical structure of some amino acids on the corrosion of armco iron in molar HCl, *Int. J. Electrochem. Sci.*, 2013, **8**, 5925–5943.
38. S. Kaya, B. Tüzün, C. Kaya and I. B. Obot, Determination of corrosion inhibition effects of amino acids: quantum chemical and molecular dynamic simulation study, *J. Taiwan Inst. Chem. Eng.*, 2016, **58**, 528–535.
39. A. Yurt, G. Bereket and C. Ogretir, Quantum chemical studies on inhibition effect of amino acids and hydroxy carboxylic acids on pitting corrosion of aluminium alloy 7075 in NaCl solution, *J. Mol. Struct.: THEOCHEM*, 2005, **725**, 215–221.
40. J. Haque, V. Srivastava, C. Verma and M. A. Quraishi, Experimental and quantum chemical analysis of 2-amino-3-((4-((S)-2-amino-2-carboxyethyl)-1H-imidazol-2-yl) thio) propionic acid as new and green corrosion inhibitor for mild steel in 1M hydrochloric acid solution, *J. Mol. Liq.*, 2017, **225**, 848–855.

41. D.-Q. Zhang, B. Xie, L.-X. Gao, Q.-R. Cai, H. G. Joo and K. Y. Lee, Intramolecular synergistic effect of glutamic acid, cysteine and glycine against copper corrosion in hydrochloric acid solution, *Thin Solid Films*, 2011, **520**, 356–361.
42. V. Srivastava, J. Haque, C. B. Verma, P. Singh, H. Lgaz, R. Salghi and M. A. Quraishi, Amino acid based imidazolium zwitterions as novel and green corrosion inhibitors for mild steel: experimental, DFT and MD studies, *J. Mol. Liq.*, 2017, **244**, 340–352.
43. G. K. Gomma and M. H. Wahdan, Schiff bases as corrosion inhibitors for aluminium in hydrochloric acid solution, *Mater. Chem. Phys.*, 1995, **39**, 209–213.
44. J. Haque, V. Srivastava, D. S. Chauhan, H. Lgaz and M. A. Quraishi, Microwave-induced synthesis of chitosan Schiff bases and their application as novel and green corrosion inhibitors: Experimental and theoretical approach, *ACS Omega*, 2018, **3**, 5654–5668.
45. P. Singh, D. S. Chauhan, S. S. Chauhan, G. Singh and M. A. Quraishi, Chemically modified expired Dapsone drug as environmentally benign corrosion inhibitor for mild steel in sulphuric acid useful for industrial pickling process, *J. Mol. Liq.*, 2019, **286**, 110903.
46. N. K. Gupta, C. Verma, M. Quraishi and A. Mukherjee, Schiff's bases derived from l-lysine and aromatic aldehydes as green corrosion inhibitors for mild steel: experimental and theoretical studies, *J. Mol. Liq.*, 2016, **215**, 47–57.
47. S. Vikneshvaran and S. Velmathi, Adsorption of L-Tryptophan-derived chiral Schiff bases on stainless steel surface for the prevention of corrosion in acidic environment: Experimental, theoretical and surface studies, *Surf. Interfaces*, 2017, **6**, 134–142.
48. H. M. Abd El-Lateef, M. Ismael and I. M. Mohamed, Novel Schiff base amino acid as corrosion inhibitors for carbon steel in CO_2-saturated 3.5% NaCl solution: experimental and computational study, *Corros. Rev.*, 2015, **33**, 77–97.
49. S. Satpati, A. Suhasaria, S. Ghosal, A. Saha, S. Dey and D. Sukul, Amino acid and cinnamaldehyde conjugated Schiff bases as proficient corrosion inhibitors for mild steel in 1 M HCl at higher temperature and prolonged exposure: Detailed electrochemical, adsorption and theoretical study, *J. Mol. Liq.*, 2021, **324**, 115077.
50. N. K. Gupta, M. Quraishi, C. Verma and A. Mukherjee, Green Schiff's bases as corrosion inhibitors for mild steel in 1 M HCl solution: experimental and theoretical approach, *RSC Adv.*, 2016, **6**, 102076–102087.
51. M. A. Amin, K. Khaled, Q. Mohsen and H. Arida, A study of the inhibition of iron corrosion in HCl solutions by some amino acids, *Corros. Sci.*, 2010, **52**, 1684–1695.
52. J.-j. Fu, S.-n. Li, Y. Wang and X.-d. Liu, Computational and electrochemical studies on the inhibition of corrosion of mild steel by l-Cysteine and its derivatives, *J. Mater. Sci.*, 2011, **46**, 3550–3559.

53. K. Khaled, Monte Carlo simulations of corrosion inhibition of mild steel in 0.5 M sulphuric acid by some green corrosion inhibitors, *J. Solid State Electrochem.*, 2009, **13**, 1743–1756.
54. K. Khaled, Corrosion control of copper in nitric acid solutions using some amino acids–A combined experimental and theoretical study, *Corros. Sci.*, 2010, **52**, 3225–3234.
55. M. Yadav, S. Kumar and L. Gope, Experimental and theoretical study on amino acid derivatives as eco-friendly corrosion inhibitor on mild steel in hydrochloric acid solution, *J. Adhes. Sci. Technol.*, 2014, **28**, 1072–1089.
56. Q. Deng, H.-W. Shi, N.-N. Ding, B.-Q. Chen, X.-P. He, G. Liu, Y. Tang, Y.-T. Long and G.-R. Chen, Novel triazolyl bis-amino acid derivatives readily synthesized via click chemistry as potential corrosion inhibitors for mild steel in HCl, *Corros. Sci.*, 2012, **57**, 220–227.
57. Q. Deng, X.-P. He, H.-W. Shi, B.-Q. Chen, G. Liu, Y. Tang, Y.-T. Long, G.-R. Chen and K. Chen, Concise CuI-catalyzed azide–alkyne 1, 3-dipolar cycloaddition reaction ligation remarkably enhances the corrosion inhibitive potency of natural amino acids for mild steel in HCl, *Ind. Eng. Chem. Res.*, 2012, **51**, 7160–7169.
58. E. Kowsari, S. Arman, M. Shahini, H. Zandi, A. Ehsani, R. Naderi, A. PourghasemiHanza and M. Mehdipour, In situ synthesis, electrochemical and quantum chemical analysis of an amino acid-derived ionic liquid inhibitor for corrosion protection of mild steel in 1M HCl solution, *Corros. Sci.*, 2016, **112**, 73–85.
59. J. Haque, V. Srivastava, M. A. Quraishi, D. S. Chauhan, H. Lgaz and I.-M. Chung, Polar group substituted imidazolium zwitterion as eco-friendly corrosion inhibitors for mild steel in acid solution, *Corros. Sci.*, 2020, **172**, 108665.
60. A. Ramírez-Estrada, V. Y. Mena-Cervantes, I. Elizalde, A. Manzo-Robledo, L. S. Zamudio-Rivera, D. A. Nieto-Álvarez, F. Farelas and R. l. Hernández-Altamirano, Development of a zwitterionic compound derived from β-amino acid as a green inhibitor for CO_2 corrosive environments, *ACS Sustainable Chem. Eng.*, 2017, **5**, 10396–10406.
61. A. Al-Sabagh, N. Nasser, O. E. El-Azabawy and A. E. El-Tabey, Corrosion inhibition behavior of new synthesized nonionic surfactants based on amino acid on carbon steel in acid media, *J. Mol. Liq.*, 2016, **219**, 1078–1088.
62. R. D. Rogers, and K. R. Seddon, *Ionic Liquids: Industrial Applications for Green Chemistry*, ACS Publications, 2002.
63. P. Wasserscheid, and T. Welton, *Ionic Liquids in Synthesis*, John Wiley & Sons, 2008.
64. S. Zhang, J. Zhang, Y. Zhang and Y. Deng, Nanoconfined ionic liquids, *Chem. Rev.*, 2016, **117**, 6755–6833.
65. C. Verma, E. E. Ebenso and M. A. Quraishi, Ionic liquids as green and sustainable corrosion inhibitors for metals and alloys: an overview, *J. Mol. Liq.*, 2017, **233**, 403–414.

66. J. D. Holbrey, M. B. Turner, and R. D. Rogers, Selection of ionic liquids for green chemical applications, in *Ionic Liquids as Green Solvents*, ACS Publications, 2003, pp. 2–12.
67. S. Cao, D. Liu, H. Ding, J. Wang, H. Lu and J. Gui, Task-specific ionic liquids as corrosion inhibitors on carbon steel in 0.5 M HCl solution: An experimental and theoretical study, *Corros. Sci.*, 2019, **153**, 301–313.
68. Y. Qiang, S. Zhang, L. Guo, X. Zheng, B. Xiang and S. Chen, Experimental and theoretical studies of four allyl imidazolium-based ionic liquids as green inhibitors for copper corrosion in sulfuric acid, *Corros. Sci.*, 2017, **119**, 68–78.
69. R. Aslam, M. Mobin, I. B. Obot and A. H. Alamri, Ionic liquids derived from α-amino acid ester salts as potent green corrosion inhibitors for mild steel in 1M HCl, *J. Mol. Liq.*, 2020, **318**, 113982.
70. X. D. Chen, M. Gong, Q. S. Fu, X. W. Zheng, and X. S. Feng, [Omim][Pro] amino acid ionic liquids as a corrosion inhibitor for copper in 3.5% NaCl solution, in *Applied Mechanics and Materials*, Trans Tech Publ, 2014, pp. 47–50.
71. M. A. Mazumder, H. A. Al-Muallem and S. A. Ali, The effects of N-pendants and electron-rich amidine motifs in 2-(p-alkoxyphenyl)-2-imidazolines on mild steel corrosion in CO_2-saturated 0.5 M NaCl, *Corros. Sci.*, 2015, **90**, 54–68.
72. Y. Zhu, M. L. Free, R. Woollam and W. Durnie, A review of surfactants as corrosion inhibitors and associated modeling, *Prog. Mater. Sci.*, 2017, **90**, 159–223.
73. D. Asefi, N. M. Mahmoodi and M. Arami, Effect of nonionic co-surfactants on corrosion inhibition effect of cationic gemini surfactant, *Colloids Surf., A*, 2010, **355**, 183–186.
74. Q. Zhang, Z. Gao, F. Xu and X. Zou, Adsorption and corrosion inhibitive properties of gemini surfactants in the series of hexanediyl-1, 6-bis-(diethyl alkyl ammonium bromide) on aluminium in hydrochloric acid solution, *Colloids Surf., A*, 2011, **380**, 191–200.
75. M. H. Sliem, M. Afifi, A. B. Radwan, E. M. Fayyad, M. F. Shibl, F. E.-T. Heakal and A. M. Abdullah, AEO7 surfactant as an eco-friendly corrosion inhibitor for carbon steel in HCl solution, *Sci. Rep.*, 2019, **9**, 1–16.
76. S. A. Umoren and U. M. Eduok, Application of carbohydrate polymers as corrosion inhibitors for metal substrates in different media: a review, *Carbohydr. Polym.*, 2016, **140**, 314–341.
77. S. P. Sitaram, J. O. Stoffer and T. J. O'Keefe, Application of conducting polymers in corrosion protection, *J. Coat. Technol.*, 1997, **69**, 65–69.
78. S. A. Umoren and M. M. Solomon, Protective polymeric films for industrial substrates: A critical review on past and recent applications with conducting polymers and polymer composites/nanocomposites, *Prog. Mater. Sci.*, 2019, **104**, 380–450.
79. L. Goni, M. J. Mazumder, S. Ali, M. Nazal and H. Al-Muallem, Biogenic amino acid methionine-based corrosion inhibitors of mild steel in acidic media, *Int. J. Miner., Metall. Mater.*, 2019, **26**, 467–482.

80. K. Zhang, W. Yang, X. Yin, Y. Chen, Y. Liu, J. Le and B. Xu, Amino acids modified konjac glucomannan as green corrosion inhibitors for mild steel in HCl solution, *Carbohydr. Polym.*, 2018, **181**, 191–199.
81. M. B. Freeman, W. M. Hann, Y. H. Paik, and G. Swift, Method of inhibiting corrosion in aqueous systems using poly (amino acids), *United States Patent* 5531934, United States, 1996.
82. D. Silverman, D. Kalota and F. Stover, Effect of pH on corrosion inhibition of steel by polyaspartic acid, *Corrosion*, 1995, **51**, 818–825.
83. H. Wang, Q. Wu, C. Li and N. Gu, Copper corrosion inhibition by polyaspartic acid and imidazole, *Mater. Corros.*, 2013, **64**, 347–352.
84. B. Qian, J. Wang, M. Zheng and B. Hou, Synergistic effect of polyaspartic acid and iodide ion on corrosion inhibition of mild steel in H_2SO_4, *Corros. Sci.*, 2013, **75**, 184–192.
85. A. Zeino, I. Abdulazeez, M. Khaled, M. W. Jawich and I. B. Obot, Mechanistic study of polyaspartic acid (PASP) as eco-friendly corrosion inhibitor on mild steel in 3% NaCl aerated solution, *J. Mol. Liq.*, 2018, **250**, 50–62.

CHAPTER 8

Plant Extracts as Green Corrosion Inhibitors

8.1 Introduction

Natural extracts derived from plants present one of the foremost categories of green corrosion inhibitors. Having a natural origin, these extracts are considered green, and their application in corrosion inhibition is in agreement with environmental regulations. Extracts from almost all the parts of a plant, *e.g.*, leaves, fruits (including pulp, skin, shell, seed), stem, bark, flower, roots, *etc.*, are useful as corrosion inhibitors.[1] In addition, the whole plant biomass itself has been utilized for inhibition purposes. The different sections of a plant are composed of various phytochemicals such as alkaloids, terpenoids, flavonoids, catechins, co-enzymes, *etc.*, including carbohydrates, amino acids, vitamins, proteins, *etc.* Accordingly, plant extract-based corrosion inhibitors have been utilized in a wide range of corrosive media on different metal surfaces.[2]

This chapter presents a brief account of the significance of plant extracts as corrosion inhibitors. The importance of natural extracts as inhibitors, major steps in extract preparation, some of the limitations, and some of the literary examples on the use of these inhibitors are discussed.

8.2 Significance of Plant Extracts as Corrosion Inhibitors

The global market of plant extracts was estimated to be USD 43.32 billion in 2019 and is expected to attain USD 66.81 billion by 2027, at a compound annual growth rate of 6% from 2020 to 2027. The major reason is that in different

Green Chemistry Series No. 73
Green Corrosion Inhibition: Fundamentals, Design, Synthesis and Applications
By Chandrabhan Verma, Mumtaz A. Quraishi and Dheeraj Singh Chauhan

Published by the Royal Society of Chemistry, www.rsc.org

areas of consumer goods, the preferred choices are shifting towards green and healthier alternatives to the conventionally used synthetic products. These areas include spices, cosmetics, packaging, pharmaceuticals, phytomedicines, herbal extracts, nutrition, flavours, fragrances, food, and beverages, *etc.* For corrosion inhibition, the major requirements are green nature, ease of availability, scalability of synthesis, cheap cost, effective performance *etc.* The plant extracts are considered environmentally benign and sustainable. The plant extracts are biodegradable, and the leftover extracts could be easily degraded in the environment. The abundant availability of plants facilitates the availability. The extract preparation procedures are comparatively cheaper and simpler compared to the synthesis of the conventionally used corrosion inhibitors. The natural extracts can be obtained in the powder and/or solution forms depending upon the application, resulting in a great extent of solubility in the corrosive environment. The phytochemical constituents contain various heteroatoms (N, S, O, P), phenyl rings, heterocycles, alkyl chains, π-bonds, functional groups *etc.*, that can interact with the target metal surface and afford a protective film formation.[3,4] Plant extracts are complex mixtures of various kinds of phytochemicals that allows cooperative synergistic corrosion inhibition.[1,5–12]

8.3 Key Points in Extract Preparation

The major types of phytochemical components found in plant extracts that have been utilized as corrosion inhibitors include alkaloids, terpenoids, flavonoids, phenolic acids, lignans, stilbenes, *etc.* (Figure 8.1). In corrosion inhibition, the natural extracts derived from different parts of plants have been applied in a wide variety of metal surfaces and corrosive solutions. Accordingly, several methods are reported to prepare plant extracts to obtain the same in the form of solid powder or liquid dissolved in polar and non-polar solvents as per requirement. The important points considered for the preparation of extracts are given below.

8.3.1 Extraction Solvents

Solvents diffuse into the tissues of plants and dissolve the existing phytochemicals.[13,14] Therefore, the choice of a suitable solvent for the process of extraction is very important. Some of the extracts' preparation requires the use of organic solvents such as ethanol, acetone, *n*-hexane *etc.* However, due to its simplicity, abundant availability, cost-effectiveness, non-flammability, non-toxicity, biocompatibility, and wide applicability, in most cases, water is generally considered the best solvent for this purpose.[15–18]

8.3.2 Temperature for Extraction

Temperature is an important parameter controlling the effective extraction of plant extracts. While a low temperature puts a limit on the solubility of the phytochemicals, high temperatures may lead to the decomposition

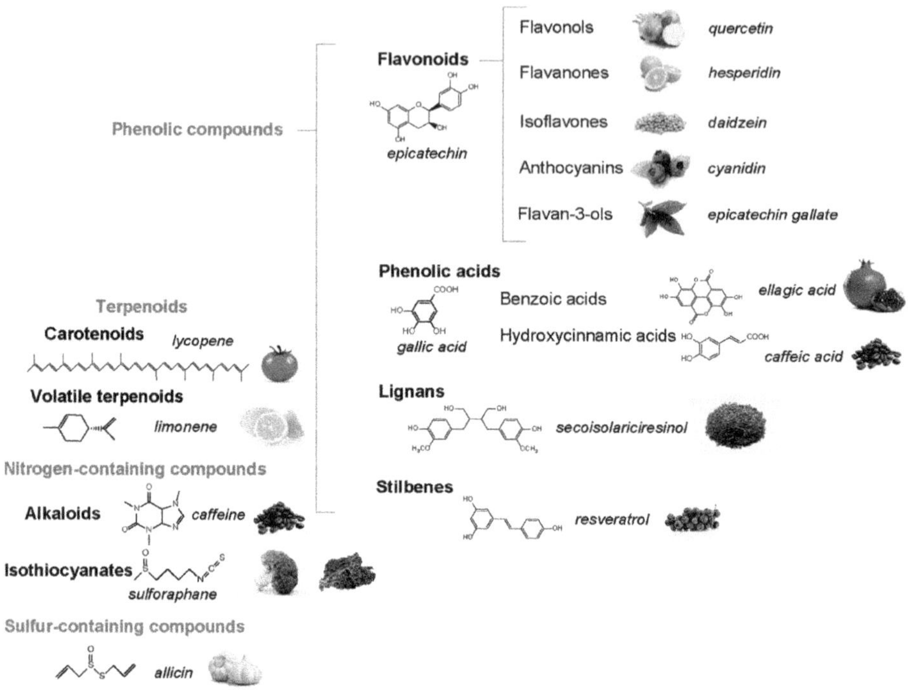

Figure 8.1 Structures of commonly occurring plant phytochemicals.[12] Reproduced from ref. 12 with permission from the Royal Society of Chemistry.

of the phytochemical constituents. In general, the range of temperature within 60–80 °C is considered acceptable to achieve optimum yield during extraction.[19,20]

8.3.3 Drying Temperature

Generally, the plant/plant materials used for extract preparation are allowed to dry at room temperature; however, this type of drying may require several days to months. Hence, drying at a slightly elevated temperature under an oven is considered generally acceptable.

8.4 Issues in Plant Extracts as Corrosion Inhibitors

Plant extracts are widely acknowledged as green corrosion inhibitors. However, there are certain associated limitations that restrict the application of natural extracts as corrosion inhibitors. In fact, due to these issues, the usage of extracts has received criticism. Over prolonged durations of storage, the extracts exhibit a natural tendency for microbial and fungal growth. In such a situation, the corrosion inhibition performance of extracts becomes different from what was observed in a fresh extract.

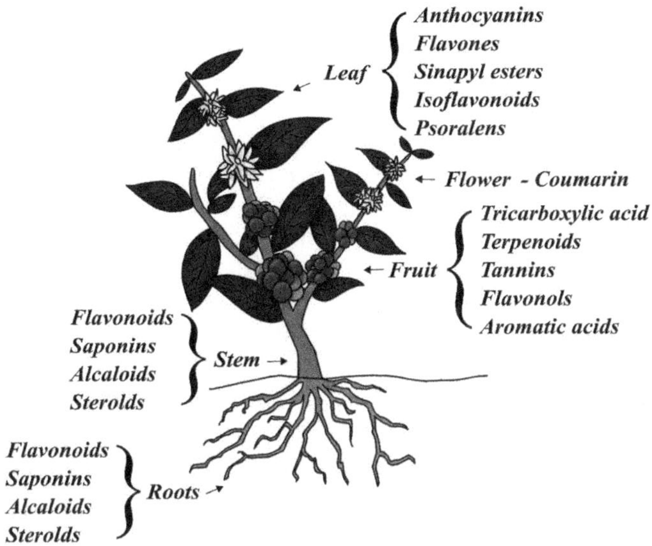

Figure 8.2 Different plant parts showing general active constituents.[22] Reproduced from ref. 22 with permission from MDPI, Copyright 2020.

Also, a plant part having varying stages of ripening, when taken for extract preparation, can have a variation in the chemical composition, which can lead to an inconsistency in inhibition performance. In fact, one study on *Musa parasidia* (banana) peels has reported this phenomenon.[21] The extract preparation condition requires a great deal of standardization and optimization, in addition to a thorough analysis of the phytochemical constituents of a plant extract. For the same reason, it is tedious to perform a thorough computational analysis of each and every phytochemical constituent to understand the adsorption behaviour. Notwithstanding the above issues, a large number of studies have reported the application of plant extracts as environmentally benign corrosion inhibitors. Figure 8.2 shows the various phytochemical constituents found in different parts of plants.

8.5 Plant Extracts Reported as Inhibitors

8.5.1 Leaf Extract

The leaf is an important part of a plant and is the centre for photosynthesis reactions. Therefore, the leaves are rich in a wide range of phytochemicals that can provide effective adsorption and corrosion. For an aluminium surface in 1 M NaOH solution, *Kalmegh* leaf extract was evaluated.[23] Andrographolide was the major phytochemical constituent of the extract. A moderately high inhibition performance of 82.4% was noticed at 1 g L^{-1} from the electroanalytical studies. In another study by the same group, the four leaf extracts from the plants *Cannabis*, *Rauwolfia*, *Cymbopogon*, *Annona*, and *Adhatoda* were evaluated for an Al surface in 1 M NaOH employing weight loss tests and electrochemical experiments (Figure 8.3).[9] Major phytochemical constituents present were ajmalicine, citral, tetrahydrocannabinol, sodium benzoate, vasicine and vasicinone. The extract of *Rauwolfia* sp. provided the best performance with 97% protection efficiency at 200 mg L^{-1}. The leaf extract of the *Calotropis* plant (major phytochemicals: calotropin, uscharin, voruscharin, and 2-oxovoruscharin) was analysed in a sweet corrosion environment on the surface of mild steel.[24] A physical mode of adsorption was revealed in alignment with Langmuir isotherm with an efficiency of 92% at 200 mg L^{-1}.

Ficus leaf extract was also studied by the same group on a mild steel surface in sweet medium in conditions simulating a practical field environment.[25] The inhibitor provided anodic shift of the corrosion potential, with 95% efficiency at 200 mg L^{-1}. *Murraya koenigii* leaf extract was studied for corrosion inhibition of mild steel in 1 M HCl and 0.5 M H_2SO_4 solutions.[26] The major phytochemicals present include β-caryophyllene, β-gurjunene, β-trans-ocimene, β-thujene, β-selinene, β-bisabolene, limonene, β-elemene, β-phellandrene and β-cadinene. Other constituents are pyrayafoline-D, murrafoline-I and mahabinine-A (Figure 8.4).[27] In both environments, quite high efficiencies were obtained at a concentration of 600 mg L^{-1}. A charge transfer control of the adsorption and inhibition behaviour with a mixed-type inhibition was noticed by the electrochemical studies. *Ginkgo biloba* is one of the most commonly studied plants for corrosion inhibition. For X70 steel, it was evaluated in 1 M HCl solution with high efficiency of 90%.[28] SEM and AFM analyses confirmed the inhibitor adsorption on metal surfaces and the formation of protective film. The major phytochemical constituents were isorhamnetin, sciadopitysin, 6-hydroxykyurenic acid, and 4-*O*-methylpyridoxine.

The measurements of the potential of zero charge and computational study shed further light on the mechanism of adsorption and inhibition. Natural olive leaf extract was studied by Pustaj *et al.* in a sweet environment, at ambient temperature.[29] At 300 mg L^{-1} concentration, an efficiency of 98.2% was achieved *via* electrochemical measurements. SEM analysis evidenced the formation of inhibitor film on the metallic substrate. Subsequently, the same extract was analysed in another study in a slightly modified sweet medium on carbon steel corrosion.[30] Bamboo leaf extract was studied for cold rolled steel and zinc surface in 0.5 M citric acid solution, wherein respectively 85.1% and 92% protection was achieved at 2 g L^{-1}.[31] Flavones were attributed to the major constituent of the extract responsible for inhibition performance. *Ananas sativum* leaves extract was reported for corrosion inhibition of Al surface in 0.1 M HCl solution.[32] The inhibitor adsorption on the metal surface obeyed the Langmuir isotherm, with the free energy values in the low and negative range suggesting the spontaneity of the adsorption.

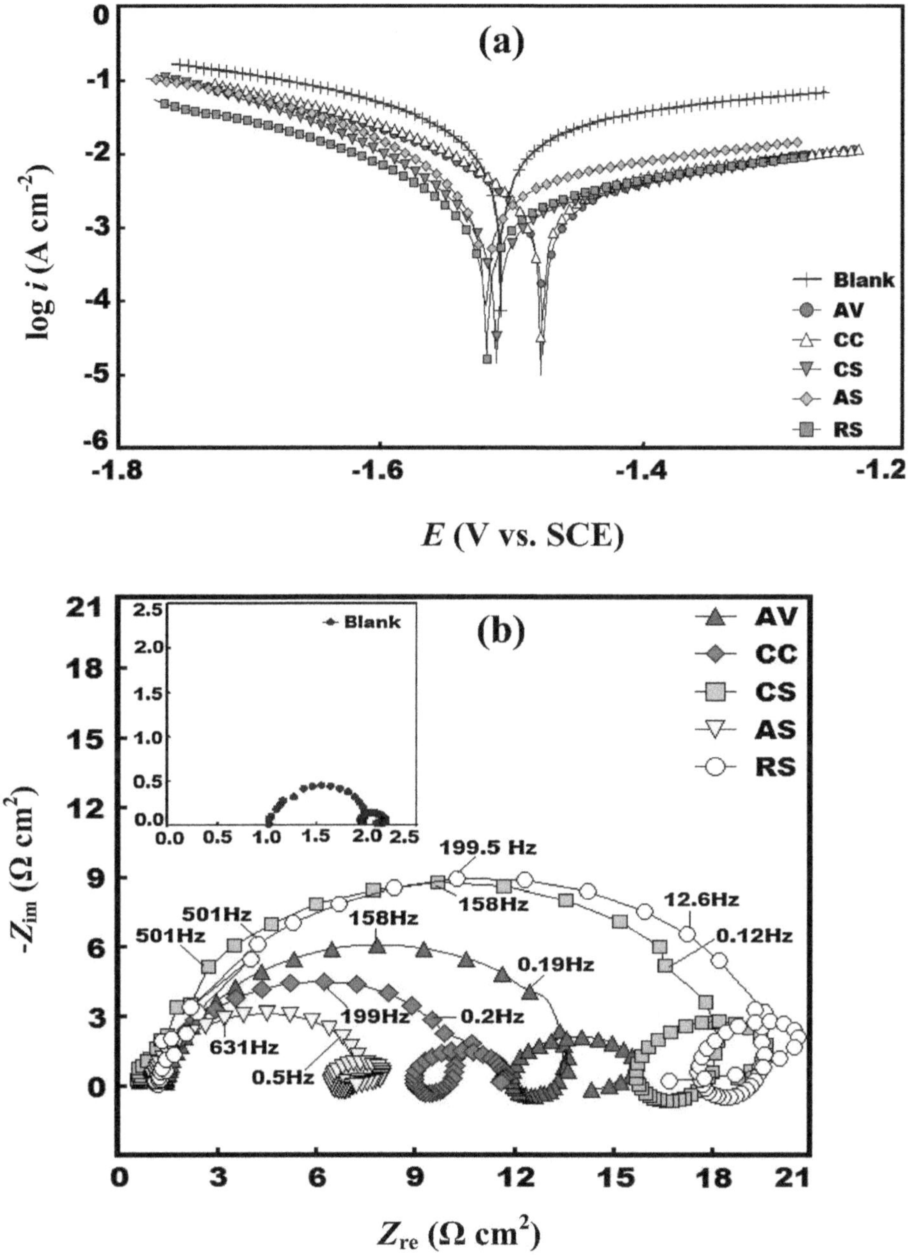

Figure 8.3 (a) PDP, and (b) EIS results (Nyquist plots) of Al surface in 1 M NaOH without and in the presence of 0.2 g L^{-1} of *Adhatoda vasica* (AV), *Cymbopogon citratus* (CC), *Cannabis sativa* (CS), *Annona squamosa* (AS) and *Rauwolfia serpentina* (RS) at 303 K. Reproduced from ref. 9 with permission from Elsevier, Copyright 2017.

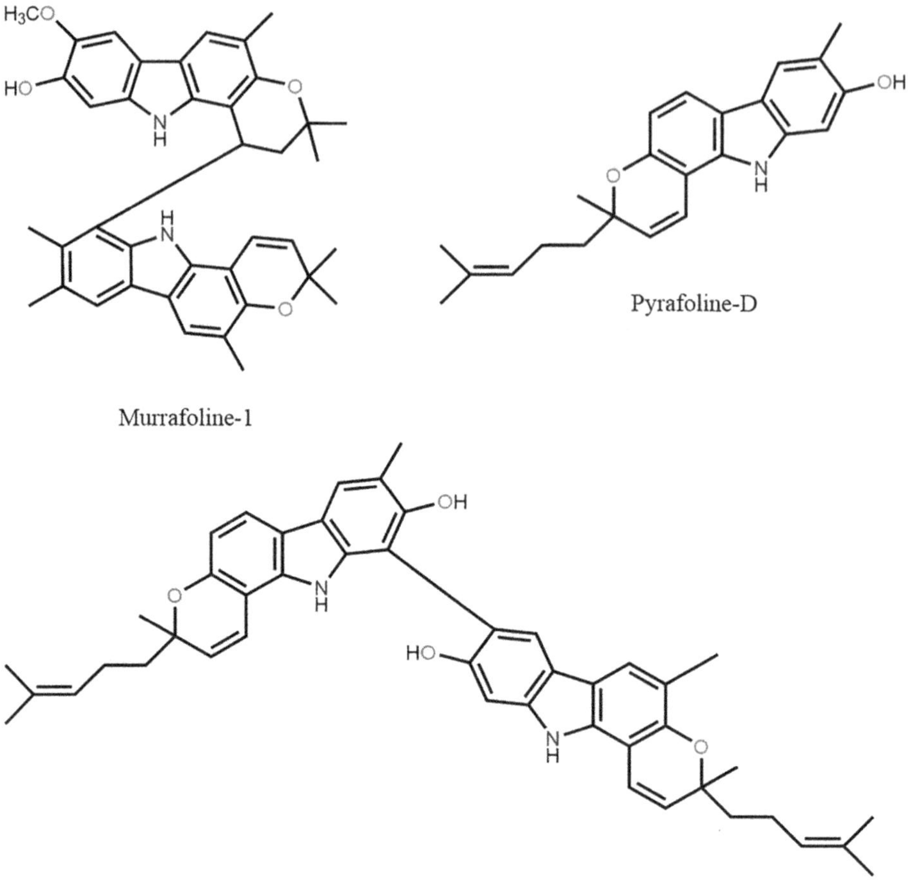

Figure 8.4 Structure of the major constituents of *Murraya koenigii* leaves. Reproduced from ref. 26 with permission from Elsevier, Copyright 2010.

8.5.2 Flower Extract

Most flowers are coloured and are rich in fragrance. This section of plants is also rich in various heteroatoms and functional groups that can aid in adsorption on the metal surface. Flower extracts from the marigold plant were studied for mild steel in 0.5 M H_2SO_4 using weight loss tests and electrochemical studies.[33] Mixed-type inhibition was noticed with predominantly physical adsorption of the major extract component lutein. Inhibitor adsorption followed the Langmuir isotherm. Borage flower aqueous extract was evaluated for mild steel in 1 M HCl.[10] The adsorption of the inhibitor on the steel surface resulted in an increase in the hydrophobic nature of the metal surface, as elucidated by the contact angle measurements. A mixed type of inhibition with cathodic predominance was noticed. An increase in the immersion time led to improvement in the adsorption and protection behaviour. The major active components responsible for protection were lactic acid, nicotinic acid, and carotene. The flower extract from *Calotropis* was studied for metal surfaces using computational studies.[34] The phytochemical constituents 4-methyl-2-phenylindole and 5,12-naphthacenedione provided the best adsorption behaviour. Table 8.1 presents the corrosion inhibition performance of some plant extracts as corrosion inhibitors.

Flower extract derived from the *Canna indica* plant was tested for mild steel in 1 M HCl solution using weight loss and electrochemical studies.[45] The extract provided a mixed type of behaviour, and the electrochemical adsorption process was charge transfer controlled. The extract contains alkaloids, glycosides, saponins, terpene, carbohydrate, steroid, protein, cholesterol, flavonoid, phylobatinin and tannin. These molecules contain fused benzene rings and O heteroatom in the rings. On mild steel and in 1 M HCl solution, in a subsequent study, researchers analysed the adsorption and inhibition effect of chamomile flower extracts.[36] After a prolonged immersion period of 8 h, a high efficiency of 98% was attained. Both experimental and computational studies verified the existence of physical and chemical modes of adsorption. Extracts of chamomile flower were also studied on S235JR steel surface in neutral chloride environment.[37] A high efficiency of 98.90% was observed at 15% v/v concentration of the extract for 2 h. The major components of the extract are phenethyl alcohol, bisabolol oxide B, coumarin, bisabolol oxide A, 2-[2.4-Hexadiynylidene]-1.6-dioxaspiro[4.4]non-3-ene, pyrrolo[2.1-*a*]isoquinoline, and 2-dodecylthiophene. The extract of *Aerva lanata* flowers was studied for low carbon steel in 1 M HCl solution as a green corrosion inhibitor.[38] These extracts contain catechin, catechol, and 3-(3-Hydroxy-3-methyl-4-penten-1-yl)-2,5,5,8*a*-tetramethyldecahydro-2-naphthalenol as the major phytochemicals. At 600 mg L^{-1} of inhibitor concentration, the protection of 88% was achieved with mixed-type behaviour. An interesting study was conducted on the effects of different growth stages of olive flowers. The extracts of olive flower inflorescence at pre-anthesis, anthesis, and grain pollen were used as eco-friendly inhibitors for steel in 1 M HCl solution.[39] Inhibition efficiency rose with the rise in inhibitor dose, with a physical mode of adsorption. The anthesis stage showed a maximum efficiency of 86.4%. Floral extract of *Sonchus oleraceus* was evaluated for mild steel in 0.5 M H_2SO_4 solution at various temperatures.[40] The phytochemical screening revealed the presence of steroids, phenolic compounds, flavonoids, tannins, glycosides, coumarin, and quinones. The inhibition efficiency decreased marginally from 96% to 89% when the temperature was raised from 298 K to 318 K, with the inhibitor adsorption showing accordance with Langmuir isotherm.

8.5.3 Fruit and Pulp Extract

The fruits are the major produce of plants that humans utilize for various purposes. The fruits and their pulp have been used for extract preparation and inhibition studies. Fruit extracts from the *Gingko biloba* plant were studied as corrosion inhibitors for N80 steel and J55 steel solutions in a sweet

Table 8.1 Performance of extracts derived from leaves, flowers, fruits, and fruit skins from different plants as corrosion inhibitors.

Name of plant	Part of plant	Metal surface/ medium	Inhibition efficiency/ concentration	Reference
Kalmegh (*Andrographis paniculata*)	Leaf	Al/1 M NaOH	82.4%/1 g L^{-1}	23
Cannabis sativa	Leaf	Al/1 M NaOH	96.7%/200 mg L^{-1}	9
Rauwolfia serpentine			97.1%/200 mg L^{-1}	
Cymbopogon citratus			87.1%/200 mg L^{-1}	
Annona squamosa			83%/200 mg L^{-1}	
Adhatoda vasica			91.2%/200 mg L^{-1}	
Calotropis procera	Leaf	Mild steel/CO_2 saturated 3.5% NaCl	91.2%/200 mg L^{-1}	24
Ficus carica	Leaf	Mild steel/CO_2 saturated 3.5% NaCl	95%/200 mg L^{-1}	25
Murraya koenigii	Leaf	Mild steel/1 M HCl	96.66%/600 mg L^{-1}	26
		Mild steel/0.5 M H_2SO_4	94.66%/600 mg L^{-1}	
Gingko biloba	Leaf	X70 steel/1 M HCl	90%/200 mg L^{-1}	28
Olive	Leaf	Carbon steel/CO_2 saturated chloride enriched water	98.2%/300 mg L^{-1}	29
Bamboo (*Phyllostachys nigra Munro*)	Leaf	Cold rolled steel/0.5 M citric acid	85.1%/2 g L^{-1}	31
		Zinc/0.5 M citric acid	92%/2 g L^{-1}	
Ananas sativum	Leaf	Al/0.1 M HCl	93.10/0.5 g L^{-1}	32
Marigold (*Tagetes erecta*)	Flower	Mild steel/0.5 M H_2SO_4	96.10%/1 g L^{-1}	33
Borage	Flower	Mild steel/1 M HCl	91%/800 mg L^{-1}	10
Calotropis procera	Flower	Computational studies	—	34
Canna indica	Flower	Mild steel/1 M HCl	95.45%/0.7%	35
Chamomile	Flower	Mild steel/1 M HCl	98%/600 mg L^{-1}	36
Matricaria recutita (Chamomile)	Flower	S235JR steel/0.5 M NaCl	98.90/15% v/v	37
Aerva lanata	Flower	Mild steel/1 M HCl	87.67/600 mg L^{-1}	38
Olea europaea. L (Olive)	Flower	Mild steel/1 M HCl	82.64%/0.02 g L^{-1}	39
Pre-anthesis			86.42%/0.02 g L^{-1}	
Anthesis			83.86%/0.02 g L^{-1}	
Grain pollen				
Sonchus oleraceus	Flower	Mild steel/0.5 M H_2SO_4	96.62%/500 mg L^{-1}	40
Gingko biloba	Fruit	N80 steel/CO_2 saturated 3.5% NaCl	92%/1000 mg L^{-1}	41
Gingko biloba	Fruit	J55 steel/CO_2 saturated 3.5% NaCl	97%/1000 mg L^{-1}	42
Chamaerops humilis L.	Fruit	Mild steel/1 M HCl	88%/1 g L^{-1}	43
Dardagan	Fruit	Mild steel/1 M HCl	92%/3 g L^{-1}	44
Rosa canina	Fruit	Mild steel/1 M HCl	85%/800 mg L^{-1}	45
Chinese gooseberry	Fruit shell	Mild steel/1 M HCl	94%/1 g L^{-1}	46

Table 8.1 (*continued*)

Name of plant	Part of plant	Metal surface/medium	Inhibition efficiency/concentration	Reference
Garlic	Peel	Carbon steel/1 M HCl	90%/400 mg L^{-1}	47
Musa parasidica (banana)	Peel	Mild steel/1 M HCl		21
Raw			90%/300 mg L^{-1}	
Ripe			82%/300 mg L^{-1}	
Over ripe			78%/300 mg L^{-1}	
Orange	Peel	Carbon steel/0.1 M HCl	95%/10% v/v	48
Pomelo	Peel	N80 steel/3.5% NaCl saturated with CO_2	86%/400 mg L^{-1}	49
Tangerine	Peel	J55 steel/3.5% NaCl saturated with CO_2	82.1%/4% v/v	50

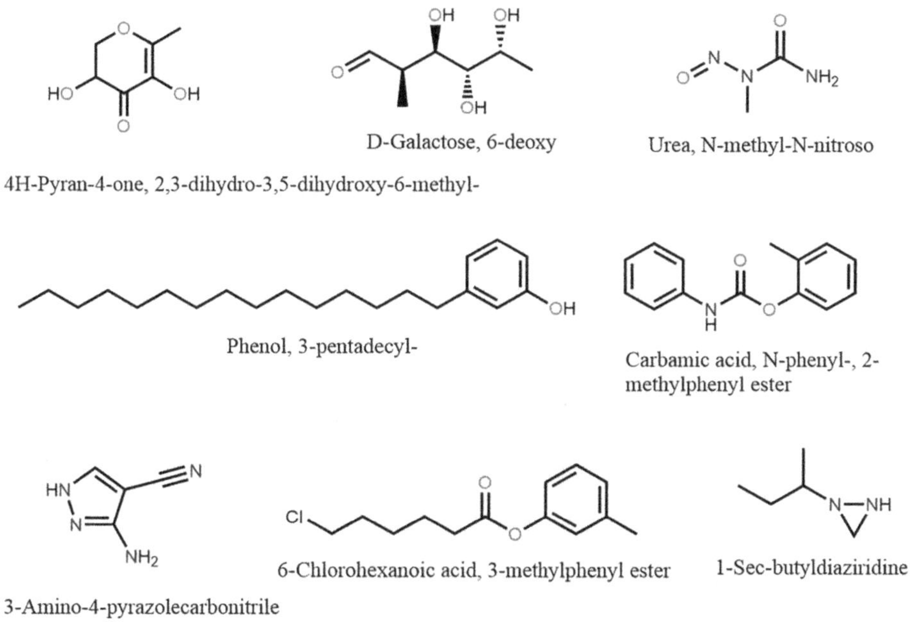

Figure 8.5 Chemical constituents present in *Gingko biloba* fruit extract. Reproduced from ref. 42 with permission from Elsevier, Copyright 2015.

corrosion medium.[41,42] The major phytochemicals existing in this extract were 1-sec-butyldiaziridine, 3-amino-4-pyrazolecarbonitrile, carbamic acid, *N*-phenyl-, 2-methylphenyl ester, 6-chlorohexanoic acid, 3-amino-4-pyrazolecarbonitrile, and 3-methylphenyl ester (Figure 8.5). Inhibitor adsorption on steel substrate was in accordance with Langmuir isotherm, and mixed-type behaviour was noted. Contact angle studies indicated an elevation in the water contact angle with a rise in the inhibitor concentration suggesting improvement in the surface hydrophobicity of the metallic substrate. Ethanol and *n*-hexane dissolved *Chamaerops humilis*, a medicinal plant, was taken and its fruit extract was studied for mild steel in 1 M HCl solution.[43] The extracts from both the solvents, at 1 g L^{-1} concentration, produced high efficiencies of 88% and 80%, respectively. Adsorption followed the Langmuir isotherm, and standard adsorption free energy (ΔG^0_{ads}) values indicated a predominance of physical adsorption. Fruit extracts from the Dardagan plant were analysed for mild steel in 1 M HCl solution, and at a high concentration (3 g L^{-1}), 92% protection was achieved.[44] A homogeneous protective film of the inhibitor was formed on the metal surface that provided excellent protection. Extracts derived from the *Rosa canina* fruit were analysed for mild steel in 1 M HCl solution.[45] An increase in the water contact angles was observed with a rise in the inhibitor dosage, indicating an improvement in the hydrophobicity with adsorbed inhibitor. The important phytochemicals present in the extract are flavonoid (marein), ascorbic acid, pectin and tannin.

8.5.4 Fruit Peel, Skin and Shell Extracts

Chinese gooseberry fruit shell extracts were analysed for mild steel in 1 M HCl solution.[46] The studies were conducted using a detailed experimental evaluation supported by DFT, MC, and MD simulations. The extract contains sucrose, maltose and folate as the major phytochemicals. The inhibitor adsorption was of a mixed-type nature with cathodic predominance. The metal-inhibitor interaction was attributed to taking place *via* polar functional groups and N atoms. Aqueous extract of garlic peel was studied for corrosion inhibition for mild steel in 1 M HCl.[47] GC study revealed the existence of an S atom containing phytochemicals. Allyl methyl disulphide, diallyl disulphide, methyl allyl trisulphide, diallyl trisulphide and 2-thiophenecarboxaldehyde were the major phytochemical components. The peel extracts of *Musa parasidica* (banana) plant were evaluated for mild steel corrosion in 1 M HCl solution.[21] A comparison of the raw, ripe, and overripe banana peels was performed, which revealed the best performance for the raw peels. Analysis by HPLC revealed the existence of gallocatechin and catechin phytochemicals in the extracts that were responsible for adsorption and corrosion inhibition. The extract of orange peel was analysed for carbon steel surface in 0.1 M HCl solution and resulted in a 95% protection performance.[48] The inhibition behaviour was ascribed to the antioxidant properties of flavonoids present in the extract. Pomelo peel extract was studied for N80 steel in 3.5% NaCl saturated with CO_2.[49] Adsorption of inhibitor resulted in a rise of the energy of activation, indicating the creation of an energy barrier against the corrosion and dissolution of steel. The extract contains naringin molecules, which were also studied *via* DFT analysis in both neutral and protonated forms. The protonated form showed a greater tendency to adsorb on a metal surface. The influence of tangerine peel extract was studied for corrosion of J55 steel in a sweet corrosion environment.[50] A mixed type of inhibition behaviour was revealed with PDP analysis. The adsorption of extract followed EL-Awady isotherm. The extract contains flavonoids, organic acids, pectin, dietary fibre, polysaccharides and trace elements that can provide effective adsorption.

8.5.5 Seed Extract

Momordica charantia (Karela) seed extracts were studied for sweet corrosion of P110SS steel substrate.[51] A mixed type of performance was observed at a concentration of 1 g L^{-1} with 89% efficiency, as revealed by the electrochemical studies. The plant extract consists of momordicin I and momordicin II, and cucurbitacin B as the main bioactive components. Other phytochemicals include glycosides (*e.g.* momordin, charantin, charantosides, goyaglycosides, momordicosides), terpenoids (*e.g.* momordicin-28, momordicinin, momordicilin, momordenol, and momordol), and cytotoxic (ribosome-inactivating) proteins *e.g.* momorcharin and momordin. The seed extracts of Azwain plant were evaluated for protection from corrosion in 0.5 M NaOH solution.[52] Mixed-type behaviour and a charge transfer controlled electrochemical mechanism was noticed with high efficiency at a moderate dosage of inhibitor. In another study, on an Al surface in 1 M HCl solution, fennel seed extracts were studied as inhibitors.[53] Among the six studied phytochemical constituents present in the extract, *E*-anethole provided the best adsorption behaviour, as revealed by the theoretical studies. The extracts of Jamun seeds were studied as environmentally-friendly corrosion inhibitors for mild steel in HCl and H_2SO_4 media.[54] At the same dose, a greater inhibition efficiency was observed in the HCl solution compared to that in the H_2SO_4. The extract provided a mixed type of inhibition performance in both the acid solutions. The extract contains an alkaloid, jambosine, and glycoside jambolin or antimellin. It also contains ellagic acid, which has the capability to reduce blood pressure by 34.6%. Seed extracts for the mustard plant were also evaluated for HCl and H_2SO_4 environment on X60 pipeline steel substrate.[55] A rise in the inhibitor concentration produced an incline in the protection performance, whereas a reduced protection performance was observed at elevated temperatures. Primarily physical mode of adsorption was observed by the kinetic/thermodynamic parameters. The major active components responsible for corrosion protection were sinapic acid, and sinigrin. A comparison of the inhibition performance of different plant extracts is given in Table 8.2.

8.5.6 Bark and Stem Extracts

Alkaloid extracts of *Geissospermum laeve* were utilized for the protection of C38 steel substrate in HCl solution.[3] Quantum chemical studies revealed that the existence of pyrrole and piperidine moieties was responsible for efficient adsorption on the steel substrate. A physical mode of adsorption was noticed that followed the Langmuir model. *Gmelina arborea* bark extracts were studied for mild steel in HCl solution.[56] The introduction of extract to corrosive solution retarded the corrosion rate from 0.6 mmpy to 0.04 mmpy. The extract contains alkaloids, carbohydrate, cardiac glycosides, tannins, phenolic compounds, saponins and flavonoids. Spontaneous adsorption was noticed on steel substrate, which was in accordance with Langmuir isotherm. *Lantana camara* bark extracts were studied for 1 M HCl as a green corrosion inhibitor.[57] At 200 mg L^{-1} concentration, a high efficiency was recorded that

Table 8.2 Performance of extracts derived from seed, stem, and root from different plants as corrosion inhibitors.

Name of plant	Part of plant	Metal surface/ medium	Inhibition efficiency/ concentration	Reference
Karela (*Momordica charantia*)	Seed	P110SS steel/3.5% NaCl CO_2 saturated	89.0%/1 g L^{-1}	51
Azwain (*Trachyspermum copticum*)	Seed	Al/0.5 M NaOH	90%/500 mg L^{-1}	52
Fennel	Seed	Al/1 M HCl	92.90%/800 mg L^{-1}	53
Jamun (*Syzygium cumini*)	Seed	Mild steel/1 M HCl	94.28%/600 mg L^{-1}	54
		Mild steel/0.5 M H_2SO_4	81.66%/600 mg L^{-1}	
Mustard	Seed	X60/2 M HCl	56.44%/1 g L^{-1}	55
		X60/1 M H_2SO_4	76%/1 g L^{-1}	
Geissospermum laeve	Bark	C38 steel/1 M HCl	90%/100 mg L^{-1}	3
Gmelina arborea	Bark	Mild steel/1 M HCl	96%/500 mg L^{-1}	56
Lantana camara	Bark	Mild steel/1 M HCl	97.33%/1 g L^{-1}	57
Tamarix aphylla	Bark	Carbon steel/ Seawater	83.5%/1.5 g L^{-1}	58
Ziziphus joazeiro	Stem bark	Mild steel/1 M HCl	90.5/800 mg L^{-1}	59
Azadirachta Indica	Leaf	Mild steel/2 M H_2SO_4	60.4%/4 g L^{-1}	60
	Root		66.8%/4 g L^{-1}	
	Seed		81.8%/4 g L^{-1}	
Beet + Zn^{2+} (50 mg L^{-1})	Root	Carbon steel/Well water	98%/4 mL	61
Ginseng + KI (5 mM)	Root	Al alloy AA1060/1 M HCl	96%/50% v/v	62
Licorice	Root	Tin/0.6 M NaCl	66.18%/3 g L^{-1}	63
Olive	Root	Mild steel/1 M HCl	89.24%/1.4 × 10^{-5} mL L^{-1}	64
	Stem		88.84%/1.4 × 10^{-5} mL L^{-1}	
	Leave		85.13%/1.4 × 10^{-5} mL L^{-1}	

increased further to 97.33% upon raising the solution concentration to 1 g L^{-1}. The extract consists of alkaloids, flavonoids, terperoids, saponins and tannins. PDP study revealed mixed-type inhibitor adsorption. Experimental, DFT, and MD studies were carried out on *Tamarix aphylla* plant extracts for carbon steel in seawater medium.[58] A moderate efficiency of 83.5% at 298 K was noted, which reduced slightly to 70.7% with a rise in the test solution temperature to 328 K. SEM studies revealed a considerable improvement in the surface morphology of steel substrate in the presence of adsorbed inhibitor film. The GC-MS analysis revealed the existence of ten phytochemicals in the extracts wherein tamarixetin was the dominant compound. Hence the DFT and MD studies were carried out on this molecule which revealed the electron donation tendency to the metal surface with a high adsorption affinity. Bark was collected from stem of *Ziziphus joazeiro* plant and studied for mild steel in 1 M HCl solution.[59] Chemical analysis by LC-MS revealed the presence of 18 different phytochemicals in the extract wherein, saponins was attributed to the major constituent responsible for adsorption behaviour.

8.5.7 Root Extract

An interesting comparative study was carried out on the leaf, seeds, and roots extracts of the *Azadirachta Indica* plant as an inhibitor for mild steel in 2 M H_2SO_4.[60] In all three cases, the inhibitor adsorption was correlated to the Freundlich isotherm. The order of protection performance of the extracts was leaves > roots > stem. Over 300 phytochemical constituents have been isolated from this plant, which include azadirachtin, azadirone, nimbin, margolone, gedunin, nimbandiol, nimbinene, nimbolin, nimbolide, nimonol, salannin, melianol, vilasanin, flavonoids, and related compounds. The formation of the Fe-extract complex was attributed to protection performance. Beetroot extract was studied for carbon steel in well water in synergistic combination with Zn^{2+} ions.[61] It was observed that the addition of surfactant CTAB did not improve inhibition. The protective film formed on the surface was found to have UV-fluorescent properties and consisted of Fe^{2+}-betanin complex and $Zn(OH)_2$. The root of the ginseng plant was studied for Al alloy in HCl environment.[62] The addition of KI as a synergistic agent improved the inhibition performance considerably. The extract is known to contain glycoside, saponin, phenolic acids, alkaloids and lignin. Licorice root extract was analysed for tin in 0.6 M NaCl solution.[63] Glycyrrhizin was the major extract component, which aided in the significant inhibition of general as well as pitting corrosion of tin metal in NaCl. The root extract contains sucrose which was attributed to the adsorption behaviour. A comparative analysis of inhibition performance of aqueous extracts of olive roots, stems and leaves was conducted for a steel surface in 1 M HCl solution.[64] All three extracts exhibited mixed-type inhibition with physical adsorption wherein the root extract gave the best performance.

8.6 Conclusions and Prospects

This chapter presents the significance and application of natural extracts of different parts of plants as corrosion inhibitors. Having natural origins, the plants and plant parts are available in abundance and are considered green. The major parameters in the preparation of plant extracts are discussed. Further, a glimpse of the literature is provided on the application of different parts of plants *e.g.* root, stem, leaves, flower, seed, fruit, peels *etc.* in corrosion protection. We have also tried to provide the major phytochemical constituents present in each of the plant extracts reported for corrosion protection. It is shown that although the plant extract-based inhibitors are green and eco-friendly, these inhibitors pose the issues of poor shelf life, and the requirement of detailed and tedious analysis of phytochemical composition. On the other hand, the presence of more than one phytochemical in the same extract can lead to synergistic improvement in the corrosion inhibition behaviour. The researchers have also studied the influence of different growth stages and different ripening stages of the same plant extract on inhibition behaviour. In addition, comparative study on inhibition performance of different parts of the same plant is also reported.

Further studies on plant extracts as corrosion inhibitors should focus on improving the shelf life of extract-based corrosion inhibitors. Cytotoxicity profiles of the extracts should also be considered. The studies should also be undertaken on detailed computational analysis of multiple phytochemical constituents using DFT, and molecular simulations. Development of corrosion inhibitors should be targeted on medicinal plants and plants thriving easily in the environment without the requirement of any special growth support. Also, a comparison of the plant extract-based inhibitors should be undertaken with commercially available inhibitors.

References

1. C. Verma, E. E. Ebenso, I. Bahadur and M. A. Quraishi, An overview on plant extracts as environmental sustainable and green corrosion inhibitors for metals and alloys in aggressive corrosive media, *J. Mol. Liq.*, 2018, **266**, 577–590.
2. N. Chaubey, A. Qurashi, D. S. Chauhan and M. Quraishi, Frontiers and advances in green and sustainable inhibitors for corrosion applications: A critical review, *J. Mol. Liq.*, 2020, **321**, 114385.
3. M. Faustin, A. Maciuk, P. Salvin, C. Roos and M. Lebrini, Corrosion inhibition of C38 steel by alkaloids extract of *Geissospermum laeve* in 1 M hydrochloric acid: electrochemical and phytochemical studies, *Corros. Sci.*, 2015, **92**, 287–300.
4. J. Buchweishaija, Phytochemicals as green corrosion inhibitors in various corrosive media: a review, *Tanz. J. Sci.*, 2009, **35**, 77–92.
5. P. E. Alvarez, M. V. Fiori-Bimbi, A. Neske, S. A. Brandán and C. A. Gervasi, Rollinia occidentalis extract as green corrosion inhibitor for carbon steel in HCl solution, *J. Ind. Eng. Chem.*, 2018, **58**, 92–99.
6. N. Asadi, M. Ramezanzadeh, G. Bahlakeh and B. Ramezanzadeh, Utilizing Lemon Balm extract as an effective green corrosion inhibitor for mild steel in 1 M HCl solution: a detailed experimental, molecular dynamics, Monte Carlo and quantum mechanics study, *J. Taiwan Inst. Chem. Eng.*, 2019, **95**, 252–272.
7. M. P. Casaletto, V. Figà, A. Privitera, M. Bruno, A. Napolitano and S. Piacente, Inhibition of Cor-Ten steel corrosion by "green" extracts of Brassica campestris, *Corros. Sci.*, 2018, **136**, 91–105.
8. F. S. de Souza and A. Spinelli, Caffeic acid as a green corrosion inhibitor for mild steel, *Corros. Sci.*, 2009, **51**, 642–649.
9. N. Chaubey, D. K. Yadav, V. K. Singh and M. A. Quraishi, A comparative study of leaves extracts for corrosion inhibition effect on aluminium alloy in alkaline medium, *Ain Shams Eng. J.*, 2017, **8**, 673–682.
10. A. Dehghani, G. Bahlakeh, B. Ramezanzadeh and M. Ramezanzadeh, Potential of Borage flower aqueous extract as an environmentally

sustainable corrosion inhibitor for acid corrosion of mild steel: electrochemical and theoretical studies, *J. Mol. Liq.*, 2019, **277**, 895–911.
11. N. El Hamdani, R. Fdil, M. Tourabi, C. Jama and F. Bentiss, Alkaloids extract of *Retama monosperma* (L.) Boiss. seeds used as novel eco-friendly inhibitor for carbon steel corrosion in 1 M HCl solution: Electrochemical and surface studies, *Appl. Surf. Sci.*, 2015, **357**, 1294–1305.
12. C. G. Fraga, K. D. Croft, D. O. Kennedy and F. A. Tomás-Barberán, The effects of polyphenols and other bioactives on human health, *Food Funct.*, 2019, **10**, 514–528.
13. M. Nasrollahzadeh, S. M. Sajadi and M. Khalaj, Green synthesis of copper nanoparticles using aqueous extract of the leaves of Euphorbia esula L and their catalytic activity for ligand-free Ullmann-coupling reaction and reduction of 4-nitrophenol, *RSC Adv.*, 2014, **4**, 47313–47318.
14. C. Capello, U. Fischer and K. Hungerbühler, What is a green solvent? A comprehensive framework for the environmental assessment of solvents, *Green Chem.*, 2007, **9**, 927–934.
15. H. Sharghi, R. Khalifeh and M. M. Doroodmand, Copper nanoparticles on charcoal for multicomponent catalytic synthesis of 1, 2, 3-Triazole derivatives from benzyl halides or alkyl halides, terminal alkynes and sodium azide in water as a "Green" solvent, *Adv. Synth. Catal.*, 2009, **351**, 207–218.
16. D. S. Bose, L. Fatima and H. B. Mereyala, Green chemistry approaches to the synthesis of 5-alkoxycarbonyl-4-aryl-3, 4-dihydropyrimidin-2 (1 H)-ones by a three-component coupling of one-pot condensation reaction: Comparison of ethanol, water, and solvent-free conditions, *J. Org. Chem.*, 2003, **68**, 587–590.
17. R. S. Varma, Greener and sustainable trends in synthesis of organics and nanomaterials, *ACS Sustainable Chem. Eng.*, 2016, **4**, 5866–5878.
18. H. Duan, D. Wang and Y. Li, Green chemistry for nanoparticle synthesis, *Chem. Soc. Rev.*, 2015, **44**, 5778–5792.
19. J. Seo, S. Lee, M. L. Elam, S. A. Johnson, J. Kang and B. H. Arjmandi, Study to find the best extraction solvent for use with guava leaves (Psidium guajava L.) for high antioxidant efficacy, *Food Sci. Nutr.*, 2014, **2**, 174–180.
20. N. A. N. Mohamad, N. A. Arham, J. Jai and A. Hadi, Plant extract as reducing agent in synthesis of metallic nanoparticles: a review, *Adv. Mater. Res.*, 2014, **832**, 350–355.
21. G. Ji, S. Anjum, S. Sundaram and R. Prakash, Musa paradisica peel extract as green corrosion inhibitor for mild steel in HCl solution, *Corros. Sci.*, 2015, **90**, 107–117.
22. A. Miralrio and A. Espinoza Vázquez, Plant extracts as green corrosion inhibitors for different metal surfaces and corrosive media: A review, *Processes*, 2020, **8**, 942.
23. N. Chaubey, V. K. Singh and M. A. Quraishi, Electrochemical approach of Kalmegh leaf extract on the corrosion behavior of aluminium alloy in alkaline solution, *Int. J. Ind. Chem.*, 2017, **8**, 75–82.
24. T. Ibrahim, E. Gomes, I. B. Obot, M. Khamis and M. Abou Zour, Corrosion inhibition of mild steel by *Calotropis procera* leaves extract in a CO_2 saturated sodium chloride solution, *J. Adhes. Sci. Technol.*, 2016, **30**, 2523–2543.

25. T. H. Ibrahim, E. E. Gomes, I. B. Obot, M. Khamis and M. A. Sabri, Mild steel green inhibition by *Ficus carica* leaves extract under practical field conditions, *J. Adhes. Sci. Technol.*, 2017, **31**, 2697–2718.
26. M. A. Quraishi, A. Singh, V. K. Singh, D. K. Yadav and A. K. Singh, Green approach to corrosion inhibition of mild steel in hydrochloric acid and sulphuric acid solutions by the extract of *Murraya koenigii* leaves, *Mater. Chem. Phys.*, 2010, **122**, 114–122.
27. C. Ito, M. Itoigawa, K. Nakao, T. Murata, M. Tsuboi, N. Kaneda and H. Furukawa, Induction of apoptosis by carbazole alkaloids isolated from *Murraya koenigii*, *Phytomedicine*, 2006, **13**, 359–365.
28. Y. Qiang, S. Zhang, B. Tan and S. Chen, Evaluation of Ginkgo leaf extract as an eco-friendly corrosion inhibitor of X70 steel in HCl solution, *Corros. Sci.*, 2018, **133**, 6–16.
29. G. Pustaj, F. Kapor and Z. Veinovic, Olive leaf extract as a corrosion inhibitor of carbon steel in CO_2-saturated chloride–carbonate solution, *Int. J. Electrochem. Sci.*, 2016, **11**, 7811–7829.
30. G. Pustaj, F. Kapor and S. Jakovljević, Carbon dioxide corrosion of carbon steel and corrosion inhibition by natural olive leaf extract: Die Kohlendioxid-Korrosion von Kohlenstoffstahl und die Korrosionshemmung durch natürlichen Olivenblattextrakt, *Materialwissenschaft und Werkstofftechnik*, 2017, **48**, 122–138.
31. X. Li, S. Deng, X. Xie and H. Fu, Inhibition effect of bamboo leaves' extract on steel and zinc in citric acid solution, *Corros. Sci.*, 2014, **87**, 15–26.
32. E. Ating, S. Umoren, I. Udousoro, E. Ebenso and A. Udoh, Leaves extract of *Ananas sativum* as green corrosion inhibitor for aluminium in hydrochloric acid solutions, *Green Chem. Lett. Rev.*, 2010, **3**, 61–68.
33. P. Mourya, S. Banerjee and M. Singh, Corrosion inhibition of mild steel in acidic solution by *Tagetes erecta* (Marigold flower) extract as a green inhibitor, *Corros. Sci.*, 2014, **85**, 352–363.
34. M. H. Abdellattif, S. H. Alrefaee, O. Dagdag, C. Verma and M. Quraishi, *Calotropis procera* extract as an environmental friendly corrosion Inhibitor: Computational demonstrations, *J. Mol. Liq.*, 2021, **337**, 116954.
35. A. Mathina and R. Rajalakshmi, Corrosion inhibition of mild steel in acid medium using *Canna Indica* as green corrosion inhibitor, *Rasayan J. Chem.*, 2016, **9**, 56–66.
36. M. Shahini, M. Keramatinia, M. Ramezanzadeh, B. Ramezanzadeh and G. Bahlakeh, Combined atomic-scale/DFT-theoretical simulations & electrochemical assessments of the chamomile flower extract as a green corrosion inhibitor for mild steel in HCl solution, *J. Mol. Liq.*, 2021, **342**, 117570.
37. K. Nasr, M. Fedel, K. Essalah, F. Deflorian and N. Souissi, Experimental and theoretical study of *Matricaria recutita* chamomile extract as corrosion inhibitor for steel in neutral chloride media, *Anti-Corros. Methods Mater.*, 2018, **65**, 292–309.
38. N. Hynes, R. M. Selvaraj, T. Mohamed, A. M. Mukesh, K. Olfa and M. P. Nikolova, *Aerva lanata* flowers extract as green corrosion inhibitor of low-carbon steel in HCl solution: an *in vitro* study, *Chem. Pap.*, 2021, **75**, 1165–1174.

39. D. Bouknana, B. Hammouti, S. Jodeh, M. Sbaa and H. Lgaz, Extracts of olive inflorescence flower pre-anthesis, at anthesis and grain pollen as eco-friendly corrosion inhibitor for steel in 1 M HCl medium, *Anal. Bioanal. Electrochem.*, 2018, **10**, 751–777.
40. A. Saxena and J. Kumar, Phytochemical screening, metal-binding studies and applications of floral extract of *Sonchus oleraceus* as a corrosion inhibitor, *J. Bio- Tribo-Corros.*, 2020, **6**, 1–10.
41. A. Singh, Y. Lin, W. Liu, D. Kuanhai, J. Pan, B. Huang, C. Ren and D. Zeng, A study on the inhibition of N80 steel in 3.5% NaCl solution saturated with CO_2 by fruit extract of *Gingko biloba*, *J. Taiwan Inst. Chem. Eng.*, 2014, **45**, 1918–1926.
42. A. Singh, Y. Lin, E. E. Ebenso, W. Liu, J. Pan and B. Huang, *Gingko biloba* fruit extract as an eco-friendly corrosion inhibitor for J55 steel in CO_2 saturated 3.5% NaCl solution, *J. Ind. Eng. Chem.*, 2015, **24**, 219–228.
43. G. Fekkar, F. Yousfi, H. Elmsellem, M. Aiboudi, M. Ramdani, I. Abdel-Rahman, B. Hammouti and L. Bouyazza, Eco-friendly *Chamaerops humilis L.* fruit extract corrosion inhibitor for mild steel in 1 M HCl, *Int. J. Corros. Scale Inhib.*, 2020, **9**, 446–459.
44. A. Sedik, D. Lerari, A. Salci, S. Athmani, K. Bachari, İ. Gecibesler and R. Solmaz, Dardagan Fruit extract as eco-friendly corrosion inhibitor for mild steel in 1 M HCl: Electrochemical and surface morphological studies, *J. Taiwan Inst. Chem. Eng.*, 2020, **107**, 189–200.
45. Z. Sanaei, M. Ramezanzadeh, G. Bahlakeh and B. Ramezanzadeh, Use of *Rosa canina* fruit extract as a green corrosion inhibitor for mild steel in 1 M HCl solution: A complementary experimental, molecular dynamics and quantum mechanics investigation, *J. Ind. Eng. Chem.*, 2019, **69**, 18–31.
46. A. Dehghani, G. Bahlakeh and B. Ramezanzadeh, A detailed electrochemical/theoretical exploration of the aqueous Chinese gooseberry fruit shell extract as a green and cheap corrosion inhibitor for mild steel in acidic solution, *J. Mol. Liq.*, 2019, **282**, 366–384.
47. S. S. d. A. A. Pereira, M. M. Pêgas, T. L. Fernández, M. Magalhães, T. G. Schöntag, D. C. Lago, L. F. de Senna and E. D'Elia, Inhibitory action of aqueous garlic peel extract on the corrosion of carbon steel in HCl solution, *Corros. Sci.*, 2012, **65**, 360–366.
48. N. M'hiri, D. Veys-Renaux, E. Rocca, I. Ioannou, N. M. Boudhrioua and M. Ghoul, Corrosion inhibition of carbon steel in acidic medium by orange peel extract and its main antioxidant compounds, *Corros. Sci.*, 2016, **102**, 55–62.
49. Z. Sun, A. Singh, X. Xu, S. Chen, W. Liu and Y. Lin, Inhibition effect of pomelo peel extract for N80 steel in 3.5% NaCl saturated with CO_2 solution, *Res. Chem. Intermed.*, 2017, **43**, 6719–6736.
50. S. Wang, B. Wu, L. Qiu, Y. Chen, J. Yuan, S. Chen, M. Bao, C. Fu and X. Wang, Inhibition effect of tangerine peel extract on J55 steel in CO_2-saturated 3.5 wt.% NaCl solution, *Int. J. Electrochem. Sci.*, 2017, **12**, 11195–11211.
51. A. Singh, Y. Lin, W. Liu, E. E. Ebenso and J. Pan, Extract of *Momordica charantia* (Karela) seeds as corrosion inhibitor for P110SS steel

in CO_2 saturated 3.5% NaCl solution, *Int. J. Electrochem. Sci.*, 2013, **8**, 12884–12893.
52. S. Ambrish and M. Quraishi, Azwain (Trachyspermum copticum) seed extract as an efficient corrosion Inhibitor for Aluminium in NaOH solution, *Res. J. Recent Sci.*, 2012, **2277**, 2502.
53. D. G. Ladha, P. M. Wadhwani, M. Y. Lone, P. C. Jha and N. K. Shah, Evaluation of fennel seed extract as a green corrosion inhibitor for pure aluminum in hydrochloric acid: an experimental and computational approach, *Anal. Bioanal. Electrochem.*, 2015, **7**, 59–74.
54. A. Singh and M. Quraishi, The extract of Jamun (*Syzygium cumini*) seed as green corrosion inhibitor for acid media, *Res. Chem. Intermed.*, 2015, **41**, 2901–2914.
55. S. A. Umoren, Biomaterials for corrosion protection: evaluation of mustard seed extract as eco-friendly corrosion inhibitor for X60 steel in acid media, *J. Adhes. Sci. Technol.*, 2016, **30**, 1858–1879.
56. L. A. Nnanna, K. O. Uchendu, F. O. Nwosu, U. Ihekoronye and P. Eti, Gmelina Arborea bark extracts as a corrosion inhibitor for mild steel in an acidic environment, *Int. J. Mater. Res.*, 2014, **4**, 34–39.
57. P. R. Shrestha, H. B. Oli, B. Thapa, Y. Chaudhary, D. K. Gupta, A. K. Das, K. B. Nakarmi, S. Singh, N. Karki and A. P. Yadav, Bark extract of *Lantana camara* in 1 M HCl as green corrosion inhibitor for mild steel, *Eng. J.*, 2019, **23**, 205–211.
58. I. H. Ali, Experimental, DFT and MD assessments of bark extract of *Tamarix aphylla* as corrosion inhibitor for carbon steel used in desalination plants, *Molecules*, 2021, **26**, 3679.
59. A. C. Mauro, B. D. Ribeiro, R. Garrett, R. M. Borges, T. U. da Silva, S. de Paula Machado, J. R. de Araujo, S. de Oliveira Massafra, F. O. R. de Oliveira Junior and E. D'Elia, *Ziziphus joazeiro* stem bark extract as a green corrosion inhibitor for mild steel in acid medium, *Processes*, 2021, **9**, 1323.
60. P. C. Okafor, E. E. Ebenso and U. J. Ekpe, *Azadirachta indica* extracts as corrosion inhibitor for mild steel in acid medium, *Int. J. Electrochem. Sci.*, 2010, **5**, 978–993.
61. J. A. Selvi, S. Rajendran, V. G. Sri, A. J. Amalraj and B. Narayanasamy, Corrosion inhibition by beet root extract, *Port. Electrochim. Acta*, 2009, **27**, 1–11.
62. I. Obot, N. Obi-Egbedi and S. Umoren, Ginseng root: a new efficient and effective eco-friendly corrosion inhibitor for aluminium alloy of type AA 1060 in hydrochloric acid solution, *Int. J. Electrochem. Sci.*, 2009, **4**, 1277–1288.
63. M. Abdallah, A. El-Etre, E. Kamar and S. Eid, Licorice root extract as corrosion inhibitor for tin in sodium chloride solutions, *J. Basic Environ. Sci.*, 2020, **7**, 191–194.
64. D. Bouknana, B. Hammouti, S. Jodeh, A. Bouyanzer, A. Aouniti and I. Warad, Aqueous extracts of olive roots, stems, and leaves as eco-friendly corrosion inhibitor for steel in 1 MHCl medium, *Int. J. Ind. Chem.*, 2015, **6**, 233–245.

CHAPTER 9

Chemical Medicines (Drugs) as Green Corrosion Inhibitors

9.1 Introduction

Medicines are organic compounds and therefore have heteroatoms (N, S, O, P), functional groups, heterocyclic rings, phenyl rings, π-bonds, *etc.*, in their molecular structure. The structural design and formulation of medicines are made with easy dissolution and absorption into human body fluids in mind. Therefore medicines can be solubilized by water and oils. In this way, medicines fulfil the structural requirements of organic molecules to be used as corrosion inhibitors. For these reasons, drug molecules are widely acknowledged as potential corrosion inhibitors for metallic substrates.[1,2] Considering this fact, a number of research articles have been published on the use of drugs or chemical medicines as corrosion inhibitors.

Almost all kinds of drugs *e.g.* antimalarial, antibacterial, antifungal, antiviral, antihypertensive, anticancer, *etc.*, have been utilized as corrosion inhibitors. These studies include almost all the different types of aqueous corrosive media, metals, and alloy surfaces. This chapter provides a brief overview of the significance of drugs as corrosion inhibitors. Some of the associated issues and literary examples on the application of drugs as inhibitors are also outlined and discussed.

9.2 Significance of Chemical Medicines as Corrosion Inhibitors

Organic corrosion inhibitors act by adsorption on the target metal surfaces, by sharing π-electrons, *via* heterocyclic or phenyl rings, by functional groups, or heteroatoms. Chemical medicines or drugs are organic molecules having

Green Chemistry Series No. 73
Green Corrosion Inhibition: Fundamentals, Design, Synthesis and Applications
By Chandrabhan Verma, Mumtaz A. Quraishi and Dheeraj Singh Chauhan

Published by the Royal Society of Chemistry, www.rsc.org

structural attributes similar to organic corrosion inhibitors. Most drugs are complex organic molecules containing imidazole, pyridine, pyrimidine *etc.* and heterocyclic rings[3] that can facilitate interaction with metallic substrates. In addition to phenyl rings, the drug molecules are composed of several five- and six-membered heterocyclic rings. The fundamental heterocyclic structures are shown in Figure 9.1. The fundamental five-membered heterocycles with N, S, and O are pyrrole, thiophene, and furan, respectively. Six-membered heterocycles include pyridine, pyrimidine, triazines, *etc.* The replacement of more than one C atom with N and/or other heteroatoms yields other heterocycles, *e.g.* pyrazoles, imidazoles, oxazoles, thiazoles, thiadiazoles, pyrazines, pyrimidines, thiazines *etc.*[4] These organic moieties also facilitate the formation of

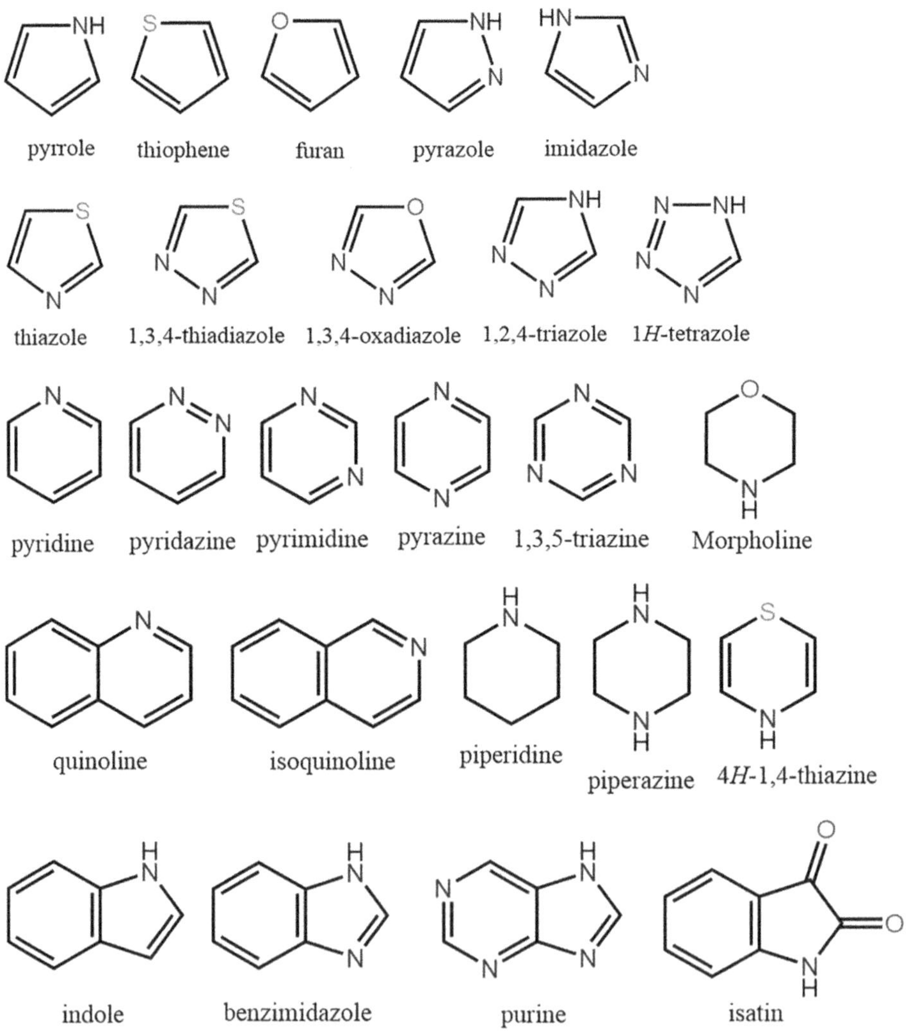

Figure 9.1 Structures of some heterocyclic molecules.

metal-inhibitor complexes and other surface-bound complexes that can aid in the formation of protective films. Therefore, over the years, medicines have been projected as a greener alternative to the otherwise used toxic imidazoline, amide, acetylenic alcohols *etc.* corrosion inhibitors. Due to high solubility in water or oil media, the preparation of organic inhibitor formulations from drugs is quite convenient, taking into account the practical application.

9.3 Issues in the Commercial Utility of Medicines as Corrosion Inhibitors

9.3.1 Issues in the Use of Chemical Medicines as Corrosion Inhibitors

Although medicines provide a green alternative to the otherwise used toxic organic corrosion inhibitors, their application for inhibition purposes has its own limitations. Chemical medicines are quite expensive; however, for human intake, governments provide these drugs at subsidized rates. For commercial application in corrosion inhibition, this would be difficult; hence the high cost of medicines is an important issue. A major issue associated with drugs as inhibitors is that most of the reports are for alloy steels in HCl and H_2SO_4 media, simulating the acid pickling application. Few reports have been undertaken for oil-well acidizing and sweet corrosion environments. In addition, the drugs are designed for miscibility in the human body at ambient temperature, whereas some industrial applications require inhibitors to function at elevated temperatures, pressures, and under flow conditions.[5–7] This requires testing drug-based corrosion inhibitors at high temperatures and high pressures. These issues require attention for the overall utilization of drugs for commercial applications.

9.3.2 Expired or Leftover Medicines as Corrosion Inhibitors

Several reports in the literature mention the application of drug molecules as corrosion inhibitors. A major limitation, as discussed above, is the high cost of the medicines. Another aspect is that the active ingredient of the medicines, *i.e.*, the major organic component retains its activity for up to one to two years. Beyond that date (expiry date), the medicines are no longer recommended for administration to patients. Although strictly in chemical terms, it has been reported that several drugs retain their activity well beyond the date of expiry.[8,9] Accordingly, several reports are available in the literature stating the use of expired drugs as corrosion inhibitors.[10–14] Therefore, this idea, in addition to cutting down the cost of fresh drugs, can provide another means of safe disposal of expired or leftover drugs. The application of expired, leftover and/or unused drugs as inhibitors is a considerably convenient, time-saving, and cheaper method compared to the use of synthetic organic molecules-based inhibitors.

9.3.3 Chemically Modified Medicines as Corrosion Inhibitors

In previous chapters we have discussed that the performance of corrosion inhibitors from natural polymers and amino acids can be enhanced using simple chemical functionalization methods. The same concept can be applied in the case of drug-based corrosion inhibitors as well. Having an organic framework, the drug molecules too present the option to undergo further chemical reactions. Another means is to perform simple chemical functionalization of the drug molecules to design effective inhibitors. Although, this is still a growing area of research requiring detailed investigations. In the sections given below, we have provided some examples of fresh, expired, and chemically functionalized drug molecules that have been utilized as inhibitors.

9.4 Chemical Medicines Reported as Corrosion Inhibitors

9.4.1 Chemical Medicines as Corrosion Inhibitors

Gokhan Gece has provided an excellent overview of the application of drugs as corrosion inhibitors.[1] Different types of chemical medicines have been reported as inhibitors, including β-lactam antibiotics, penicillin, cephalosporins, quinolones, tetracyclines, macrolides, lincosamides, sulphonamides, aminoglycosides, amphenicols, opioid analgesics, histamines, muscle relaxants, antifungal drugs, antihelminthics, antiviral drugs, antipsychotic drugs, antihypertensive drugs, amoebicidal drugs *etc.* Acyclovir, an antiviral medication, was evaluated for mild steel in a 1 M HCl environment.[15] The drug molecule consists of a purine unit in the framework, which in turn is composed of pyrimidine and imidazole moieties. The two joined heterocycles with –C=O and $-NH_2$ groups provide effective adsorption on the steel surface. The cefazolin drug (cephalosporin antibiotics) was also studied on mild steel in a 1 M HCl medium.[16] This drug contains thiadiazole, thiazine, and tetrazole rings in the molecular framework. The energy of activation for the drug's adsorption was greater than that of the blank solution, suggesting the creation of an energy barrier. Antibiotic Streptomycin was also studied for mild steel in a 1 M HCl solution, which provided a high efficiency although at a dose of 500 mg L^{-1}.[17] This drug molecule consists of plenty of –OH and $-NH_2$ functionalities that allow the adsorption and inhibition performance. Dapsone, a sulfone drug, contains two $-NH_2$ substituted phenyl rings on either side of the $-SO_2$ group was analysed on a mild steel surface in both HCl and H_2SO_4 environments.[18] This type of symmetrical structure is likely to result in a greater coverage on the metallic substrate. A superior adsorption and inhibition behaviour was observed in HCl solution at the identical concentration. Disulfiram drug, which is used to treat chronic alcoholism, was also studied for mild steel.[19] This molecule consists of 4 S atoms leading to efficient adsorption. Mixed-type inhibitor adsorption was noticed in a PDP study. The adsorption properties and inhibition effect of Ceftobiprole drug (fifth-generation cephalosporin) molecule that contains thiazole, thiadiazole and thiazine rings were also studied for mild steel in 1 M HCl solution.[20] Polarization curves indicated a mixed-type inhibitor. The inhibitor adsorption obeyed the Langmuir isotherm model. Adsorption and corrosion inhibition performance of Cefotaxime sodium, which is a cephalosporin antibiotic, was evaluated for mild steel in a 1 M HCl medium.[21] This drug molecule contains a thiazole and a thiazine ring. At a moderately high dose of 500 mg L^{-1}, high efficiency of 95.8% was noticed. EIS indicated single semicircular capacitive loops, which were consistent with a modified Randles circuit, indicating a charge transfer control of the adsorption behaviour. Corrosion inhibition performance of Cefuzonam drug, a second generation cephalosporin antibiotic, was studied for mild steel in a 1 M HCl environment.[22] This molecule contains thiazole, thiadiazole and thiazine rings. The addition of increasing inhibitor concentrations to the acid solution produced a rise in the water contact angles. This shows that the inhibitor adsorption resulted in a superior hydrophobic nature of the metal/solution interface, which supported the water repelling behaviour of the inhibitor film. Ceftazidime drug, a third-generation cephalosporin antibiotic (containing thiazole, thiazine and pyridine rings), when evaluated for steel surface in acidic medium showed a mixed mode of physical and chemical adsorption.[23] Table 9.1 summarizes the structures and performance of some drugs as corrosion inhibitors.

The adsorption process was in agreement with the Langmuir isotherm. Ceftriaxone, a third-generation cephalosporin antibiotic, reduced the rate of corrosion of steel substrate from 44.85 mmpy to 4.43 mmpy.[24] Mixed-type inhibition performance was noted for the inhibitor. This drug contains thiazole, thiazine, and triazine moieties in the molecular structure. Diethylcarbamazine was also evaluated for mild steel in 1 M HCl solution.[25] This molecule contains a piperazine ring. At a very low dose of 0.5 mM, high efficiency of >94% was noticed. The adsorption behaviour was in agreement with the Langmuir isotherm. The adsorption/corrosion inhibition of the Fexofenadine drug was also studied for mild steel in a 1 M HCl medium.[26] This molecule contains a phenyl and a piperidine ring. A very high protection performance of 97% was noticed at a low dose of 0.5 mM. FTIR study indicated the interaction of O atom of inhibitor with Fe atom of the metal surface resulting in Fe–O bond formation. Quantum chemical studies revealed an inclination of electron donation from the inhibitor molecule to the steel surface. Molecular structures and inhibition performance of some chemical medicines on ferrous and non-ferrous alloys are displayed in Table 9.2. Chloroquine, an antimalarial drug, was studied for mild steel in a 1 M HCl solution.[27] This molecule contains a quinolone ring attached to –NH– and –Cl groups. At 3.1×10^{-4} M inhibitor concentration, 99% protection was achieved. EIS indicated a charge transfer controlled mechanism with semicircular capacitive loops at high frequency and inductive behaviour at low frequencies. Computational studies show a high tendency of electron donation

Table 9.1 Structures and inhibition performance of some chemical medicines.

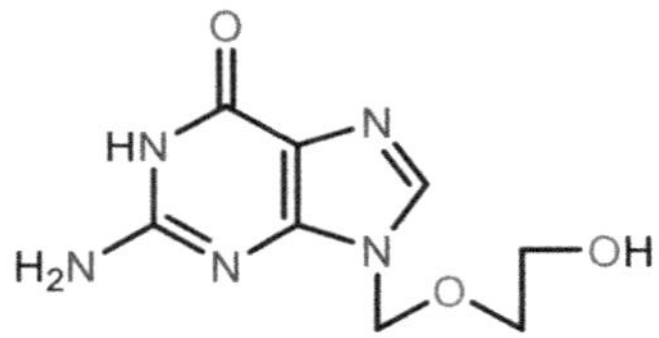	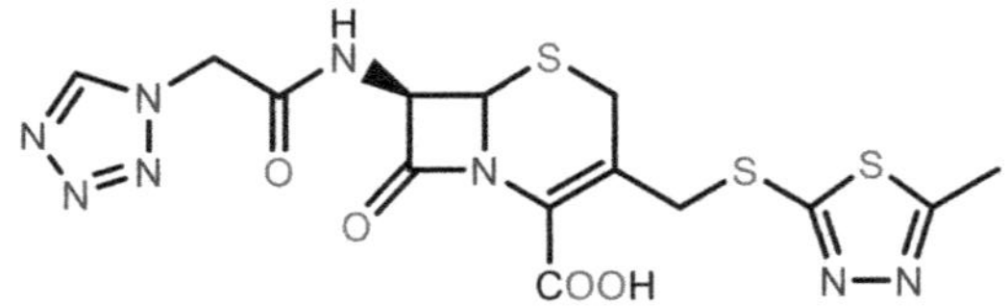	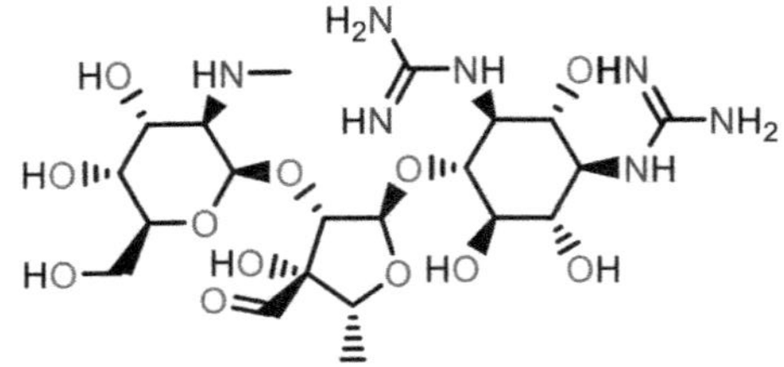
Acyclovir Mild steel/1 M HCl IE 92.6% (500 mg L^{-1})[15]	Cefazolin Mild steel/1 M HCl IE 93.9% (1.095 mM)[16]	Streptomycin Mild steel/1 M HCl IE 88.5% (500 mg L^{-1})[17]
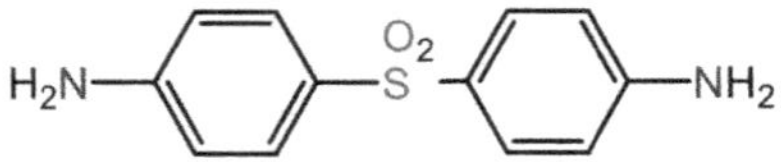	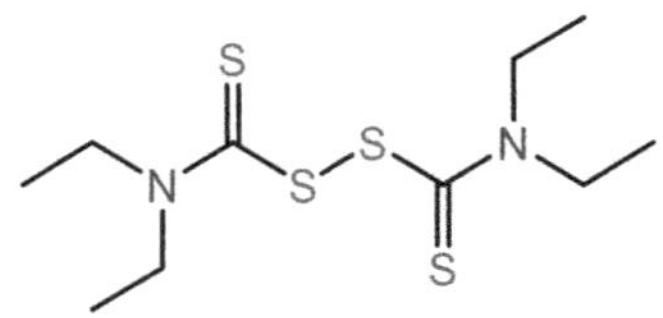	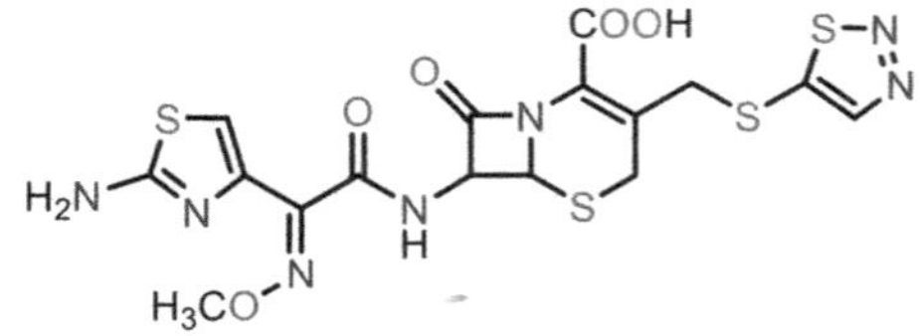
Dapsone Mild steel/1 M HCl; 0.5 M H_2SO_4 IE 93.33%; 90.66% (400 mg L^{-1})[18]	Disulfiram Mild steel/1 M HCl IE 97% (0.337 mM)[19]	Ceftobiprole Mild steel/1 M HCl IE 92.2% (0.931 mM)[20]
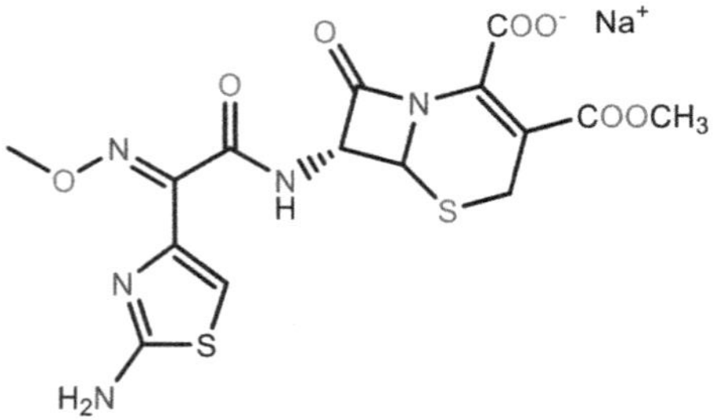	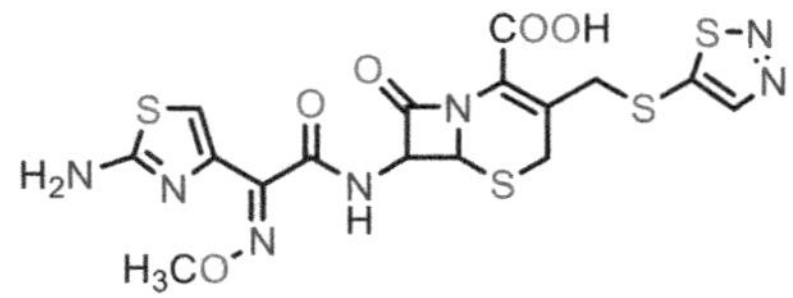	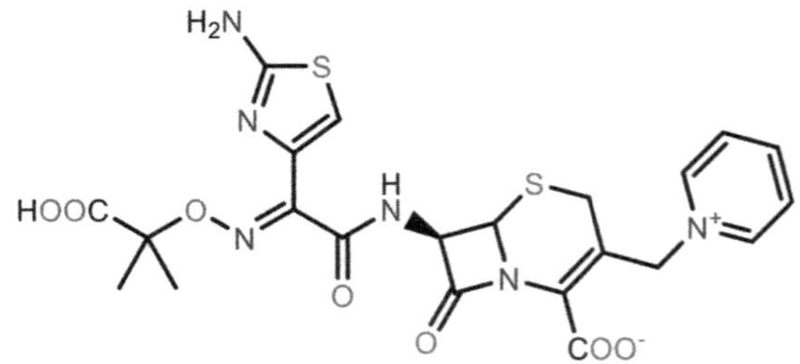

Cefotaxime sodium Mild steel/1 M HCl IE 95.8% (300 mg L^{-1})[21]	Cefuzonam Mild steel/1 M HCl IE 92.8% (400 mg L^{-1})[22]	Ceftazidime Mild steel/1 M HCl IE 96% (0.183 mM)[23]
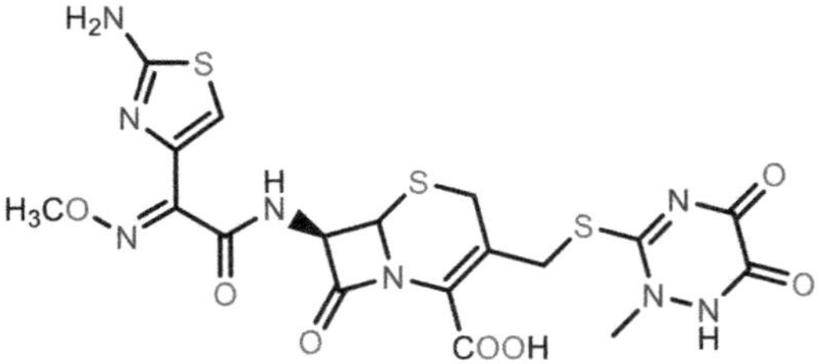	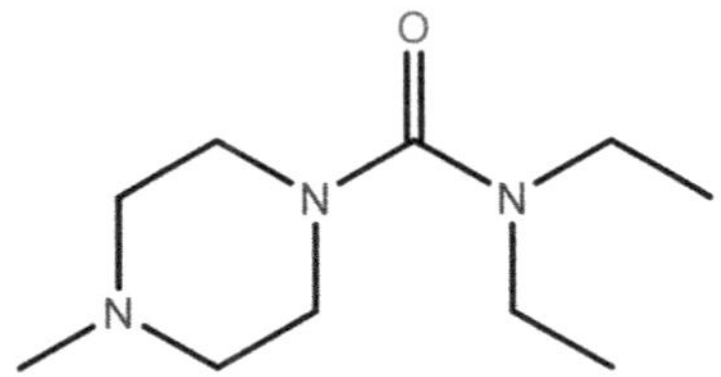	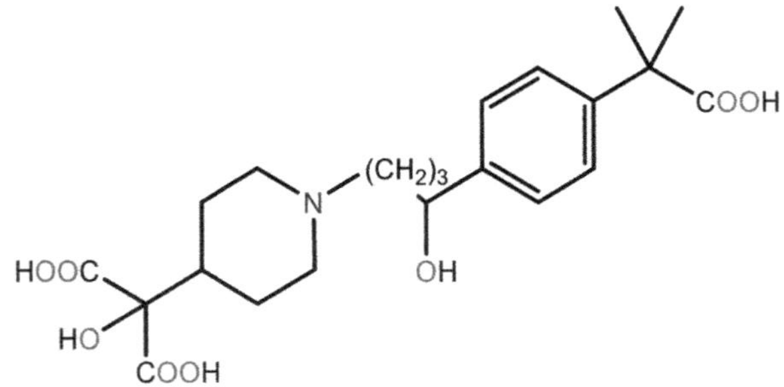
Ceftriaxone Mild steel/1 M HCl IE 90.0% (400 mg L^{-1})[24]	Diethylcarbamazine Mild steel/1 M HCl IE 94.31% (0.501 mM)[25]	Fexofenadine Mild steel/1 M HCl IE 98% (0.3 mM)[26]

Table 9.2 Structures and inhibition performance of some chemical medicines on ferrous and non-ferrous alloys.

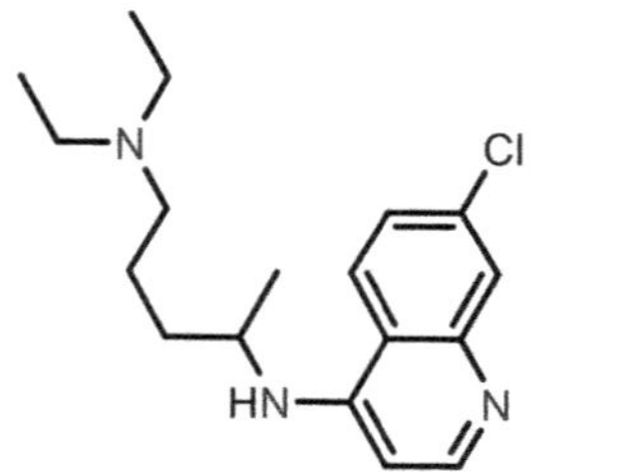 Chloroquine Mild steel/1 M HCl IE 99% (0.31 mM)[27]	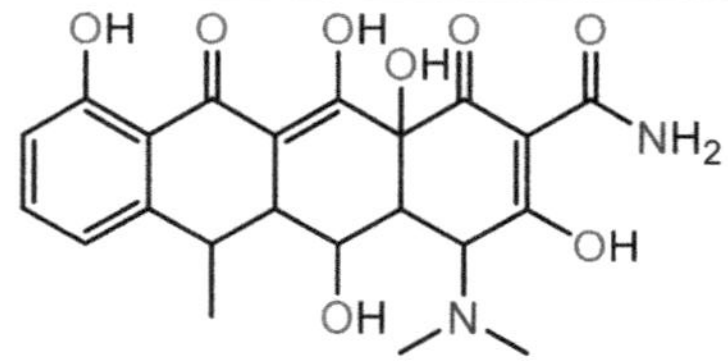 Doxycycline Mild steel/1 M HCl IE 94.8% (0.905 mM)[28]	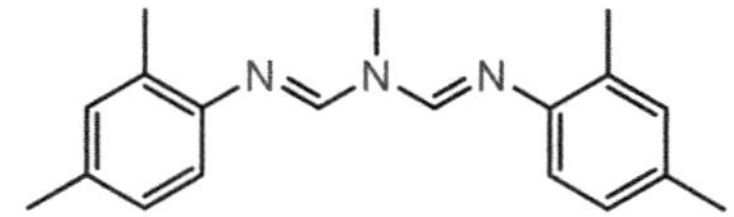 *N*,*N*′-[(methylimino)dimethylidyne] di-2,4-xylidine Mild steel/0.5 M H_2SO_4 IE 90% (0.852 mM)[29]
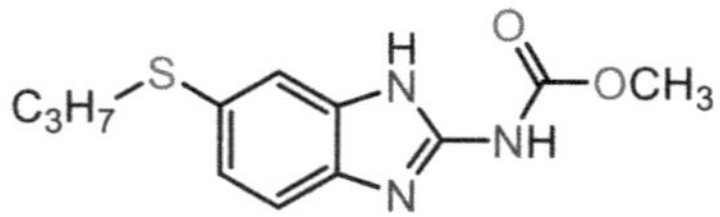 Albendazole Mild steel/1 M HCl IE 96% (0.4 mM)[30]	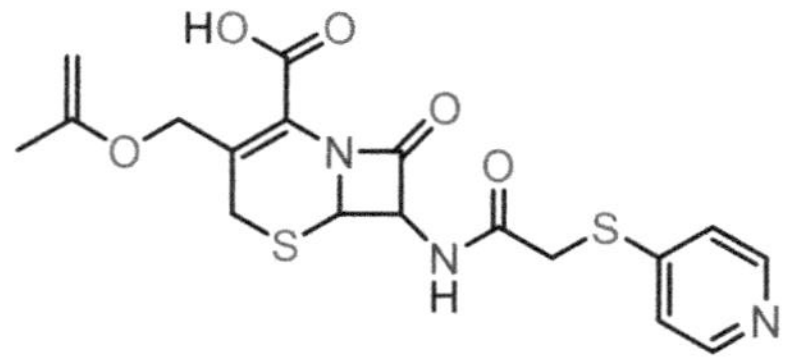 Cefapirin Mild steel/1 M HCl IE 94% (400 mg L^{-1})[31]	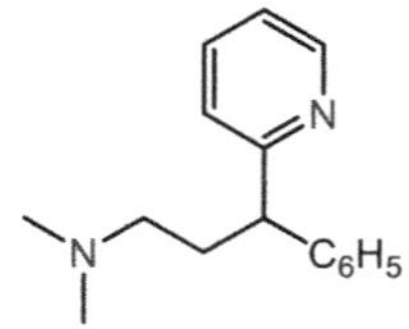 Pheniramine Mild steel/1 M HCl IE 92.2% (0.931 mM)[32]
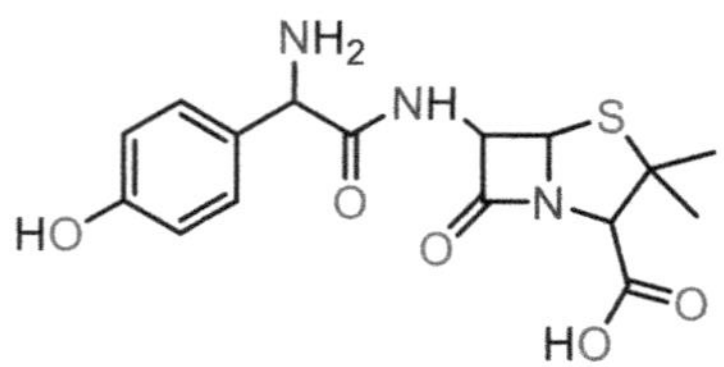 Amoxicillin 99.99% Al/2 M HCl IE 90.00% (2 g L^{-1})[33]	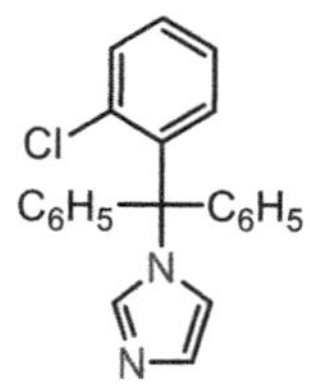 Clotrimazole AA1060/0.5 M H_3PO_4 IE 90.00% (2 g L^{-1})[34]	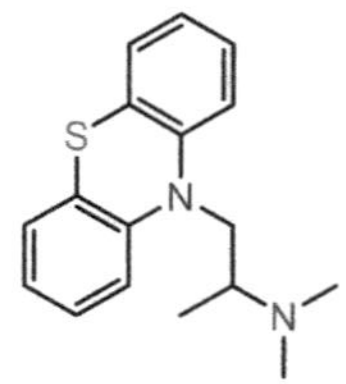 Promethazine hydrochloride Cu/0.5 M H_2SO_4 IE 96.98% (400 mg L^{-1})[35]

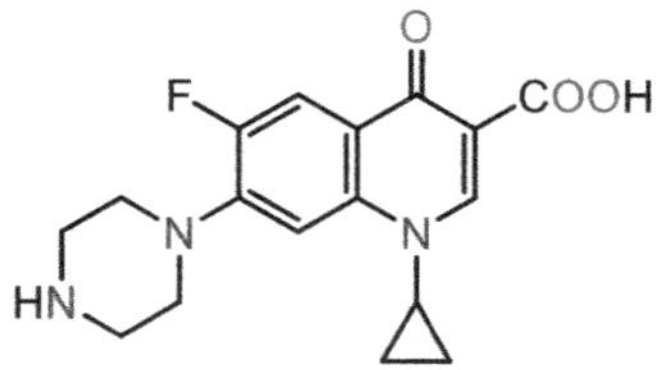 Ciprofloxacillin Bronze/Simulated acid rain (pH 4) IE 56.6% (2 g L^{-1})[36]	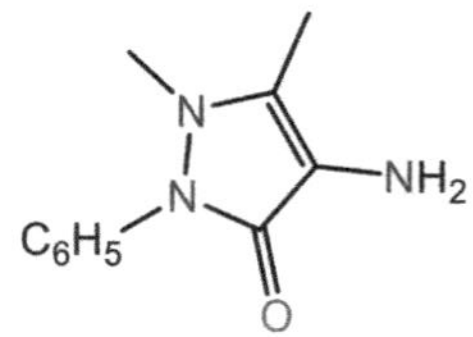 4-Amino antipyrine Cu/3% NaCl IE 88.1% (10 mM)[37]	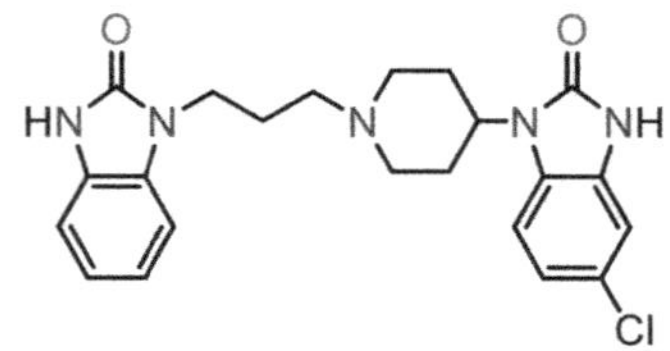 Domperidone Cu/3% NaCl IE 93.6% (20 mg L^{-1})[38]

from inhibitor to metal surface. The doxycycline drug was evaluated for mild steel in a 1 M HCl solution.[28] This molecule contains several –C=O and –OH groups and six member rings. PDP analyses suggested mixed-type behaviour. The inhibitor adsorption on steel surface followed Langmuir adsorption isotherm. *N,N′*-[(methylimino)dimethylidyne] di-2,4-xylidine (MIDX) was evaluated on mild steel in 0.5 M H_2SO_4.[29] This drug contains two phenyl rings and three N atoms in connection with two π-bonds. The authors proposed that the inhibitor shows electrostatic interaction, chemical adsorption and back-donation from the metal surface. Albendazole was studied for mild steel in a 1 M HCl solution *via* gravimetric and electrochemical analyses supported by DFT studies.[30] The molecule contains a benzimidazole moiety. Mixed-type behaviour was obtained with >90% protection at 0.4 mM. An increase in temperature from 308 to 338 K resulted in a slight lowering in the protection efficiency. Cefapirin drug containing a pyridine moiety was studied for mild steel in 1 M HCl solution.[31] Water contact angles were recorded for various concentrations of the drug on the steel surface and showed an increase with inhibitor concentration supporting the formation of a hydrophobic film of inhibitor molecules. The pheniramine drug that contains a phenyl ring and a pyridine ring was studied for mild steel in a 1 M HCl solution using chemical, electrochemical, surface and computational studies.[32] The drug showed mixed-type of inhibition behaviour with adsorption following the Langmuir isotherm.

Four antibacterial drugs, ampicillin, cloxacillin, flucloxacilin, and amoxicillin, were reported for an Al surface in a 2 M HCl solution.[33] The increase in the electron donor characteristics resulted in an improvement in the protection performance. The order of inhibition was amoxicillin > ampicillin > cloxacillin > flucloxacillin. Flucloxacillin, dicloxacillin, cloxacillin, cephalexin, and cefadroxil were investigated on an Al alloy AA6063 in 0.5 M H_3PO_4.[39] Cloxacillin and cefadroxil showed better performance compared to other molecules. Synergistic improvement in the protection behaviour was also investigated using KI. Gentamicin, kanamycin, and amikacin were studied for 99.99% Al in 1 M HCl.[40] The protection behaviour was in the order of amikacin > kanamycin > gentamicin. The change in the molecular weight and the number of heteroatoms was attributed to the adsorption and the protection behaviour of the inhibitors. Two azole category drugs, namely Clotrimazole and Fluconazole, were studied on an Al alloy in 0.1 M HCl solution at 30 °C and 50 °C.[34] The high value of E_{HOMO} and the lower ΔE of Clotrimazole were attributed to better adsorption and inhibition performance. Adsorption of Fluconazole was also studied on Al in another study,[41] and it was observed that the drug adsorbed *via* cationic species on the metal surface. In addition, the drug also underwent adsorption *via* O, N heteroatoms and benzylic C atoms as the active centres. The inhibitive action of Clotrimazole on Al in 0.1 M HCl in other studies was attributed to physical adsorption.[42,43] Miconazole nitrate was investigated on a 96.9–97.8% Al surface in 1 M HCl.[44] The N atoms of the imidazole ring were attributed to the adsorption and protection performance. Ketoconazole was also studied as an inhibitor for an Al substrate in a 2 M HCl environment.[45] Synergistic improvement in the protection was observed by the introduction of KI in the solution. Two anti-histamine drugs, namely Dioxopromethaxine hydrochloride (DPZ) and Promethazine hydrochloride (PZ), were studied on a Cu substrate in 0.5 M H_2SO_4.[35] Electrochemical studies revealed high protection efficiencies of >90% for both. The inhibitor adsorption was in accordance with the Langmuir isotherm. Computational studies *via* DFT and MD simulations supported the experimentally obtained results. Amoxicillin, Ciprofloxacin, Doxycycline and Streptomycin were investigated for bronze in a simulated acid rain solution.[36] A polluted acid rain medium was prepared (pH 4) in which moderate efficiencies of 28.8% to 68.5% were observed at a concentration of 800 mg L^{-1}. Quantum chemical calculations indicated that the most effective inhibitor showed the lowest energy gap. These drugs contain phenyl rings and $-NH_2$, –COOH and –OH groups that were attributed to the inhibition performance. A pharmaceutical intermediate 4-Amino-antipyrine was analysed on Cu surface in 3% NaCl solution employing gravimetric and electrochemical measurements.[37] The drug adsorption was in agreement with Langmuir isotherm. Computational studies indicated that the N and O atoms were the major active sites responsible for adsorption and protection behaviour. On a Cu surface in 3% NaCl, Domperidone was also evaluated *via* experimental and computational studies.[38] A maximum efficiency of 94.2% was obtained at a concentration of 20 mg L^{-1}. This drug molecule contains two benzimidazole moieties with one piperidine ring.

9.4.2 Expired Chemical Medicines as Corrosion Inhibitors

Vaszilcsin *et al.* proposed using expired drug molecules as corrosion inhibitors by exemplifying paracetamol and carbamazepine.[14] Both of these drugs (after 12 months from the date of expiry) are commonly used and can be a cost-effective alternative to expensive corrosion inhibitors. Paracetamol contains a phenyl ring, and carbamazepine contains two phenyl rings on either side of a *N*-containing seven-membered ring (azepine) attached to an amide group. Electrochemical studies indicated a stable adsorbed film of inhibitors formed at the metal-solution interface. Subsequently, the application of expired drugs as corrosion inhibitors has been comprehensively reviewed.[46] On a carbon steel surface in a 2 M HCl solution, an expired Tobramycin (ETo) drug was evaluated.[47] The drug molecule consists of several –OH and $-NH_2$ groups, and oxygen-containing six-membered rings. Moderate efficiencies were recorded at the dose of 500 mg L^{-1}. Expired atorvastatin was evaluated for mild steel in a 1 M HCl solution.[12] The atorvastatin molecule contains a pyrrole ring and three phenyl rings in addition to –OH and –COOH functional groups. A comparison of the fresh and expired drugs was conducted, which revealed that the drug molecule retained stability and activity even after the expiry date. Figure 9.2 shows the adsorption and protection mechanism of the atorvastatin drug on a metal surface.

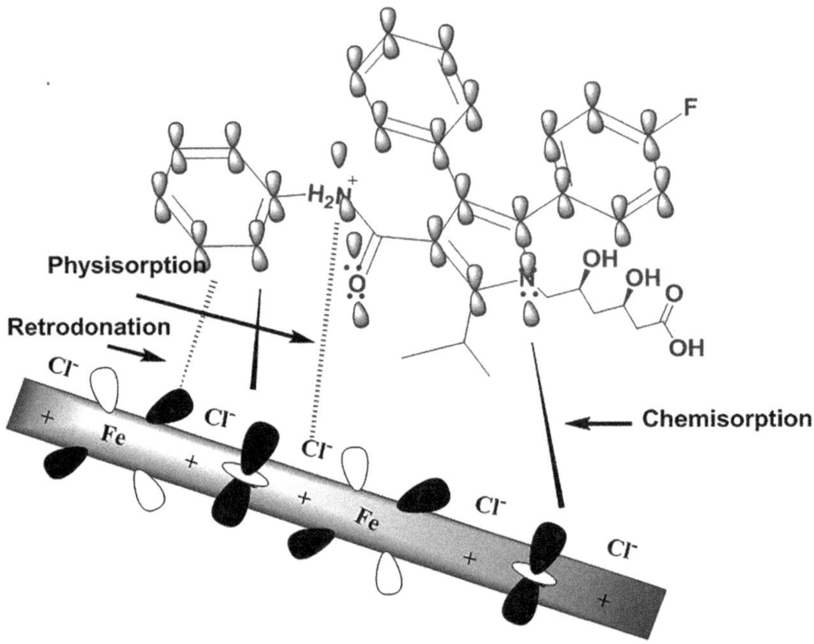

Figure 9.2 Adsorption and protection mechanism of expired drug atorvastatin on a mild steel surface.[12] Reproduced from ref. 12, https://doi.org/10.1007/s40090-017-0120-5, under the terms of the CC BY 4.0 licence https://creativecommons.org/licenses/by/4.0/.

The inhibition effect of the fresh and expired drugs was almost the same, supporting the utility of expired drugs for corrosion inhibition. In subsequent studies, expired tramadol[10] and Podocip[13] drugs were studied for mild steel in a 1 M HCl solution, and efficient performance was observed. Tramadol contains benzene and cyclohexane rings, while Podocip contains thiazole and thiazine rings. In the case of tramadol, the quantum chemical investigations showed that the inhibitor showed better activity in protonated form compared to that in the neutral form. The same group further studied expired metformin drugs for the sweet corrosion of C1018 steel surface[48] and 15% HCl.[49] In fact, these two were the first examples of the use of expired drugs for sweet corrosion and oil-well acidizing. The metformin molecule has a simple chemical structure and contains five N atoms that can allow efficient adsorption. For application in sweet corrosion, a comprehensive study using DFT simulations was undertaken that supported the action of the drug in protonated form. In the case of 15% HCl, the drug showed efficient adsorption and corrosion inhibition that improved further with the incorporation of KI, indicating synergism. In a study by Geethamani and Kasthuri, expired Asthalin drug was studied for mild steel in 1 M HCl and 1 M H_2SO_4 environments.[50] Asthalin molecule contains a phenyl ring with three –OH groups and an –NH group. A detailed study on the influence of various concentrations and temperatures was carried out. Spontaneous adsorption of inhibitor on the steel surface was elucidated, which was in agreement with Langmuir isotherm. Better inhibition efficiency was observed in the HCl solution, with mixed-type inhibitor adsorption in both media. Good inhibition performance was obtained in both media even at elevated temperatures. Three anti-diabetic drugs, namely acarbose, voglibose and miglitol, were evaluated for mild steel in NaCl medium.[51] The miglitol molecule contains a six-membered ring with an N atom and five –OH groups. The introduction of Zn ions produced a remarkable improvement in the adsorption and inhibition performance. The studied inhibitors showed anodic predominance for adsorption and obeyed the Langmuir isotherm. Table 9.3 presents the structures and performance of some expired drugs as corrosion inhibitors.

Expired etoricoxib drug was utilized as a green corrosion inhibitor for carbon steel in 0.5 M H_3PO_4 solution using electrochemical methods.[52] This molecule contains a phenyl ring with two pyridine rings and an $-SO_2-$ group. Moderated inhibition performance was achieved at 30 °C, which decreased with elevation in temperature to 60 °C. Surface examination using SEM-EDX and FTIR studies supported the adsorption and film formation of expired drug molecules. The inhibition effect of expired amoxicillin was studied on mild steel in 1 M HCl solution using gravimetric and electrochemical studies.[53] High efficiency of >94% was observed although at a high dose of 1800 mg L^{-1}. The inhibition of the drug followed the physical mode while obeying the Langmuir isotherm. Expired drugs atenolol and nifedipine were studied for mild steel in 1 M HCl solution.[54] Nifedipine molecule contains a phenyl ring and a pyridine moiety in the structure with a $-NO_2$ group and two $-COOCH_3$ groups. Both molecules acted as mixed-type inhibitors, and the inhibition mechanism's charge transfer control was evidenced. SEM study revealed the formation of protective inhibitor film. The DFT studies indicated high reactivity in the form of low values of global softness for both molecules. Expired metronidazole (MNZ) was studied for Cu surface in 1 M HCl solution.[55] UV-vis studies indicated a decrease in the corrosion product formation in the presence of an inhibitor. Cyclic voltammetry studies supported the protective action of an MNZ inhibitor by suppression of the oxidation currents of Cu. DFT study showed the possibility of inhibitor adsorption *via* surface complexes on Cu, and *via* electrostatic interaction by the formation of chloride bridges.

9.4.3 Chemically Modified Medicines as Corrosion Inhibitors

The isoniazid drug was subjected to a two-step chemical reaction to develop four derivatives, which were tested as inhibitors for mild steel in 1 M HCl environment.[56] Among the four tested inhibitors, the derivatives prepared using piperazine, and 2-mercaptobenzimidazole provided the best performance at a moderate dosage of 200 mg L^{-1}. EIS indicated a charge transfer control of the inhibitor adsorption, whereas PDP study revealed mixed-type behaviour.

Table 9.3 Structures and inhibition performance of some expired medicines.

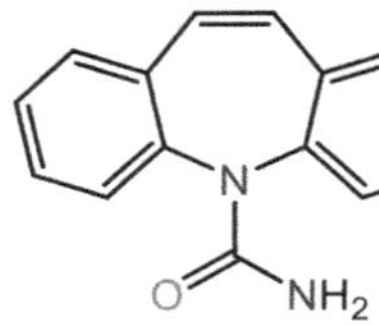	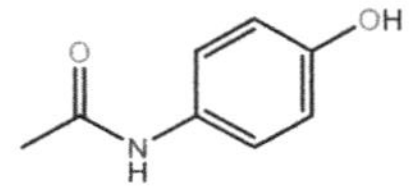	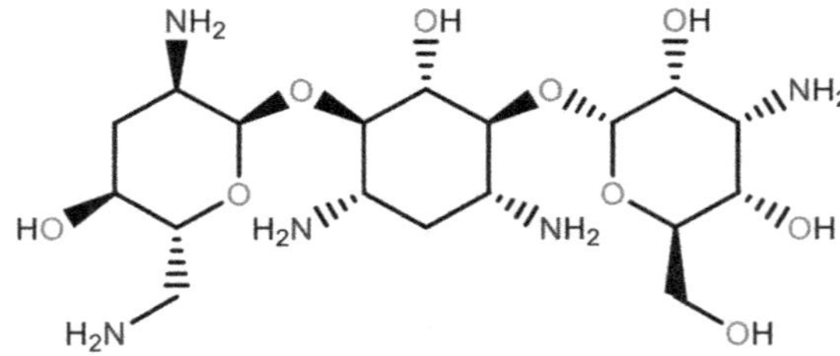
Carbamazepine Carbon steel/0.1 M H_2SO_4 IE 90% (5 mM)[14]	Paracetamol Carbon steel/0.25 M acetic acid/0.25 M sodium acetate IE 85% (5 mM)[14]	Tobramycin Carbon steel/2 M HCl IE 80% (500 mg L^{-1})[47]
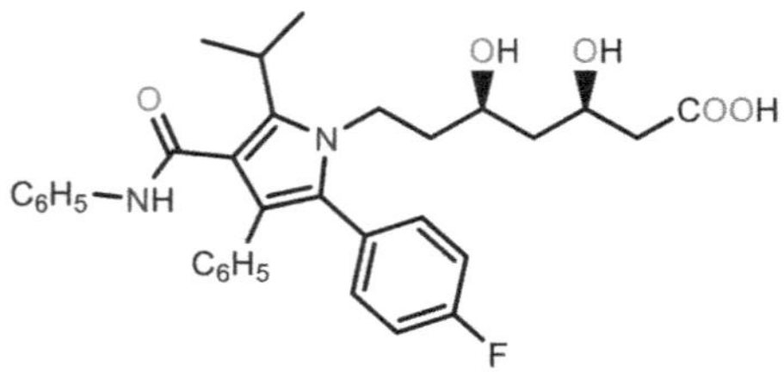	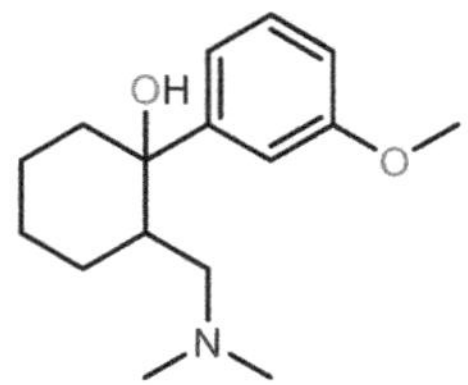	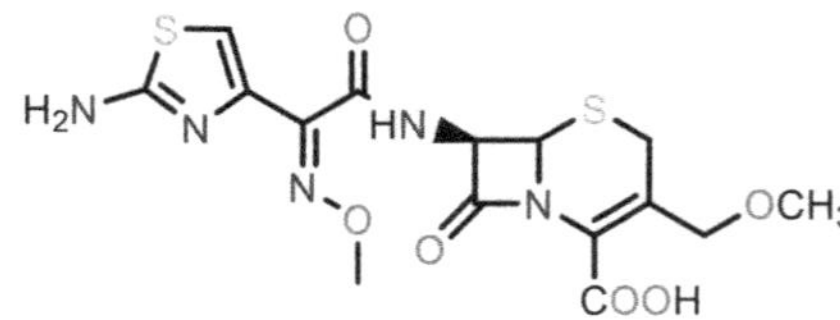
Atorvastatin Mild steel/1 M HCl IE 99.08% (150 mg L^{-1})[12]	Tramadol Mild steel/1 M HCl IE 97.2% (100 mg L^{-1})[10]	Podicip Mild steel/1 M HCl IE 97.93% (100 mg L^{-1})[13]
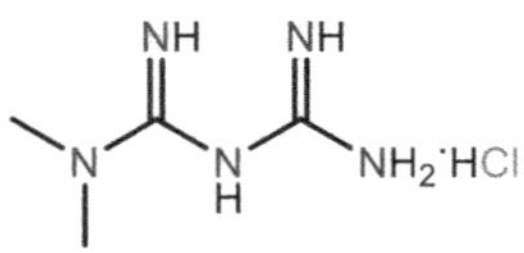	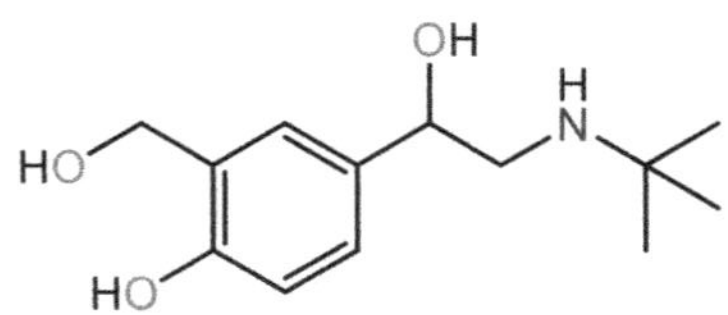	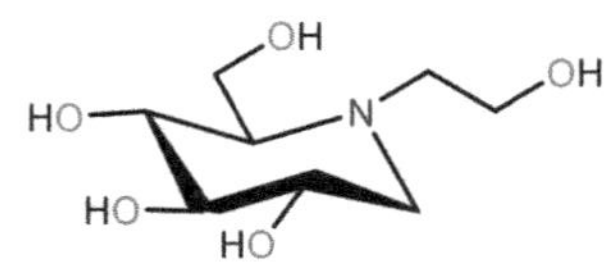

Metformin C1018 steel/CO_2 saturated 3.5% NaCl + 340 mg L^{-1} acetic acid IE 90% (200 mg L^{-1})[48] Carbon steel/15% HCl IE 68.79% (1000 mg L^{-1})[49]	Asthalin Mild steel/1 M HCl Mild steel/1 M H_2SO_4 IE 94.76%; 82.52% (9% v/v)[50]	Miglitol Mild steel/700 mg L^{-1} NaCl IE 78.25% (100 mg L^{-1})[51]
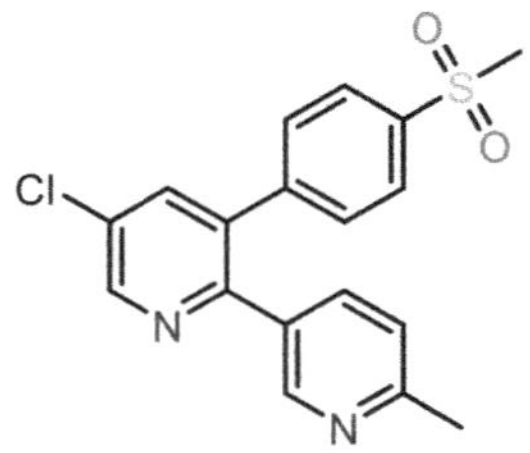	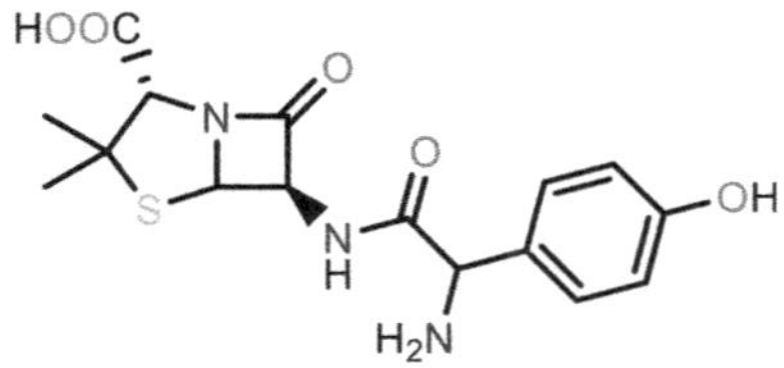	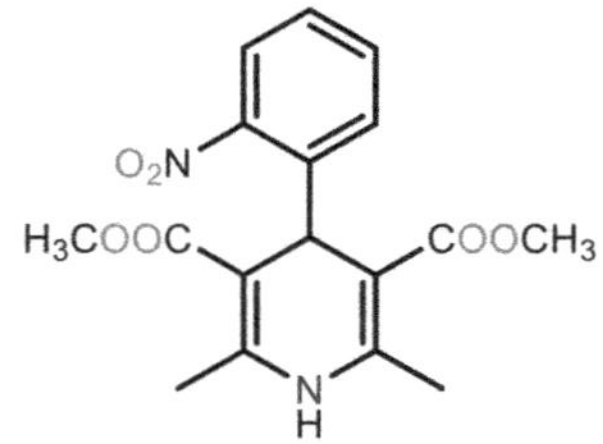
Etoricoxib Carbon steel/0.5 M H_3PO_4 IE 80.627% (225 mg L^{-1})[52]	Amoxicillin Mild steel/1 M HCl IE 94.47% (1.8 g L^{-1})[53]	Nifedipine Mild steel/1 M HCl IE 93.91% (200 mg L^{-1})[54]

As discussed in the previous chapters, one of the simplest strategies for chemical modification that has been utilized to develop novel inhibitor molecules is the Schiff base formation. Table 9.4 shows the corrosion protection performance of some chemically modified drugs. In another investigation, Schiff bases were prepared from expired isoniazid drug *via* refluxing in ethanol.[57] FTIR studies of the inhibitor adsorbed films on the metal surface supported the inhibitor adsorption. The inhibitor, containing an $-OCH_3$ group, provided the best performance. Schiff bases of the aminoantipyrine drug were synthesized *via* simple room temperature stirring. The developed inhibitors were evaluated for P110 steel in a simulated oil-well acidizing environment.[58] Computational analysis suggested the adsorption of the inhibitors *via* the imine linkage (–CH=N–). High inhibition efficiencies were afforded for both the Schiff bases in acid solution. The expired drug dapsone was chemically modified using benzaldehyde and salicylaldehyde to afford Schiff bases and analysed for mild steel in 0.5 M H_2SO_4.[59] KI was also introduced in a trace amount to explore synergistic improvement in corrosion inhibition. Synergistic action of KI on adsorption and protection of the inhibitor on steel surface is schematically displayed in Figure 9.3. Surface studies using FTIR supported inhibitor adsorption and film formation. The theoretical studies *via* DFT computations provided support for the involvement of the imine linkage in the adsorption. In an earlier study, Schiff base of dapsone with salicylaldehyde was utilized for the protection of mild steel in 1 M HCl.[60] High efficiency of 95% was observed at 500 mg L^{-1} concentration. Physical adsorption was noted as the major mode of metal-inhibitor interaction, wherein the inhibitor obeyed the Langmuir isotherm.

9.5 Conclusions and Prospects

This chapter describes the significance of chemical medicines in corrosion inhibition. It is well-established that organic molecules, especially heterocycles, show effective adsorption and protection performance on metal surfaces in a variety of corrosive media. The drug molecules also contain the same structural aspects *e.g.* polar functional groups, heteroatoms, heterocycles, phenyl rings, *etc.* Several researchers have explored drug molecules as environmentally sustainable corrosion inhibitors over the years. The drug molecules are biocompatible, have well-known and stable structures and high solubility in aqueous media, and show effective adsorption on metal surfaces. In different media, *viz.* acid pickling, acidizing, sweet corrosion, saline environment, alkaline media, *etc.*, drugs have been reported as inhibitors. Different metal and alloy surfaces, *viz.* Al, Cu, ferrous alloys, have been investigated. In recent years, researchers have also explored expired or leftover medicines as inhibitors for a cost-effective alternative. Moreover, simple chemical modifications have been performed on both fresh and expired chemical medicines to develop effective inhibitor molecules.

Further studies should target the cost-effective and cheaply available drug molecules to develop effective inhibitors. The newly introduced areas such

Table 9.4 Structures and inhibition performance of some chemically modified medicines.

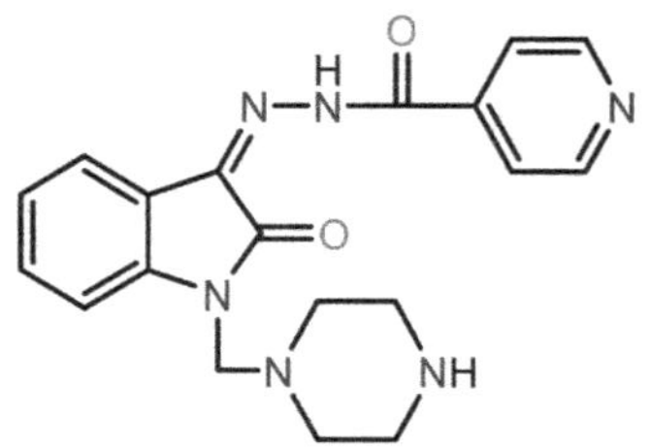	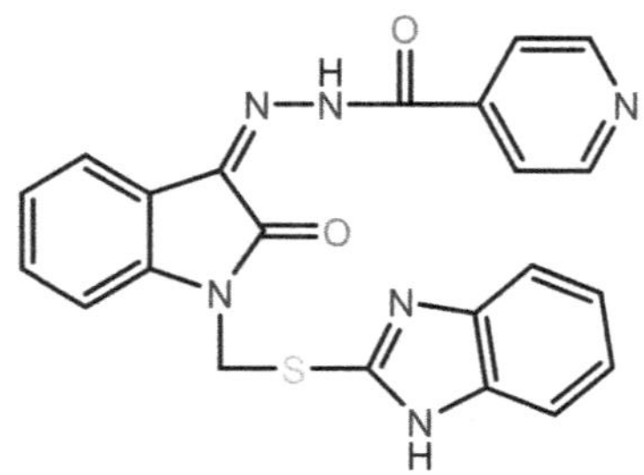	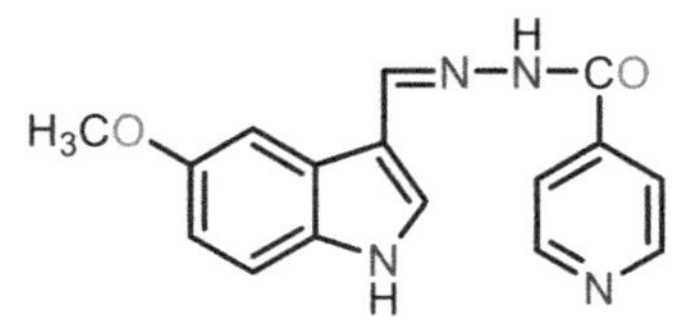
Isoniazid derivative Mild steel/1 M HCl IE 97.61% (200 mg L^{-1})[56]	Isoniazid derivative Mild steel/1 M HCl IE 98.52% (200 mg L^{-1})[56]	Isoniazid derivative Mild steel/1 M HCl IE 96.00% (200 mg L^{-1})[57]
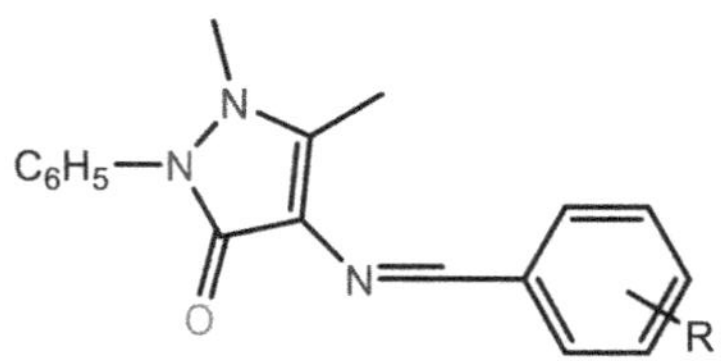	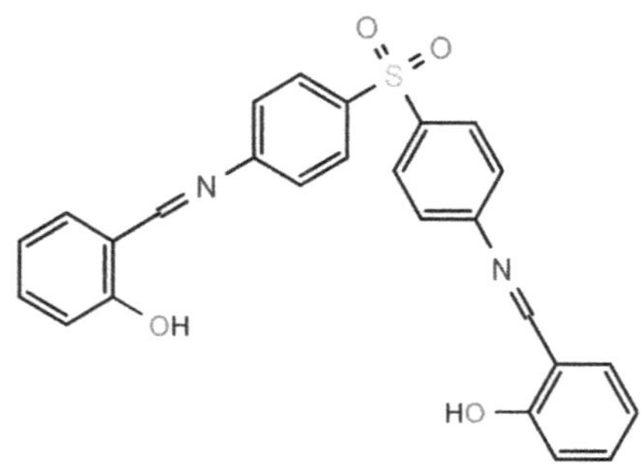	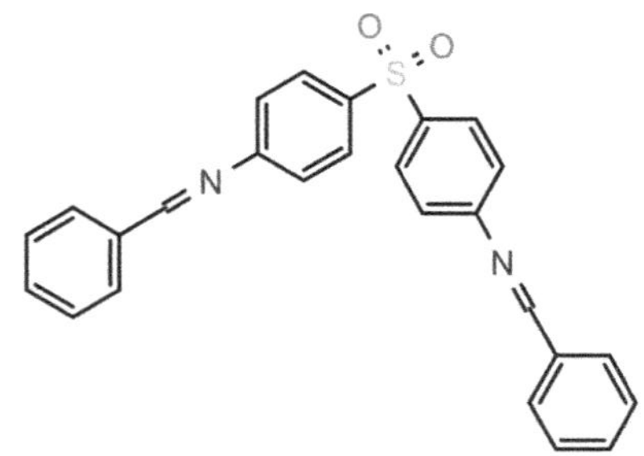
Aminoantipyrine Schiff bases P110 steel/15% HCl IE 95% (500 mg L^{-1})[58]	Dapsone Schiff base-1 Mild steel/0.5 M H_2SO_4 IE 99% (0.21 mM) + 0.6 mM KI[59]	Dapsone Schiff base-2 Mild steel/0.5 M H_2SO_4 IE 97% (0.21 mM) + 0.6 mM KI[59]

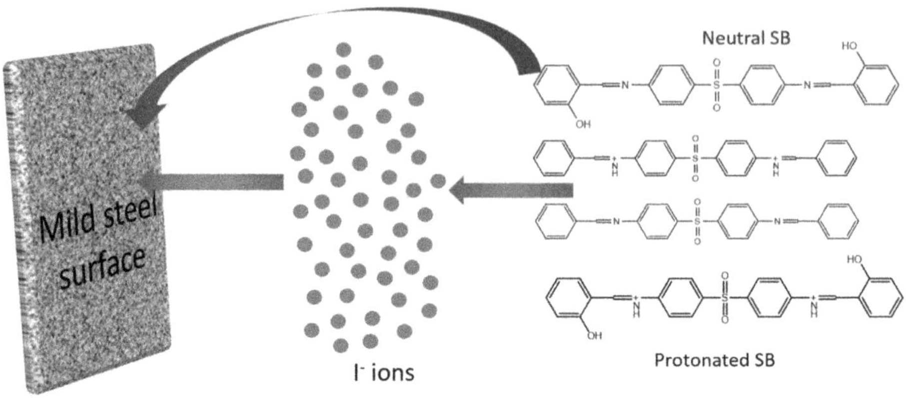

Figure 9.3 Schematic of the adsorption and protection mechanism of neutral and protonated forms of Schiff bases prepared from expired drug Dapsone.[59] Reproduced from ref. 59 with permission from Elsevier, Copyright 2019.

as expired drugs and chemically modified fresh/expired drugs should be emphasized. A realistic estimate of the effectiveness of drugs after the date of expiry should be determined, and simple methods of chemical functionalization of drug-based inhibitors should be carried out. Thorough computational studies on structure-activity correlation should be performed. The drug molecules should be further explored with other chemical additives to develop corrosion inhibitor formulations that can be compared with commercially available corrosion inhibitors.

References

1. G. Gece, Drugs: A review of promising novel corrosion inhibitors, *Corros. Sci.*, 2011, **53**, 3873–3898.
2. C. Verma, D. S. Chauhan and M. A. Quraishi, Drugs as environmentally benign corrosion inhibitors for ferrous and nonferrous materials in acid environment: An overview, *J. Mater. Environ. Sci.*, 2017, **8**, 4040–4051.
3. T. Eicher, S. Hauptmann and A. Speicher, *The Chemistry of Heterocycles: Structures, Reactions, Synthesis, and Applications*, John Wiley & Sons, 2013.
4. M. A. Quraishi, D. S. Chauhan and V. S. Saji, Heterocyclic biomolecules as green corrosion inhibitors, *J. Mol. Liq.*, 2021, **341**, 117265.
5. G. Schmitt, Application of inhibitors for acid media: report prepared for the European federation of corrosion working party on inhibitors, *Br. Corros. J.*, 1984, **19**, 165–176.
6. K. R. Ansari, D. S. Chauhan, A. Singh, V. S. Saji and M. A. Quraishi, Corrosion inhibitors for acidizing process in oil and gas sectors, *Corrosion Inhibitors in the Oil and Gas Industry*, ed. V. S. Saji and S. A. Umoren, Wiley-VCH Verlag GmbH & Co. KGaA, 2020, pp. 153–176, ISBN 978-3-527-34618-9.

7. D. S. Chauhan, M. Quraishi and A. Qurashi, Recent trends in environmentally sustainable Sweet corrosion inhibitors, *J. Mol. Liq.*, 2021, **326**, 115117.
8. J. Sandford-Smith, Outdated drugs may be useful, *BMJ*, 2003, **326**, 51.
9. W.H. Organization, I.P. Association, I.S.W. Association and UNICEF, *Guidelines for Safe Disposal of Unwanted Pharmaceuticals in and after Emergencies*, World Health Organization, Geneva, 1999.
10. P. Dohare, D. S. Chauhan, A. A. Sorour and M. A. Quraishi, DFT and experimental studies on the inhibition potentials of expired Tramadol drug on mild steel corrosion in hydrochloric acid, *Mater. Discovery*, 2017, **9**, 30–41.
11. P. Dohare, D. S. Chauhan, B. Hammouti and M. A. Quraishi, Experimental and DFT investigation on the corrosion inhibition behavior of expired drug lumerax on mild steel in hydrochloric acid, *Anal. Bioanal. Electrochem.*, 2017, **9**, 762.
12. P. Singh, D. S. Chauhan, K. Srivastava, V. Srivastava and M. A. Quraishi, Expired atorvastatin drug as corrosion inhibitor for mild steel in hydrochloric acid solution, *Int. J. Ind. Chem.*, 2017, **8**, 363–372.
13. P. Dohare, D. S. Chauhan and M. A. Quraishi, Expired Podocip drug as potential corrosion inhibitor for carbon steel in acid chloride solution, *Int. J. Corros. Scale Inhib.*, 2018, **7**, 25–37.
14. N. Vaszilcsin, V. Ordodi and A. Borza, Corrosion inhibitors from expired drugs, *Int. J. Pharm.*, 2012, **431**, 241–244.
15. C. Verma, M. Quraishi and E. Ebenso, Electrochemical studies of 2-amino-1, 9-dihydro-9-((2-hydroxyethoxy) methyl)-6*H*-purin-6-one as green corrosion inhibitor for mild steel in 1.0 M hydrochloric acid solution, *Int. J. Electrochem. Sci.*, 2013, **8**, 7401–7413.
16. A. K. Singh and M. A. Quraishi, Effect of Cefazolin on the corrosion of mild steel in HCl solution, *Corros. Sci.*, 2010, **52**, 152–160.
17. S. K. Shukla, A. K. Singh, I. Ahamad and M. A. Quraishi, Streptomycin: A commercially available drug as corrosion inhibitor for mild steel in hydrochloric acid solution, *Mater. Lett.*, 2009, **63**, 819–822.
18. A. Singh, A. Kumar Singh and M. A Quraishi, Dapsone: a novel corrosion inhibitor for mild steel in acid media, *Open Electrochem. J.*, 2010, **2**, 43–51.
19. A. K. Singh and M. A. Quraishi, Investigation of the effect of disulfiram on corrosion of mild steel in hydrochloric acid solution, *Corros. Sci.*, 2011, **53**, 1288–1297.
20. A. K. Singh and M. A. Quraishi, Adsorption properties and inhibition of mild steel corrosion in hydrochloric acid solution by ceftobiprole, *J. Appl. Electrochem.*, 2011, **41**, 7–18.
21. S. K. Shukla and M. A. Quraishi, Cefotaxime sodium: a new and efficient corrosion inhibitor for mild steel in hydrochloric acid solution, *Corros. Sci.*, 2009, **51**, 1007–1011.
22. A. K. Singh, E. E. Ebenso and M. A. Quraishi, Corrosion inhibition behavior of cefuzonam at mild steel/HCl acid interface, *Res. Chem. Intermed.*, 2013, **39**, 3033–3042.

23. A. K. Singh, S. K. Shukla, M. Singh and M. A. Quraishi, Inhibitive effect of ceftazidime on corrosion of mild steel in hydrochloric acid solution, *Mater. Chem. Phys.*, 2011, **129**, 68–76.
24. S. K. Shukla and M. A. Quraishi, Ceftriaxone: a novel corrosion inhibitor for mild steel in hydrochloric acid, *J. Appl. Electrochem.*, 2009, **39**, 1517–1523.
25. A. K. Singh and M. Quraishi, Inhibitive effect of diethylcarbamazine on the corrosion of mild steel in hydrochloric acid, *Corros. Sci.*, 2010, **52**, 1529–1535.
26. I. Ahamad, R. Prasad and M. A. Quraishi, Experimental and theoretical investigations of adsorption of fexofenadine at mild steel/hydrochloric acid interface as corrosion inhibitor, *J. Solid State Electrochem.*, 2010, **14**, 2095–2105.
27. A. K. Singh, S. Khan, A. Singh, S. Quraishi, M. Quraishi and E. E. Ebenso, Inhibitive effect of chloroquine towards corrosion of mild steel in hydrochloric acid solution, *Res. Chem. Intermed.*, 2013, **39**, 1191–1208.
28. S. K. Shukla and M. Quraishi, The effects of pharmaceutically active compound doxycycline on the corrosion of mild steel in hydrochloric acid solution, *Corros. Sci.*, 2010, **52**, 314–321.
29. A. K. Singh, S. K. Shukla, M. Quraishi and E. E. Ebenso, Investigation of adsorption characteristics of *N,N'*-[(methylimino) dimethylidyne] di-2, 4-xylidine as corrosion inhibitor at mild steel/sulphuric acid interface, *J. Taiwan Inst. Chem. Eng.*, 2012, **43**, 463–472.
30. I. Ahamad, R. Prasad, E. E. Ebenso and M. Quraishi, Electrochemical and quantum chemical study of albendazole as corrosion inhibitor for mild steel in hydrochloric acid solution, *Int. J. Electrochem. Sci.*, 2012, **7**, 3436.
31. A. K. Singh, E. E. Ebenso and M. Quraishi, Adsorption behaviour of cefapirin on mild steel in hydrochloric acid solution, *Int. J. Electrochem. Sci.*, 2012, **7**, 2320–2333.
32. I. Ahamad, R. Prasad and M. Quraishi, Inhibition of mild steel corrosion in acid solution by Pheniramine drug: Experimental and theoretical study, *Corros. Sci.*, 2010, **52**, 3033–3041.
33. M. Abdallah, Antibacterial drugs as corrosion inhibitors for corrosion of aluminium in hydrochloric solution, *Corros. Sci.*, 2004, **46**, 1981–1996.
34. I. Obot, N. Obi-Egbedi and S. Umoren, Antifungal drugs as corrosion inhibitors for aluminium in 0.1 M HCl, *Corros. Sci.*, 2009, **51**, 1868–1875.
35. Y. Lu, L. Zhou, B. Tan, B. Xiang, S. Zhang, S. Wei, B. Wang and Q. Yao, Two common antihistamine drugs as high-efficiency corrosion inhibitors for copper in 0.5 M H_2SO_4, *J. Taiwan Inst. Chem. Eng.*, 2021, **123**, 11–20.
36. I. Rotaru, S. Varvara, L. Gaina and L. M. Muresan, Antibacterial drugs as corrosion inhibitors for bronze surfaces in acidic solutions, *Appl. Surf. Sci.*, 2014, **321**, 188–196.
37. S. Hong, W. Chen, H. Q. Luo and N. B. Li, Inhibition effect of 4-amino-antipyrine on the corrosion of copper in 3 wt.% NaCl solution, *Corros. Sci.*, 2012, **57**, 270–278.

38. D. Wang, B. Xiang, Y. Liang, S. Song and C. Liu, Corrosion control of copper in 3.5 wt.% NaCl solution by domperidone: experimental and theoretical study, *Corros. Sci.*, 2014, **85**, 77–86.
39. A. Fouda, A. Al-Sarawy, F. S. Ahmed and H. El-Abbasy, Corrosion inhibition of aluminum 6063 using some pharmaceutical compounds, *Prot. Met. Phys. Chem. Surf.*, 2009, **45**, 635–643.
40. M. Abdallah and B. Jahdaly, Gentamicin, kanamycin and amikacin drugs as non-toxic inhibitors for corrosion of aluminum in 1.0 M hydrochloric acid, *Int. J. Electrochem. Sci.*, 2015, **10**, 9808–9823.
41. I. Obot and N. Obi-Egbedi, Fluconazole as an inhibitor for aluminium corrosion in 0.1 M HCl, *Colloids Surf., A*, 2008, **330**, 207–212.
42. I. Obot, N. Obi-Egbedi, S. Umoren and E. Ebenso, Adsorption and kinetic studies on the inhibition potential of fluconazole for the corrosion of Al in HCl solution, *Chem. Eng. Commun.*, 2011, **198**, 711–725.
43. I. Obot, N. Obi-Egbedi and S. Umoren, Adsorption characteristics and corrosion inhibitive properties of clotrimazole for aluminium corrosion in hydrochloric acid, *Int. J. Electrochem. Sci.*, 2009, **4**, 863–877.
44. J. I. Bhat and V. D. Alva, Miconazole nitrate as a potential corrosion inhibitor for aluminium in hydrochloric acid medium, *Synth. React. Inorg., Met.-Org., Nano-Met. Chem.*, 2012, **42**, 951–957.
45. I. Obot and N. Obi-Egbedi, Inhibition of aluminium corrosion in hydrochloric acid using nizoral and the effect of iodide ion addition, *E.-J. Chem.*, 2010, 7, 837–843.
46. D. S. Chauhan, A. A. Sorour and M. A. Quraishi, An overview of expired drugs as novel corrosion inhibitors for metals and alloys, *Int. J. Chem. Pharm. Sci.*, 2016, **4**, 680–691.
47. F. Abeng, V. Anadebe, V. Idim and M. Edim, Anti-corrosion behaviour of expired tobramycin drug on carbon steel in acidic medium, *S. Afr. J. Chem.*, 2020, **73**, 125–130.
48. I. B. Onyeachu, S. Abdel-Azeim, D. S. Chauhan and M. A. Quraishi, Electrochemical and computational insights on the application of expired Metformin drug as a novel inhibitor for the Sweet corrosion of C1018 steel, *ACS Omega*, 2021, **6**, 65–76.
49. K. Haruna, T. A. Saleh and M. Quraishi, Expired metformin drug as green corrosion inhibitor for simulated oil/gas well acidizing environment, *J. Mol. Liq.*, 2020, 113716.
50. P. Geethamani and P. Kasthuri, The inhibitory action of expired asthalin drug on the corrosion of mild steel in acidic media: A comparative study, *J. Taiwan Inst. Chem. Eng.*, 2016, **63**, 490–499.
51. R. G. Sundaram, G. Vengatesh and M. Sundaravadivelu, Surface morphological and quantum chemical studies of some expired drug molecules as potential corrosion inhibitors for mild steel in chloride medium, *Surf. Interfaces*, 2021, **22**, 100841.
52. R. A. Anaee, I. H. R. Tomi, M. H. Abdulmajeed, S. A. Naser and M. M. Kathem, Expired Etoricoxib as a corrosion inhibitor for steel in acidic solution, *J. Mol. Liq.*, 2019, **279**, 594–602.

53. B. Fadila, A. Sihem, A. Sameh and G. Kardas, A study on the inhibition effect of expired amoxicillin on mild steel corrosion in 1 N HCl, *Mater. Res. Express*, 2019, **6**, 046419.
54. N. K. Gupta, C. Gopal, V. Srivastava and M. Quraishi, Application of expired drugs in corrosion inhibition of mild steel, *Int. J. Pharm. Chem. Anal.*, 2017, **4**, 8–12.
55. A. Samide, B. Tutunaru, A. Dobrițescu, P. Ilea, A.-C. Vladu and C. Tigae, Electrochemical and theoretical study of metronidazole drug as inhibitor for copper corrosion in hydrochloric acid solution, *Int. J. Electrochem. Sci.*, 2016, **11**, e5534.
56. A. K. Singh and M. Quraishi, Investigation of adsorption of isoniazid derivatives at mild steel/hydrochloric acid interface: Electrochemical and weight loss methods, *Mater. Chem. Phys.*, 2010, **123**, 666–677.
57. A. K. Singh, A. K. Pandey, P. Banerjee, S. K. Saha, B. Chugh, S. Thakur, B. Pani, P. Chaubey and G. Singh, Eco-friendly disposal of expired anti-tuberculosis drug isoniazid and its role in the protection of metal, *J. Environ. Chem. Eng.*, 2019, 7, 102971.
58. A. Singh, K. Ansari, M. Quraishi, S. Kaya and L. Guo, Aminoantipyrine derivatives as a novel eco-friendly corrosion inhibitors for P110 steel in simulating acidizing environment: Experimental and Computational studies, *J. Nat. Gas Sci. Eng.*, 2020, **83**, 103547.
59. P. Singh, D. S. Chauhan, S. S. Chauhan, G. Singh and M. A. Quraishi, Chemically modified expired Dapsone drug as environmentally benign corrosion inhibitor for mild steel in sulphuric acid useful for industrial pickling process, *J. Mol. Liq.*, 2019, **286**, 110903.
60. A. Singh, J. Avyaya, E. E. Ebenso and M. Quraishi, Schiff's base derived from the pharmaceutical drug Dapsone (DS) as a new and effective corrosion inhibitor for mild steel in hydrochloric acid, *Res. Chem. Intermed.*, 2013, **39**, 537–551.

CHAPTER 10

Oleochemicals as Green Corrosion Inhibitors

10.1 Introduction

In the previous chapters, we have provided a glimpse of green and sustainable corrosion inhibitors having a natural origin, namely, biopolymers, amino acids, and plant extracts. Oleochemicals, which means the fats and oils extracted from plant and animal sources, also represent this category. Having a natural origin, oleochemicals are environmentally benign. Oleochemicals bear the structure of long-chain fatty acids, which aid in the effective adsorption over the target metal surfaces. The presence of other surface functionalities, in the form of polar functional groups, heteroatoms, heterocycles and/or phenyl rings, π-bonds, facilitate their adsorption. Due to their ease of availability, low toxicity, and biodegradable properties, the oleochemicals present a major category of environmentally benign corrosion inhibitors.

Several research articles have come in the literature reporting the utility of oleochemicals as corrosion inhibitors. This chapter introduces oleochemicals, their significance as corrosion inhibitors, some literary examples, and discusses some bottlenecks associated with their usage for inhibition purposes.

10.2 Production/Recovery of Oleochemicals

Oils and fats are crude materials for the preparation of oleochemicals. These precursors include unsaturated fats, methyl esters, greasy alcohols and amines, and glycerols. Vegetable oils account for around 80% of the global

Green Chemistry Series No. 73
Green Corrosion Inhibition: Fundamentals, Design, Synthesis and Applications
By Chandrabhan Verma, Mumtaz A. Quraishi and Dheeraj Singh Chauhan

Published by the Royal Society of Chemistry, www.rsc.org

oil and fat production. In 2019, this was about 200 million tons.[1] Soybeans are the most significant oil crop, followed by palm, rapeseed, and sunflower oils. About 15–17 million tons of vegetable oils are utilized for the production of surfactants, coatings, oils, makeup, and other items. Fundamentally, saturated or unsaturated fatty acids can be derived from oil or fat precursors *via* hydrolytic or lipolytic parting (*via* reaction with water response at high temperature or pressure or by enzymatic reaction). It is reported that around eight or so oils and fats add to the bulk production of fatty acids, depending upon geographical production. The manufacture of oleochemicals involves two major processes, including hydrolysis and transesterification, as shown below:

(i) **Hydrolysis:** Involves the parting (or hydrolysis) of the fatty substances to yield unsaturated fats and glycerol:

$$\underset{\text{Triglyceride}}{RCO_2CH_2\text{–}CHO_2CR\text{–}CH_2O_2CR} + \underset{\text{water}}{3H_2O} \rightarrow \underset{\text{fatty acid}}{3RCOOH} + \underset{\text{glycerol}}{HOCH_2\text{–}CHOH\text{–}CH_2OH}$$

(ii) **Transesterification:** In this process, the oil/fat reacts with alcohols (R'OH) rather than with water as in hydrolysis (shown above). Glycerol is obtained along with the unsaturated fatty esters. Generally, the response involves the utilization of methanol (MeOH) to yield unsaturated fatty methyl esters:

$$\underset{\text{Triglyceride}}{RCO_2CH_2\text{–}CHO_2CR\text{–}CH_2O_2CR} + \underset{\text{Methanol}}{3MeOH} \rightarrow \underset{\text{Fatty acid methyl ester}}{3RCO_2Me} + \underset{\text{Glycerol}}{HOCH_2\text{–}CHOH\text{–}CH_2OH}$$

(iii) **Hydrogenation:** In this process, the unsaturated fats are converted to corresponding fatty alcohols by direct hydrogenation. For certain applications, unsaturated fats are changed over to fatty nitriles. Hydrogenation of these nitriles yields fatty amines having various applications.[2] A schematic of the production of oleochemicals and their derivatives is shown in Figure 10.1.

10.3 Significance of Oleochemicals as Corrosion Inhibitors

The use of oleochemicals in various applications is in alignment with the provisos of Green Chemistry principles and, in general, promotes the use of renewable, sustainable, and bio-based chemicals. Oleochemicals find applications in personal care and cosmetics, food and beverages, pharmaceuticals, lubricants, paints, coatings, corrosion inhibitors *etc.* A major source of oleochemicals is the biomass derived from plant and animal wastes. Therefore, the usage of oleochemicals from biomass aids in minimizing waste production and environmental sustainability.[3,4] The most important sources of natural oils are soybean oils, palm oil, rapeseed oil, and sunflower oil.

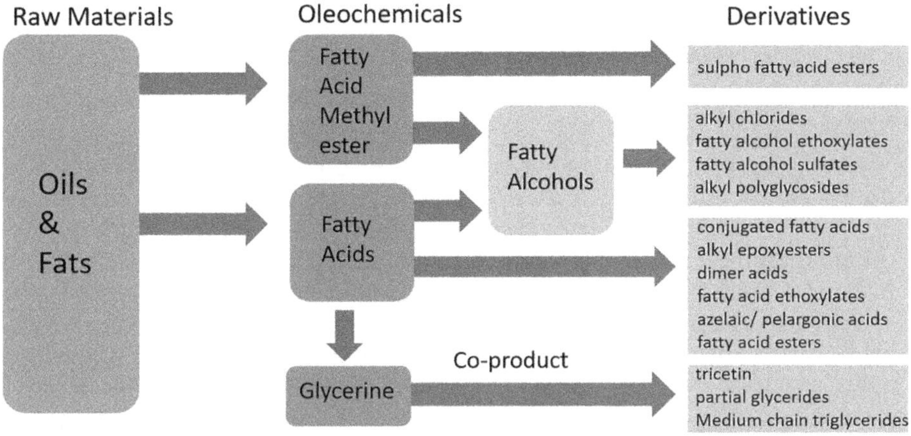

Figure 10.1 Production of oleochemicals and their derivatives.

The oils and fats are triesters of fatty acids and glycerol, having varying chemical structures and applications. Fatty acids form the basic structural units of oleochemicals, from which different derivatives such as fatty acid esters, alcohols, amine derivatives *etc.* are formed *via* chemical and enzymatic transformations. Some oleochemicals-based products include speciality esters, glycerol esters, alkoxylates, *etc.* These molecules can undergo strong adsorption with the metal surface leading to the formation of stable films, allowing high protection efficiency. For the detailed procedure of production and recovery of oleochemicals from fats and oils involving hydrolysis, transesterification, and hydrogenation, readers are referred to elsewhere.[5,6] In the domains of corrosion inhibition, the oleochemicals find attention due to the high cost of petrochemicals and the shift towards the preference for the use of eco-friendly products. Various types of oleochemicals-based corrosion inhibitors are reported in the literature, including fatty amines, long-chain fatty acids, their derivatives, including imidazolines, oxadiazoles, triazoles, amides, and their salts, as discussed in the below sections.

10.4 Issues in the Use of Oleochemicals as Corrosion Inhibitors

The long synthesis time associated with the oleochemicals application and the tedious isolation and purification from the oil/fat animal/plant source could be a major setback. For the synthesis of corrosion inhibitors, considering economic and time constraints, multicomponent reactions are preferable to allow rapid synthesis. This side of oleochemicals requires consideration with the aim of developing a commercially valuable inhibitor formulation. However, researchers have developed effective inhibitor molecules from natural oils in simple one/two-step chemical reactions.[7]

10.5 Literature Reports on Oleochemicals as Corrosion Inhibitors

10.5.1 Natural Oil Derivatives

One of the prominent sources of vegetable oils is rice bran, consisting of 47% monounsaturated and 33% polyunsaturated fats, and 10–23% fatty acid. A fatty amide was prepared from rice bran oil and used as a green inhibitor in a sweet medium for API X70 steel substrate.[7] Electrochemical measurements revealed a high efficiency of 95% with anodic predominance. Palm oil and its derivatives are also considered efficient corrosion inhibitors. Palm oil was used as a corrosion inhibitor for ductile iron and mild steel in freshwater and 1 M NaOH solution.[8] Reasonable protection was afforded in both environments. Palm kernel oil (PKO) was studied as an environmentally benign corrosion inhibitor for carbon steel SAE 1045 in 1 M HCl solution.[9] A high efficiency of 90.10% was achieved at a concentration of 4 g L^{-1}. In another study, a PKO inhibitor was investigated on the same metal surface in 1 M NaOH, wherein a high protective performance of 96.67% at 8 v/v% concentration was obtained from weight loss methods and polarization studies.[10] A palm oil-based ethanolamine derivative was synthesized and studied for the protection of carbon steel in 0.5 M HCl solution.[11] The inhibitor was prepared using a reaction between a fatty acid and ethanolamine, resulting in 2-aminoethyl fatty esters and *N*-(2-hydroxyethyl) fatty amides. The presence of –NH– and –OH groups in these inhibitors can be attributed to the high extent of adsorption over metallic substrate. Mixed-type inhibition was afforded with the inhibitor adsorption in agreement with the Langmuir isotherm. Some nonconventional oils *e.g.* neem, rice bran, and karanja were taken to prepare thiosemicarbazide derivatives that were studied as inhibitors for mild steel in HCl solution.[12] Figure 10.2 shows the performance of some oleochemicals as corrosion inhibitors.

Fatty acid derivatives from microalgae *Scenedesmus* sp. were evaluated as corrosion inhibitors of mild steel in 1 M HCl solution.[13] Surface examination *via* AFM studies revealed a reduction in the surface roughness from 138.51 nm to 45.09 nm. A high efficiency of 95.1% was attained, with a considerable decrease in hydrogen evolution. The presence of π-bonds in the long alkyl chain was attributed to the effective adsorption and protection performance. Oleic acid was found to be the major component, followed by palmitic and linoleic acid in the three oils. Further, a Schiff base was prepared from these oils, and studied as an inhibitor for mild steel corrosion in HCl solution.[19] At a concentration of 500 mg L^{-1}, a high efficiency of >92% was obtained, which had a predominance of the physical mode of inhibitor adsorption on the steel surface. Four different vegetable oils, *viz.* sunflower oil, cottonseed oil, corn oil, and palm oil, were procured and sulfated fatty acid–diethanolamine complexes were prepared.[14] The molecules contained N and O heteroatoms that could be credited with the high extent of adsorption over the metallic

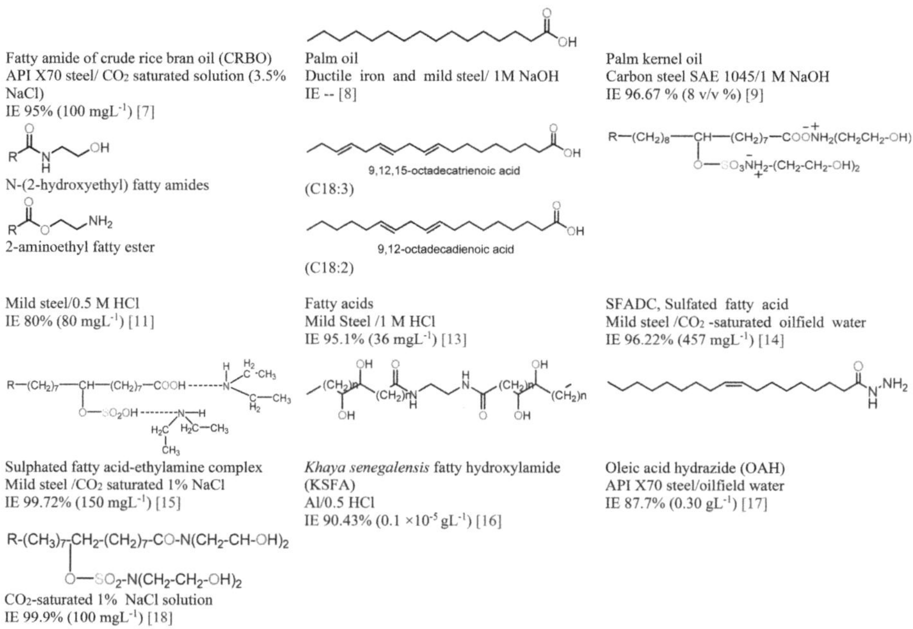

Figure 10.2 Corrosion inhibition performance of some oleochemicals-based corrosion inhibitors.

surface. A high efficiency of 99.95% was achieved at 100 mg L^{-1} from the complex prepared using corn oil. Inhibitor adsorption was in accordance with Langmuir isotherm, and further supported by X-ray fluorescence microscopy. In another study, five surfactants were prepared from palm oil and studied for carbon steel corrosion protection in CO_2-saturated brine at 50 °C.[20] An increase in the carbon chain length of the surfactants caused a lowering in the CMC of the surfactant molecules, and a corresponding improvement was visualized in the protection performance. Spontaneous adsorption of inhibitors correlating to the Langmuir isotherm was noticed. Following reflux of corn oil and ethanolamine for 7–8 h at 170–180 °C, a series of sulfonated fatty acid diethanolamides were synthesized.[21] The developed novel inhibitor molecules were analysed for a mild steel surface in a sweet corrosion environment. Polarization study revealed excellent inhibition performance of the molecules with a mixed mode of corrosion protection. Three complex surfactant inhibitors were synthesized from sunflower oil and N atom-containing molecules and studied for mild steel in the sweet environment using gravimetric and LPR analyses.[15] Around the CMC, all the surfactants provided the maximum inhibition performance. The presence of N, O and S heteroatoms can be attributed to the high inhibition performance. ΔG^0_{ads} values revealed a chemical mode of inhibitor adsorption. Energy dispersive x-ray fluorescence microscopy (EDRF) supported the formation of a surface film of inhibitor molecules. Five surfactants prepared from corn oil were evaluated for C1018 steel surface in CO_2-saturated brine at various temperatures.[18] Thermodynamic parameters suggested chemical mode of adsorption of all the studied inhibitors. Presence of N, O and S heteroatoms-containing functional groups could be attributed to protection performance. The adsorption of the studied molecules obeyed the Langmuir isotherm. SEM and EDRF studies supported the formation of a protective inhibitor film on the steel surface.

10.5.2 Fatty Amides

Amide derivatives of 13-docosenoic acid were prepared by reaction with primary and secondary aliphatic/aromatic amines. The products were utilized for the inhibition of mild steel corrosion in 1 M HCl solution.[22] The presence of long alkyl chains, π-bonds, aromatic rings along with heteroatoms provided efficient adsorption. Elevation in the temperature up to 60 °C produced a remarkable efficiency of 96.8% with mixed physical and chemical modes of adsorption. Computational studies revealed the metal-inhibitor interaction *via* electron donation from the inhibitor to the metal surface. Using the seed oil of *Khaya senegalensis* following esterification, transesterification, hydroxylation, and amidation reactions, a fatty amide was produced.[16] The inhibitor was evaluated for corrosion mitigation of aluminium substrate in 0.5 M HCl solution using weight loss tests. A physical mode of adsorption was revealed *via* the –OH and $-CONH_2$ functional groups. Oleic acid hydrazide was prepared and employed as a green inhibitor for API X70 steel surface in an Indian oilfield-produced water under hydrodynamic conditions.[17] The presence of –CONH and $-NH_2$ groups can be ascribed to the significant protection as observed in the results. A maximum efficiency of 87.7% was obtained at 0.30 g L^{-1}. A schematic of the inhibition mechanism in Figure 10.3a–c shows that with an increase in the inhibitor concentration, the surface coverage of metal by the inhibitor molecules increases, thereby impacting a better inhibition performance. The panel below shows that under the flow condition, inhibitor performance is improved, although the mass transfer of the Fe^{2+} ions to the bulk of solution is promoted, retarding the formation of a metal-inhibitor complex.

Several hydrazide and thiosemicarbazide-based molecules were developed using long-chain fatty acids and used as inhibitors in the oil-well acidizing process.[23–26] The developed molecules were cost-effective and eco-friendly compared to conventionally used inhibitors. PDP analysis carried over mild and N80 steel substrates revealed mixed-type inhibition in accordance with Temkin isotherm. Long-chain hydrazides and thiosemicarbazides were prepared from fatty acids having 11, 12, and 18 carbon atom chains and studied for mild steel in boiling 15% HCl solution.[23] At a concentration of 5 g L^{-1}, maximum efficiency was obtained by 1-decene-4-phenyl-thiosemicarbazide. Figure 10.4 shows the schematic of the synthesis of some oleochemical-based corrosion inhibitors.

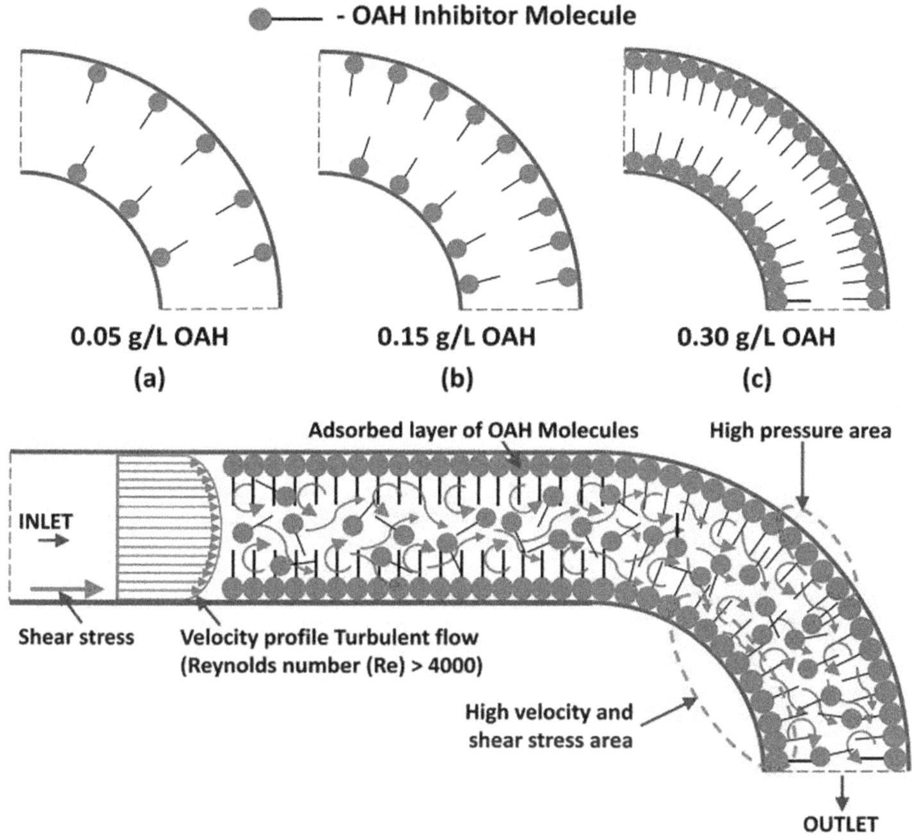

Figure 10.3 Adsorption of oleic acid hydrazide (OAH) on the elbow test section at different concentrations: (a) 0.05 g L^{-1}, (b) 0.15 g L^{-1} and (c) 0.30 g L^{-1}. The panel below shows the interaction of OAH molecules with the internal wall of the steel pipeline in hydrodynamic conditions. Reproduced from ref. 17 with permission from Taylor & Francis, Copyright 2020.

10.5.3 Heterocycles

Heterocyclic molecules are considered the primary choice of corrosion inhibitors in most corrosive media.[32] In this context, imidazolines (IM) represent one of the most commonly used heterocycle-based inhibitors in the industries. From myristic acid and diethyleneamine, myristic acid IM was prepared in a single-step reaction and utilized as a corrosion inhibitor for low carbon steel in 15% HCl.[33] A high inhibition efficiency was noted at a moderate dose of inhibitor. A mixed mode of physical and chemical adsorption was revealed from thermodynamic parameters. FTIR and XPS analyses revealed that the inhibitor adsorption resulted in a suppression of the chloride content of the metal surface. Palmitic acid IM

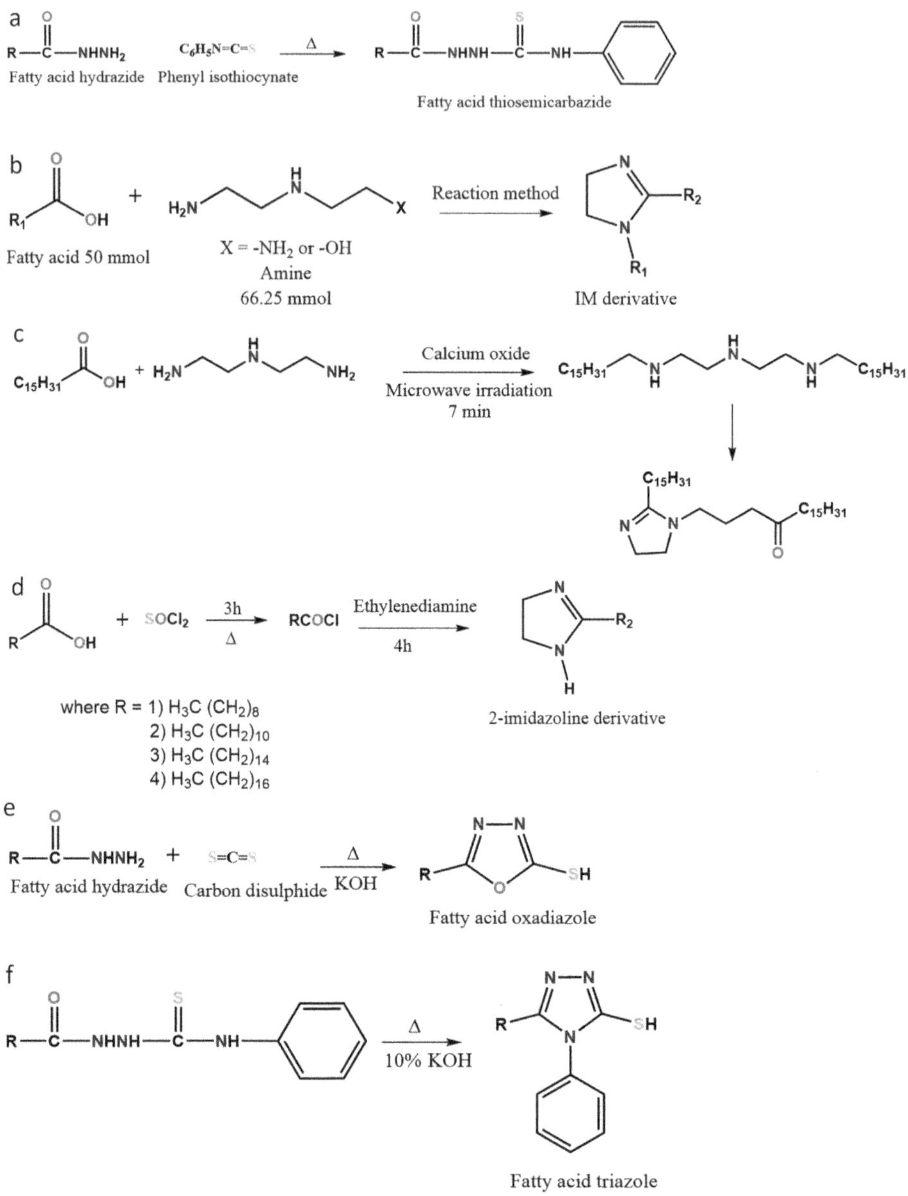

Figure 10.4 Schematic of oleochemicals-based corrosion inhibitors containing (a) thiosemicarbazide, (b) IM,[27] (c) 2-IM,[28] (d) Palmitic IM,[29] (e) oxadiazole,[30] and (f) triazole[31] moieties.

was synthesized *via* a single-step microwave method within a span of 7 min and tested for acid cleaning of desalination plants in 1 M H_2SO_4.[29] High efficiency was afforded, which increased further upon the addition of KI. Surface hydrophobicity of the metallic substrate increased in the presence of the adsorbed inhibitor film, as revealed by the static water contact angle measurements. MD-based computations indicated that the inhibitor adsorbed on the metal surface in a linear fashion. Figure 10.5 displays a schematic of the adsorption and corrosion inhibition mechanism of the inhibitor. It was proposed that while the neutral inhibitor adsorbed *via* chemical adsorption, the protonated form underwent physical adsorption *via* electrostatic interaction. Another IM-based inhibitor, namely QSI, was prepared and analysed for low carbon steel in 15% HCl medium using static and hydrodynamic methods of measurements.[34] A moderate efficiency of <50% was obtained, which improved to 90% with KI. Inhibitor adsorption on steel substrate followed the chemical adsorption. Elongation in the length of the alkyl chain attached to the inhibitor produced an enhancement in the inhibition efficiency.

Four IM-based oleochemical inhibitors, namely PDI, UDI, HDI, and NI, were studied as green inhibitors for corrosion of mild steel in 0.5 M H_2SO_4.[28] Inhibitor adsorption was in accordance with the Langmuir isotherm, while the PDP studies indicated mixed-type behaviour. For mild steel corrosion protection in 1% NaCl environment, some novel IM-based

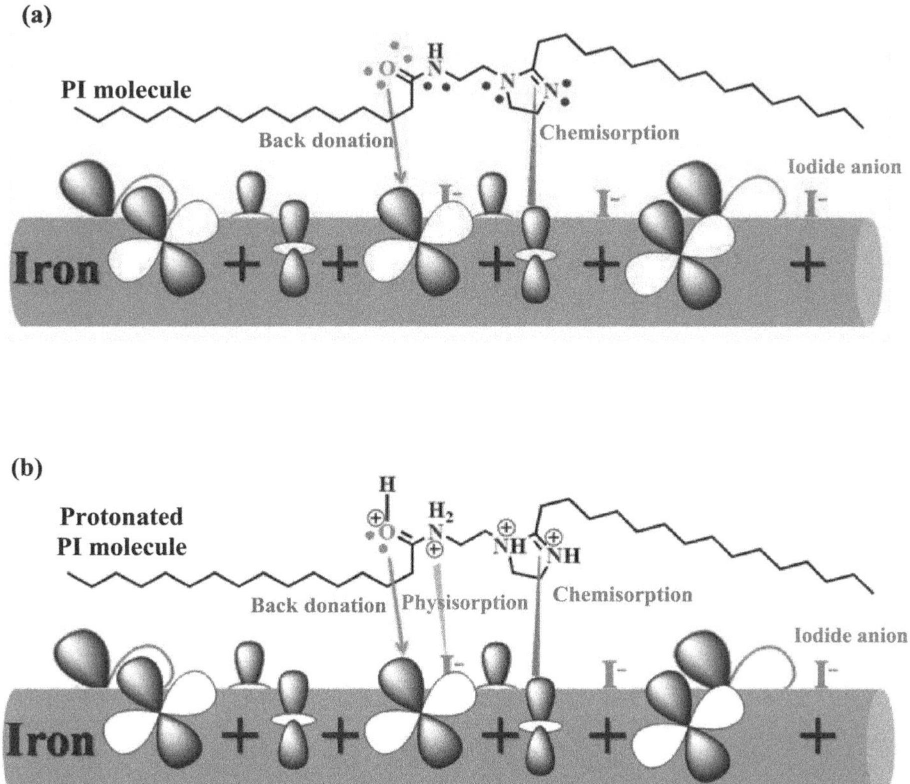

Figure 10.5 Schematic of adsorption of neutral (a) and protonated (b) forms of palmitic IM corrosion inhibitor on a mild steel surface.[29] Reproduced from ref. 29 with permission from Elsevier, Copyright 2019.

molecules prepared from oleic and stearic acids were used as inhibitors.[27] The preparation of inhibitors was achieved using conventional as well as microwave methods. Moderate efficiencies were afforded by the inhibitors containing heptadec-8-enyl and hydroxyethyl substituents. Several IM-based oleochemicals were studied as volatile corrosion inhibitors (VCI) for mild steel, brass, and copper.[35] The inhibitors were prepared from the condensation reaction of 2-dec-9-enyl-2-imidazoline with various organic acids, *viz.* maleic, orthophosphoric, nitrobenzoic, phthalic, and cinnamic acid, and corresponding salts were used as inhibitors. Corrosion testing was conducted using a gravimetric method, the Eschke test, NaCl inoculation, and SO_2 tests. The cinnamate derivative provided the best overall performance. Figure 10.6 shows the performance of some heterocycle-based oleochemicals as corrosion inhibitors.

An early study in 1960 reported Na, Ca, and Pb salts prepared using azelaic, caproic, caprylic, capric, sebacic, and suberic acids and studied as corrosion inhibitors for iron metal surfaces in the pH range of 4 to 6.[36] It

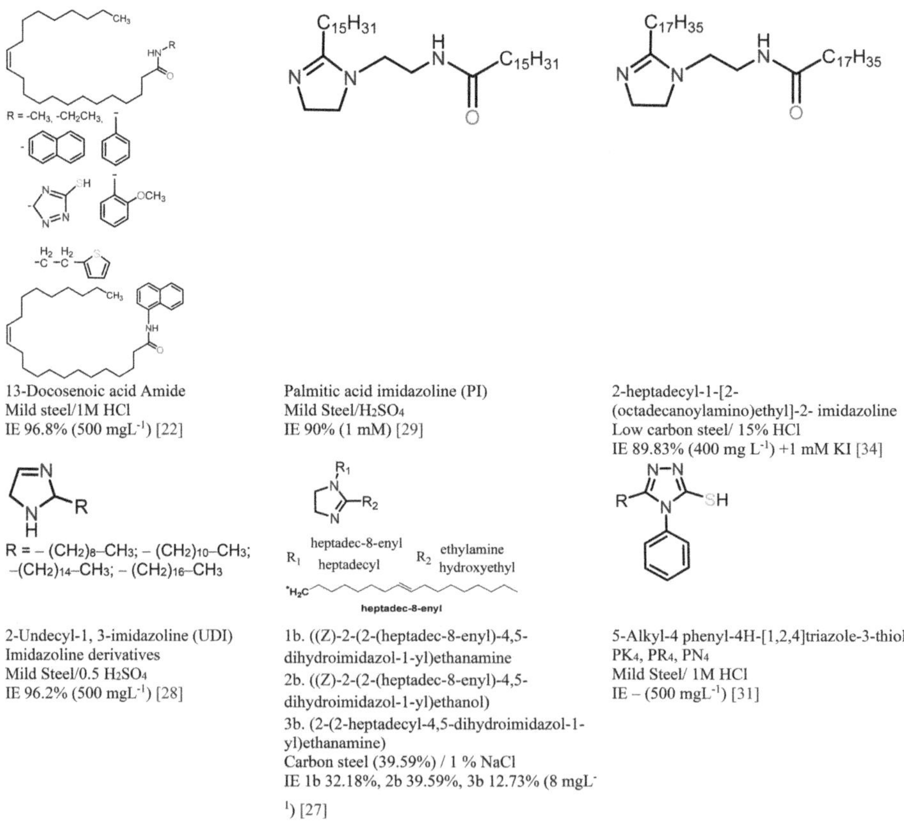

Figure 10.6 Corrosion inhibition performance of some heterocyclic-containing oleochemicals-based corrosion inhibitors.

was revealed that the increase in the alkyl chain length led to an improvement in the protection performance. Among the heterocyclic molecules, compared to oxadiazoles, the triazole-based inhibitors provided better performance. In another study, some fatty acid oxadiazoles were studied for mild steel in 1 M HCl.[25] The inhibitor 2-decane-5-mercapto-1-oxa-3,4-diazole (DMOD) provided the best performance. PDP measurements revealed a cathodic type of performance, supported by SEM investigation, which showed a smooth surface morphology of steel substrate in the presence of inhibitors. Inhibitor adsorption correlated to the Langmuir isotherm. 2-Undecane-5-mercapto-1-oxa-3,4-diazole (UMOD) showed the best performance in 15% boiling HCl on mild steel and N80 steel substrates. The inhibitor adsorption was in agreement with the Temkin isotherm.[26] All the studied inhibitors showed a mixed type of performance, suppressing both anodic and cathodic types of reactions. Using oleic, lauric and undecenoic acids, a series of hydrazides, thiosemicarbazides, oxadiazoles, and triazoles were prepared. The developed corrosion inhibitors were subjected to a series of detailed investigations in formic and acetic acid solutions using gravimetric and electrochemical measurements.[26,31,37] The fatty acid thiosemicarbazides provided the best overall performance, which was attributed to the existence of –C=S group along with an additional phenyl ring in comparison to the hydrazides.

10.6 Conclusions and Prospects

This chapter discusses the significance of oleochemicals derived from naturally occurring fats and oils and their application in corrosion inhibition. Several articles have appeared in the literature on the application of oleochemicals as corrosion inhibitors for acid pickling, acidizing, CO_2 corrosion, in saline, alkaline media, and even as volatile corrosion inhibitors. These molecules show effective surface coverage of the target metal due to large molecular structures. Moreover, having N, O, *etc.* heteroatoms facilitates strong metal-inhibitor interactions. Although from a review of the existing literature, it is evident that, compared to other environmentally benign inhibitors such as amino acids, plant extracts, natural polymers, drugs, *etc.*, the studies on oleochemicals as inhibitors are scarce. One possible issue could be the tedious production procedure of oleochemical compounds from the oil/fat source. However, considering the large-scale application, this can be addressed *via* thorough screening using experimental and computational studies. Considering the large molecular structures, the solubility of oleochemicals could be improved by the incorporation of other additives. In fact, this can also aid in the development of inhibitor formulations. Further, prospective studies should also target non-ferrous metals and alloys *e.g.* Cu, Al, Zn, *etc.* Detailed studies on the effects of temperature, flow rates, pH, *etc.* should be carried out on oleochemical-based corrosion inhibitors. Comparison with existing commercially used inhibitors should be carried out in simulated conditions.

References

1. M. Shahbandeh, Vegetable oils: production worldwide 2012/13-2019/20, by type, 2019.
2. P. Roose, K. Eller, E. Henkes, R. Rossbacher and H. Höke, *Amines, Aliphatic, Ullmann's Encyclopedia of Industrial Chemistry*, 2000, pp. 1–55.
3. J. Salimon, N. Salih and E. Yousif, Industrial development and applications of plant oils and their biobased oleochemicals, *Arabian J. Chem.*, 2012, **5**, 135–145.
4. P. Gallezot, Conversion of biomass to selected chemical products, *Chem. Soc. Rev.*, 2012, **41**, 1538–1558.
5. A. Behr and T. Seidensticker, The basics of oleochemistry-Basic oleochemicals, in *Chemistry of Renewables*, Springer, 2020, pp. 37–60.
6. W. Abdelmoez, N. Mostafa and A. Mustafa, Utilization of oleochemical industry residues as substrates for lipase production for enzymatic sunflower oil hydrolysis, *J. Cleaner Prod.*, 2013, **59**, 290–297.
7. E. Reyes-Dorantes, J. Zuñiga-Díaz, A. Quinto-Hernandez, J. Porcayo-Calderon, J. Gonzalez-Rodriguez and L. Martinez-Gomez, Fatty amides from crude rice bran oil as green corrosion inhibitors, *J. Chem.*, 2017, **2017**, 2871034.
8. A. Daniyan, O. Ogundare, B. AttahDaniel and B. Babatope, Effect of palm oil as corrosion inhibitor on ductile iron and mild steel, *Asia Pac. J .Sci. Technol.*, 2011, **12**, 45–53.
9. M. Y. Zulkafli, N. K. Othman, A. M. Lazim and A. Jalar, Effect of carboxylic acid from palm kernel oil for corrosion prevention, *Int. J. Basic Appl. Sci.*, 2013, **13**, 29–32.
10. M. Zulkafli, N. Othman, A. Lazim and A. Jalar, Inhibitive effects of palm kernel oil on carbon steel corrosion by alkaline solution, in *AIP Conference Proceedings*, American Institute of Physics, 2013, pp. 42–47.
11. M. M. Ali, T. T. Irawadi, N. Darmawan, M. Khotib and Z. A. Mas'ud, Reaction products of crude palm oil-based fatty acids and monoethanolamine as corrosion inhibitors of carbon steel, *Makara J. Sci.*, 2019, **23**, 155–161.
12. S. Toliwal and K. Jadav, Inhibition of corrosion of mild steel by phenyl thiosemicarbazides of nontraditional oils, *J. Sci. Ind. Res.*, 2009, **68**, 235–241.
13. A. Khanra, M. Srivastava, M. P. Rai and R. Prakash, Application of unsaturated fatty acid molecules derived from microalgae toward mild steel corrosion inhibition in HCl solution: a novel approach for metal–inhibitor association, *ACS Omega*, 2018, **3**, 12369–12382.
14. H. M. Abd El-Lateef, I. Ismayilov, V. Abbasov, E. Efremenko, L. Aliyeva and E. Qasimov, Green surfactants from the type of fatty acids as effective corrosion inhibitors for mild steel in CO_2-saturated NaCl solution, *Am. J. Phys. Chem.*, 2013, **2**, 16–23.
15. V. M. Abbasov, H. M. Abd El-Lateef, L. I. Aliyeva, I. T. Ismayilov, E. E. Qasimov and M. M. Narmin, Efficient complex surfactants from the

type of fatty acids as corrosion inhibitors for mild steel C1018 in CO_2-environments, *J. Korean Chem. Soc.*, 2013, **57**, 25–34.
16. A. Adewuyi and R. A. Oderinde, Synthesis of hydroxylated fatty amide from underutilized seed oil of *Khaya senegalensis*: a potential green inhibitor of corrosion in aluminum, *J. Anal. Sci. Technol.*, 2018, **9**, 1–13.
17. T. Ajmal, S. B. Arya, L. Thippeswamy, M. Quraishi and J. Haque, Influence of green inhibitor on flow-accelerated corrosion of API X70 line pipe steel in synthetic oilfield water, *Corros. Eng., Sci. Technol.*, 2020, **55**, 487–496.
18. V. Abbasov, H. M. Abd El-Lateef, L. Aliyeva, E. Qasimov, I. Ismayilov and M. M. Khalaf, A study of the corrosion inhibition of mild steel C1018 in CO_2-saturated brine using some novel surfactants based on corn oil, *Egypt. J. Pet.*, 2013, **22**, 451–470.
19. S. Toliwal, K. Jadav and T. Pavagadhi, Corrosion inhibition study of a new synthetic Schiff base derived from nontraditional oils on mild steel in 1N HCl solution, *J. Sci. Ind. Res.*, 2010, **69**, 43–47.
20. H. M. Abd El-Lateef, V. Abbasov, L. Aliyeva, E. Qasimov and I. Ismayilov, Inhibition of carbon steel corrosion in CO_2-saturated brine using some newly surfactants based on palm oil: experimental and theoretical investigations, *Mater. Chem. Phys.*, 2013, **142**, 502–512.
21. I. T. Ismayilov, H. M. Abd El-Lateef, V. M. Abbasov, S. A. Mamedxanova, U. C. Yolchuyeva and C. K. Salmanova, Anticorrosion ability of some surfactants based on corn oil and monoethanolamine, *Am. J. Phys. Chem.*, 2013, **1**, 79–86.
22. A. M. Elsharif, S. A. Abubshait, I. Abdulazeez and H. A. Abubshait, Synthesis of a new class of corrosion inhibitors derived from natural fatty acid: 13-Docosenoic acid amide derivatives for oil and gas industry, *Arabian J. Chem.*, 2020, **13**, 5363–5376.
23. M. Quaraishi, D. Jamal and M. Tariq Saeed, Fatty acid derivatives as corrosion inhibitors for mild steel and oil-well tubular steel in 15% boiling hydrochloric acid, *J. Am. Oil Chem. Soc.*, 2000, **77**, 265–268.
24. M. Ajmal, D. Jamal and M. Quraishi, Fatty acid oxadiazoles as acid corrosion inhibitors for mild steel, *Anti-Corros. Methods Mater.*, 2000, **47**, 77–82.
25. M. Rafiquee, N. Saxena, S. Khan and M. Quraishi, Some fatty acid oxadiazoles for corrosion inhibition of mild steel in HCl, *Indian J. Chem. Technol.*, 2007, **14**, 576–583.
26. M. A. Quraishi and D. Jamal, Corrosion inhibition by fatty acid oxadiazoles for oil well steel (N-80) and mild steel, *Mater. Chem. Phys.*, 2001, **71**, 202–205.
27. D. Wahyuningrum, S. Achmad, Y. M. Syah, B. Buchari and B. Ariwahjoedi, The synthesis of imidazoline derivative compounds as corrosion inhibitor towards carbon steel in 1% NaCl solution, *J. Math. Fundam. Sci.*, 2008, **40**, 33–48.
28. M. Rafiquee, S. Khan, N. Saxena and M. Quraishi, Investigation of some oleochemicals as green inhibitors on mild steel corrosion in sulfuric acid, *J. Appl. Electrochem.*, 2009, **39**, 1409–1417.

29. D. B. Tripathy, M. Murmu, P. Banerjee and M. A. Quraishi, Palmitic acid based environmentally benign corrosion inhibiting formulation useful during acid cleansing process in MSF desalination plants, *Desalination*, 2019, **472**, 114128.
30. M. Quraishi and F. A. Ansari, Fatty acid oxadiazoles as corrosion inhibitors for mild steel in formic acid, *J. Appl. Electrochem.*, 2006, **36**, 309–314.
31. M. Quraishi and D. Jamal, Fatty acid triazoles: Novel corrosion inhibitors for oil well steel (*N*-80) and mild steel, *J. Am. Oil Chem. Soc.*, 2000, **77**, 1107–1111.
32. M. A. Quraishi, D. S. Chauhan and V. S. Saji, Heterocyclic biomolecules as green corrosion inhibitors, *J. Mol. Liq.*, 2021, **341**, 117265.
33. M. M. Solomon, S. A. Umoren, M. A. Quraishi and M. Salman, Myristic acid based imidazoline derivative as effective corrosion inhibitor for steel in 15% HCl medium, *J. Colloid Interface Sci.*, 2019, **551**, 47–60.
34. M. M. Solomon, S. A. Umoren, M. A. Quraishi, D. B. Tripathy and E. Abai, Effect of akyl chain length, flow, and temperature on the corrosion inhibition of carbon steel in a simulated acidizing environment by an imidazoline-based inhibitor, *J. Pet. Sci. Eng.*, 2020, **187**, 106801.
35. M. Quraishi and D. Jamal, Synthesis and evaluation of some organic vapour phase corrosion inhibitors, *Indian J. Chem. Technol.*, 2004, **11**, 459–464.
36. J. Mayne and E. Ramshaw, Inhibitors of the corrosion of iron. II. Efficiency of the sodium, calcium and lead salts of long chain fatty acids, *J. Appl. Chem.*, 1960, **10**, 419–422.
37. M. Quraishi and F. Ansari, Corrosion inhibition by fatty acid triazoles for mild steel in formic acid, *J. Appl. Electrochem.*, 2003, **33**, 233–238.

CHAPTER 11

Surfactants and Biosurfactants as Green Corrosion Inhibitors

11.1 Introduction

11.1.1 Surfactants and Biosurfactants: Properties, Classification and Applications

The name SURFACTANT is a blend of three words, *i.e.* SURFace ACTive AgeNT, which specifies the collection of chemicals that are active at the surface or interface.[1] The surface active agents obtained from living organisms, especially microorganisms, *i.e.* of microbial origin, are known as biosurfactants.[2,3] Surfactants are amongst the most established and important categories of organic compounds. Biosurfactants play a crucial role in living systems. Surfactants have acquired various industrial and biological properties and applications, including in medicine and pharmacology. Surfactants have also been used extensively in other numerous industries such as household, personal care, dying, agrochemicals, polymers, painting, coating, foodstuffs, oils, cosmetics and petroleum production *etc.* More so, surfactants are widely employed in biotechnological and microbiological applications. They are used as materials for waste water treatment as well as catalysts for chemical transformations and micellar catalysts for micelle formation. In view of the high demand and consumption, surfactants are produced in large quantities. The current global production of surfactants is greater than 10 million tons/year.

Structurally, surfactants are amphiphlic molecules that are composed of two components, namely an oil-loving long hydrophobic hydrocarbon non-polar chain and a water-loving hydrophilic polar head.[4,5] The polar head

Green Chemistry Series No. 73
Green Corrosion Inhibition: Fundamentals, Design, Synthesis and Applications
By Chandrabhan Verma, Mumtaz A. Quraishi and Dheeraj Singh Chauhan

Published by the Royal Society of Chemistry, www.rsc.org

may be ionic or non-ionic in nature. Based on the nature of the polar head, surfactants may be classified as follows:[6,7]

(i) *Cationic surfactants*: These surfactants contain a positively charged, *i.e.* cationic, head *e.g.* hexadecyltrimethylammonium bromide ($CH_3(CH_2)_{15}N^+(CH_3)^3Br^-$).
(ii) *Anionic surfactants*: These surfactants contain a negatively charged, *i.e.* anionic, head *e.g.* sodium dodecylbenzene sulfonate ($CH_3(CH_2)_{11}C_6H_4SO_3^-Na^+$).
(iii) *Zwitterionic surfactants*: These surfactants contain both positive and negative charges, *i.e.* a zwitterionic head, *e.g.* *N*-dodecyl-*N*,*N*-dimethyl betaine ($C_{12}H_{25}N^+(CH_3)_2CH_2COO^-$).
(iv) *Non-ionic surfactants*: These surfactants do not hold any charge, *e.g.* polyoxyethylene (20) cetyl ether ($C_{16}H_{33}(CH_2CH_2O)_{20}$–OH).

Surfactants can also be classified as bolaform surfactants in which two hydrophilic heads are bound together with a long hydrophobic tail.[8] Potassium hexadecanedioate ($^-O_2C(CH_2)_{14}CO_2^-2K^+$) represents a common example of a bolaform surfactant.[9] Further, surfactants may also be classified as dimeric or gemini surfactants in which two surfactant molecules are joined together by a spacer.[9,10] Diagrammatic illustrations of different types of surfactants are presented in Figure 11.1. Previous studies show that surfactants are widely used as corrosion inhibitors. Because of their ionic or polar nature, they show a strong affinity for solubility in aqueous electrolytes. In aqueous solutions, surfactants can undergo various types of changes, such as being adsorbed on the metallic surface, precipitation, oil/water partitioning, micellization and surface interactions, *etc.*[11] Therefore, adversely affect the existence of the monomeric form of the surfactants. Nevertheless, the inhibition efficiency of the surfactants is attributed to their adsorption on a metallic surface and the formation of the protective film on the metal surface.[12–14] The adsorption of monomeric, dimeric (gemini) and micellar surfactants occurs in such a way that the polar head is directed towards the metallic surface and the hydrophobic hydrocarbon chains are directed towards the solution side (Figure 11.2).

The corrosion protection effectiveness of surfactants greatly varies with the size of the hydrophobic chain. Obviously, an increase in the size of the hydrocarbon chain causes a subsequent increase in the %IE. For example, E. A. Badr[15] synthesized, characterized and evaluated the corrosion inhibition potential of three 2-aminobenzothiazole based cationic surfactants for mild steel in 1 M HCl using various methods. The surfactants were designated as AB8, AB10 and AB12, where the numerical digits represent the number of carbon atoms in the hydrocarbon tail. Gravimetric and electrochemical studies suggest that %IE of the investigated cationic surfactants follow the sequence: AB8 < AB10 < AB12. A Tafel polarization study indicated that AB8, AB10 and AB12 behave as mixed-type inhibitors. They retard both anodic and cathodic Tafel reactions through their adsorption at the actives sites.

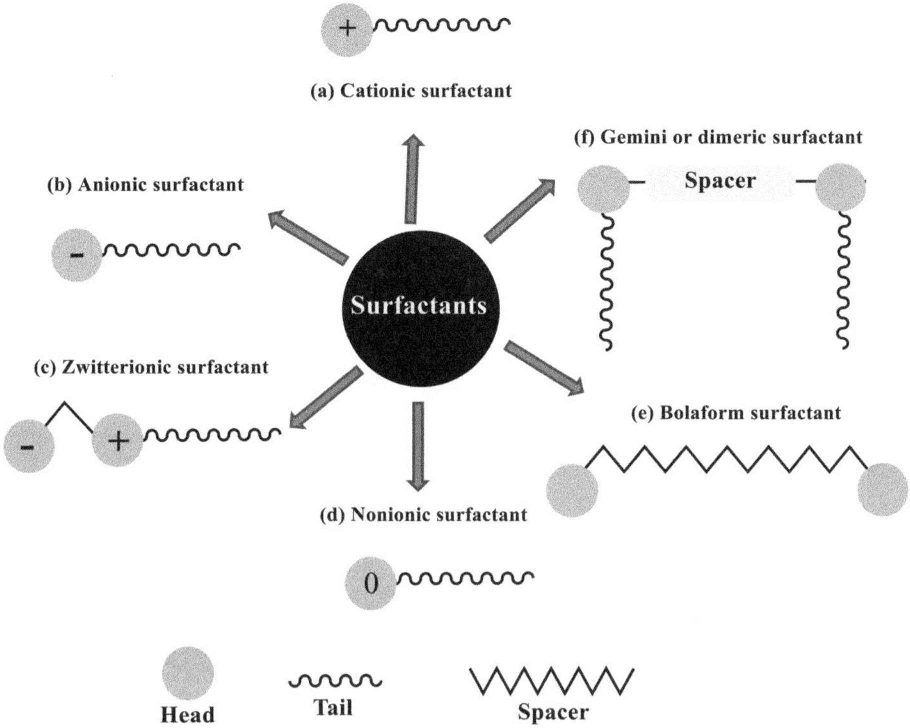

Figure 11.1 Diagrammatic illustrations of different types of surfactants.

The adsorption of AB8, AB10 and AB12 on metallic surfaces follows the Langmuir adsorption isotherm model.

In another report, Shaban *et al.*[16] synthesized, characterized and evaluated the corrosion inhibition potential of three alginate polymeric cationic surfactants having different sizes of hydrocarbon chains for carbon steel in 1 M HCl. The %IE of studied surfactants, designated as ALGOB (C-8), ALGDB (C-12) and ALGHB (C-16), was evaluated at their various concentrations and temperatures. Results suggest that ALGOB, ALGDB and ALGHB exhibit their maximum inhibition performance at 1000 ppm concentration. The %IE of ALGOB, ALGDB and ALGHB increases with their concentration and decreases with the rise in temperature. Potentiodynamic polarization studies were also carried out at different concentrations of ALGOB, ALGDB and ALGHB, and results demonstrate a similar trend of inhibition efficiencies derived from weight loss measurements. Obviously, a rise in ALGOB, ALGDB and ALGHB concentrations causes subsequent decreases in the corrosion rate and corrosion current density (i_{corr}). The maximum decrease in the i_{corr} value was derived at 1000 ppm concentration of ALGOB, ALGDB and ALGHB. This finding suggests that ALGOB, ALGDB and ALGHB become effective by blocking the surface active sites through their adsorption. They act as mixed-type inhibitors as they did

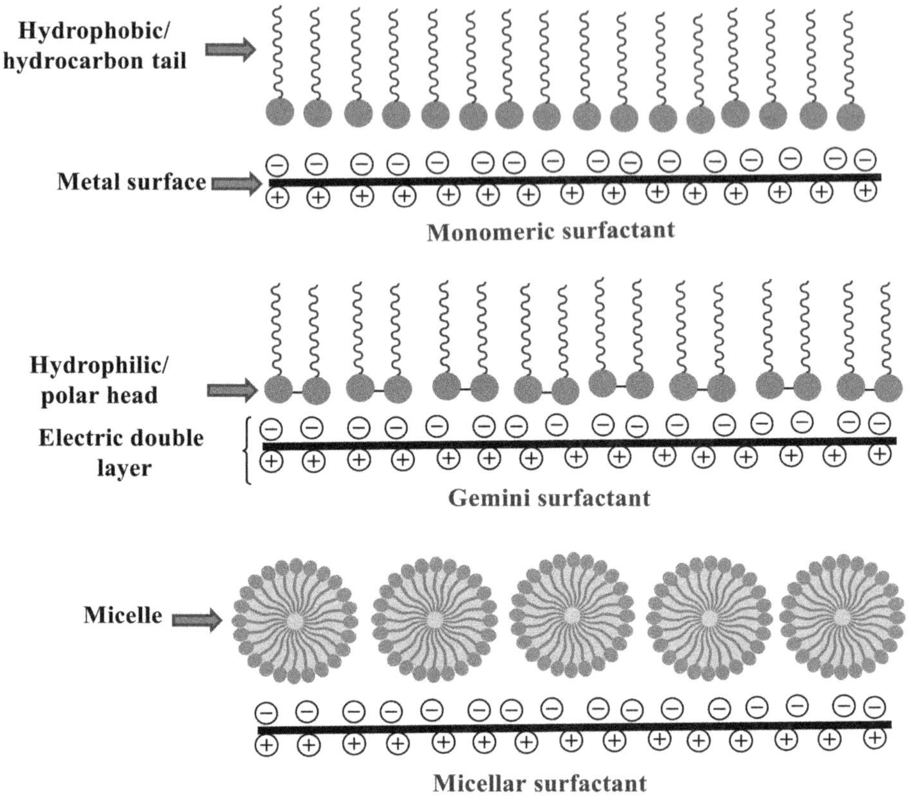

Figure 11.2 Diagrammatic illustration of adsorption behaviour of monomeric, Gemini and micellar surfactants.

not cause any appreciable shift in the corrosion potential values (E_{corr}). EIS study suggests that ALGOB, ALGDB and ALGHB act as interface-type corrosion inhibitors as their presence increases the value of charge transfer resistance. The increase in the diameter of Nyquist curves in the presence of ALGOB, ALGDB and ALGHB validates this finding. The Tafel polarization and Nyquist curves for carbon steel corrosion in 1 M HCl solution are presented in Figure 11.3.

Literature survey suggests that numerous cationic, non-ionic, gemini and ionic liquid-based surfactants are widely reported as corrosion inhibitors.[11,14] Their inhibition efficiency increases with the rise in their concentration and diminishes with a rise in the temperature. Independent to their nature and chemical structures, most surfactants become effective by adsorbing on the metallic surface, mostly following the Langmuir adsorption isotherm model. Electrochemical analyses reveal that most of the investigated surfactants behave as mixed- and interface-type corrosion inhibitors, identifying that they become effective by adsorbing at the active sites present over the metal surface. Generally, this type of

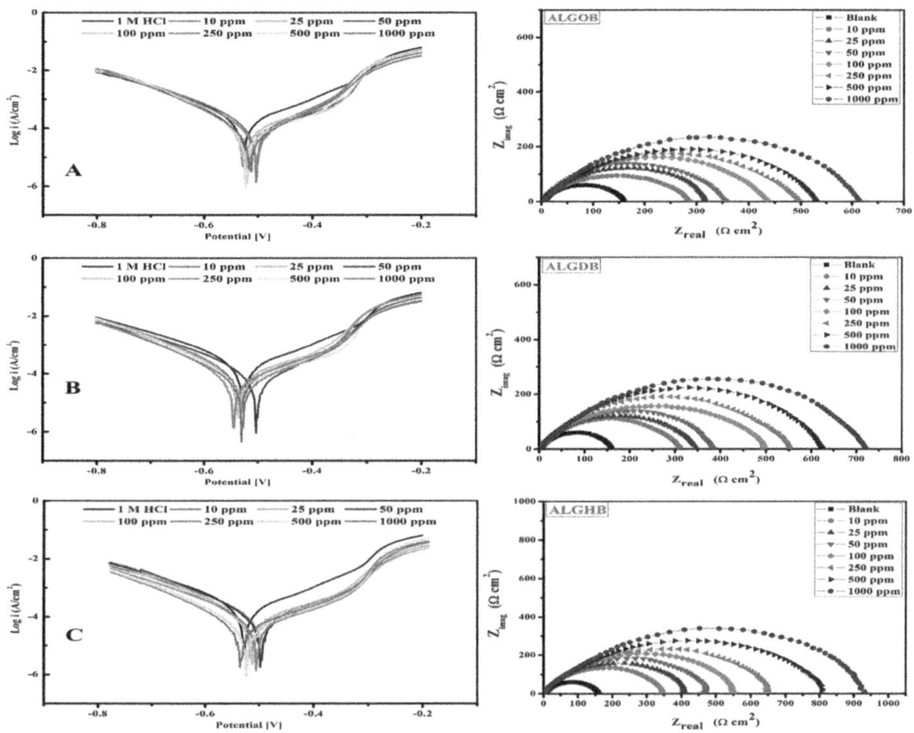

Figure 11.3 Tafel polarization (left-hand-side) and Nyquist (right-hand-side) curves for carbon steel corrosion in the absence and presence of numerous studied concentrations of (A) ALGOB, (B) ALGDB and (C) ALGHB.[16] Reproduced from ref. 16 with permission from Elsevier, Copyright 2019.

adsorption results in the formation of a hydrophobic film over the metal surface that prevents the attack of aggressive corrosive species. The adsorbed surfactant molecules impose a barrier for the charge transfer process. The adsorption of surfactants on the metal surface and their subsequent mechanism of corrosion prevention can be reported by surface analyses, including SEM, AFM, EDX, XRD and XPS, *etc.* The mode and effectiveness of interactions between the metal surface and surfactant molecules have also been demonstrated using computational studies, principally DFT, MDS and MCS methods. Most surfactant molecules interact spontaneously with the metallic surface using the charge transfer mechanism. The chemical structures of some common cationic, non-ionic, gemini and ionic liquid-based surfactants used as corrosion inhibitors and other information derived from a few major reports are presented in Tables 11.1–11.4.

The surfactants derived from biological systems are known as biosurfactants (Figure 11.4).[2,92] The need for the development of biosurfactants is

Table 11.1 Summary of some major reports on cationic surfactants as corrosion inhibitors.

Structure and abbreviation	Opt. Conc.	Medium	%IE (EIS)	Reference
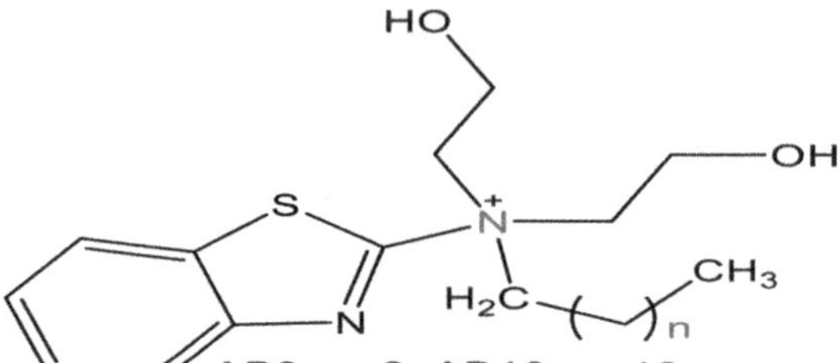 AB8: n=8; AB10: n=10 AB12: n=12	500 ppm (AB12)	CS/1 M HCl	89.3%	15
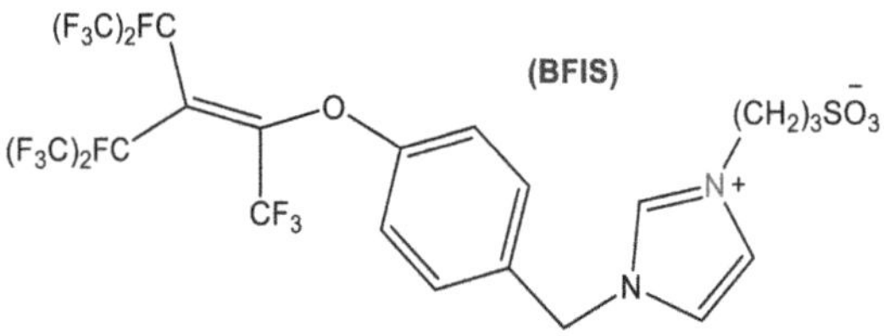 (BFIS)	5×10^{-5} M	MS/1 M HCl	95.4%	17
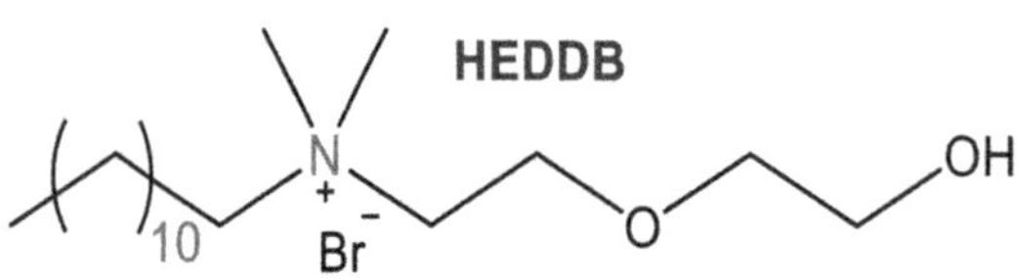 HEDDB	1×10^{-3} M	1018 CS/1 M HCl	93.4%	19
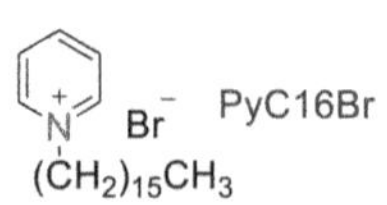 (PyC16Br)	1×10^{-3} M	MS/0.5 M H_2SO_4	—	21
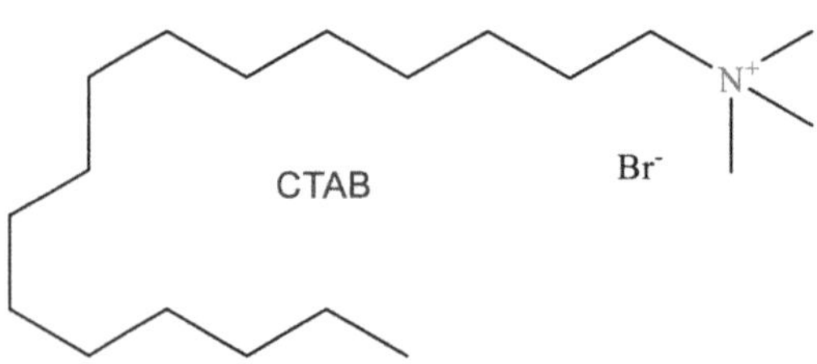 CTAB	50×10^{-3} M	MS/2 M HCl	—	23
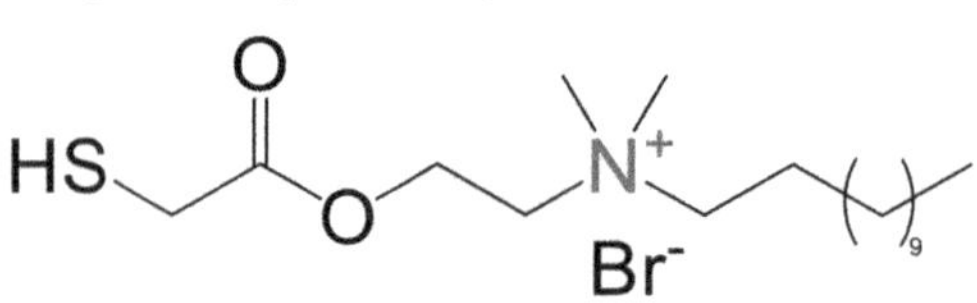	1×10^{-4} M	CS/1 M HCl	93.3%	25 and 26

Structure and abbreviation	Opt. Conc.	Medium	%IE (EIS)	Reference
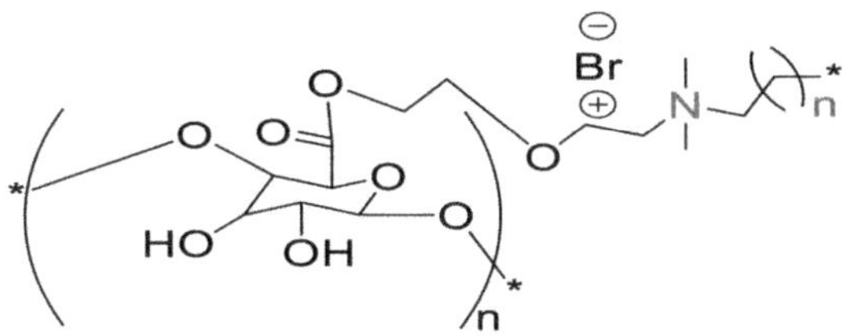 ALGOB: n=6; ALGDB: n=10 ALGHB: n=14	1000 ppm (ALGHB)	MS/0.5 M HCl	83.3%	16
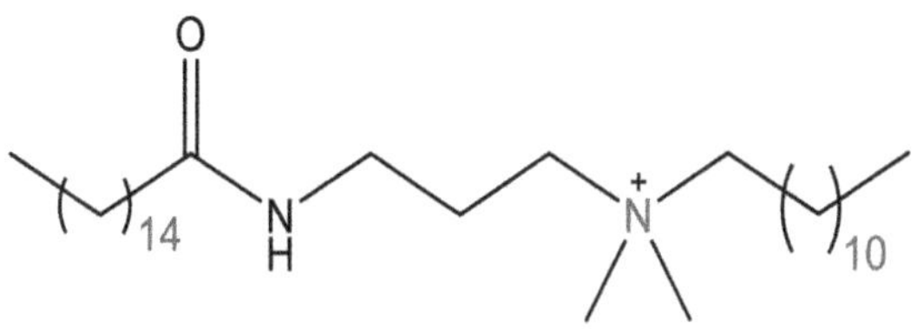	1×10^{-3} M	N80 CS/1 M HCl	98.4%	18
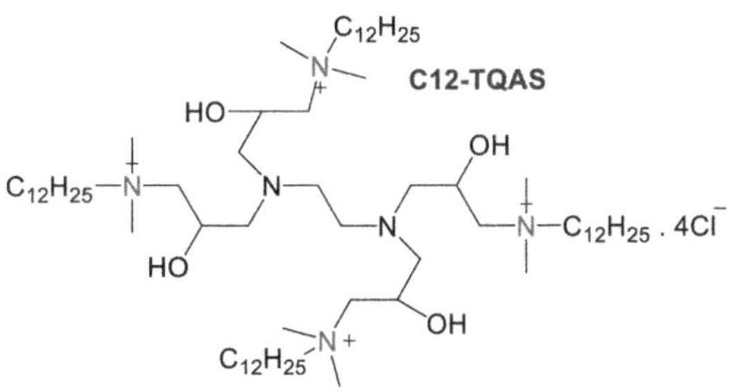	40 ppm	N80 CS/20% HCl	—	20
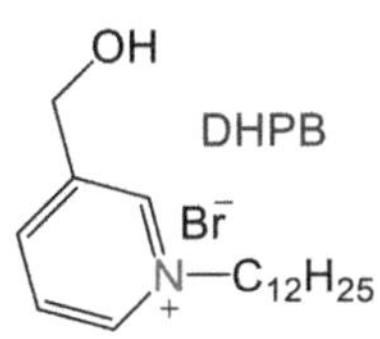	300 ppm	CS/2 M HCl	91.7%	22
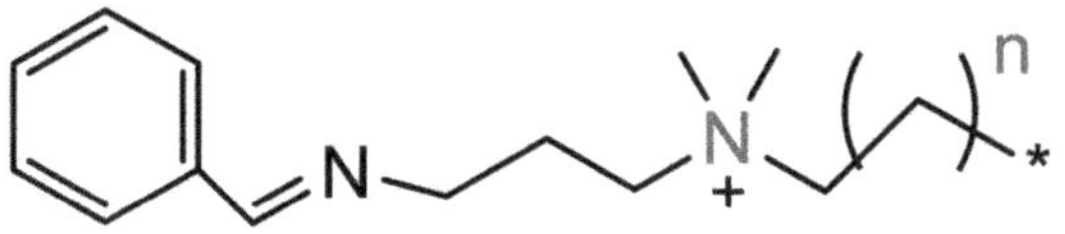 Comp I: n=6, II: n=10 III:n=14	1×10^{-2} M	1 M HCl	89.9%	24
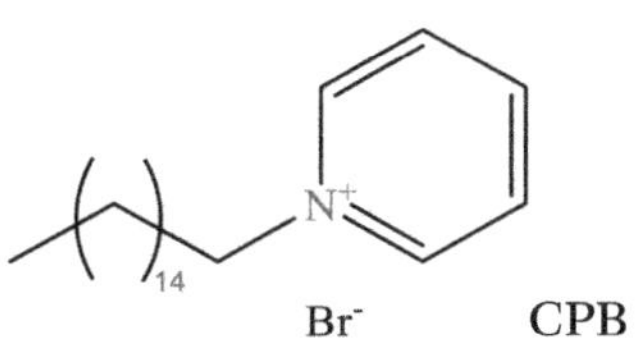	150 ppm	CS/1 M HCl	94.0%	27

(*continued*)

Table 11.1 (*continued*)

Structure and abbreviation	Opt. Conc.	Medium	%IE (EIS)	Reference
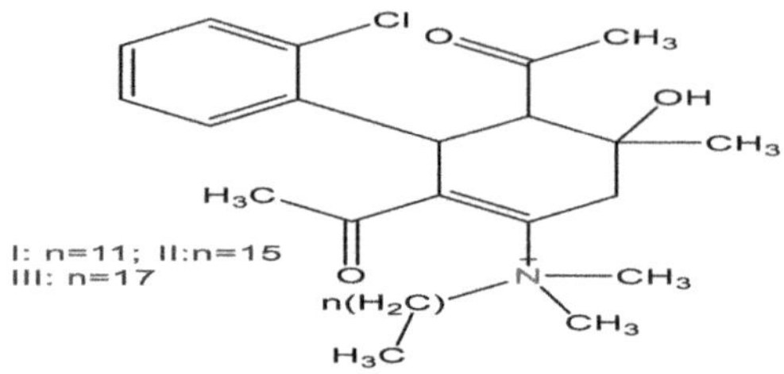	250 ppm (III)	C1018CS/15% HCl	95.5%	28
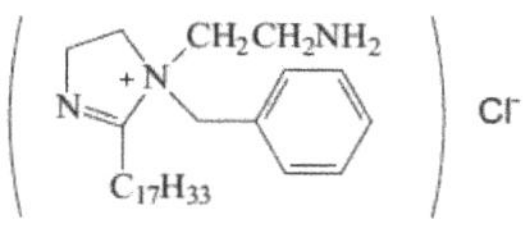	25	MS/10 vol% HCl	81.9%	30
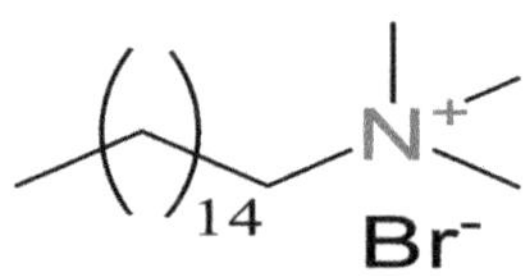 CTAB	100 ppm	MS/5 M HCl	96.2%	32
CPB 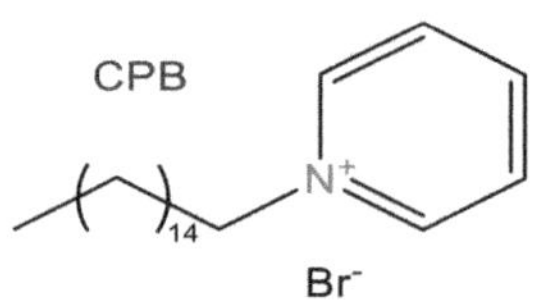	150 ppm	CS/1 M HCl	89.4%	26
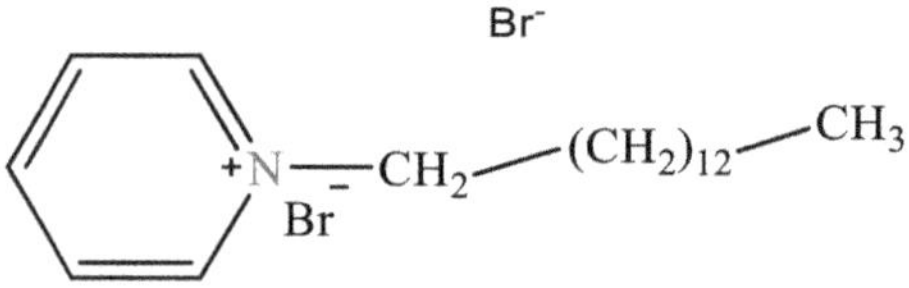	1 mM	CRS/7 M H_3PO_4	92.2%	35

Table 11.2 Summary of some major reports on non-ionic surfactants as corrosion inhibitors.

Structure and abbreviation	Opt. Conc.	Medium	%IE (EIS)	Reference
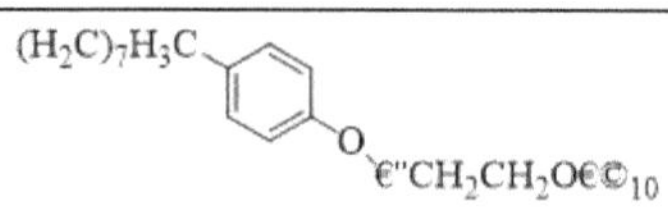	100 ppm	MS/10 vol% HCl	57.2%	31
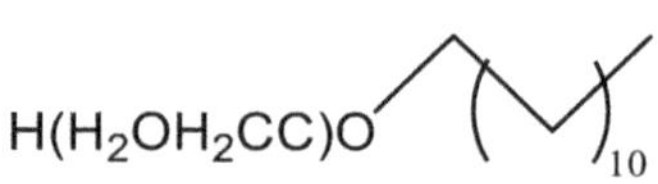 FAEO	2.48 × 10−5	CS/1 M HCl	80.2%	38

Structure and abbreviation	Opt. Conc.	Medium	%IE (EIS)	Reference
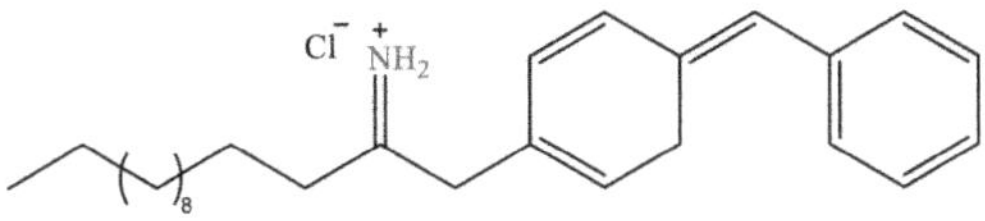	5×10^{-4} M	CS/1 M HCl	93.3%	29
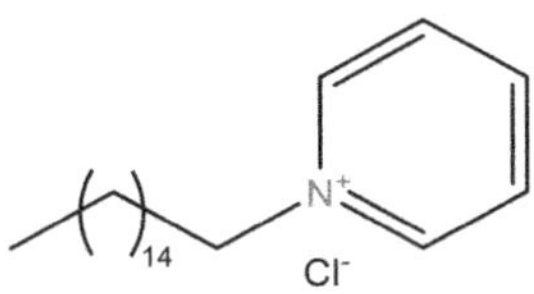	251×10^{-3} M	MS/0.5 M H_2SO_4	94.2%	31
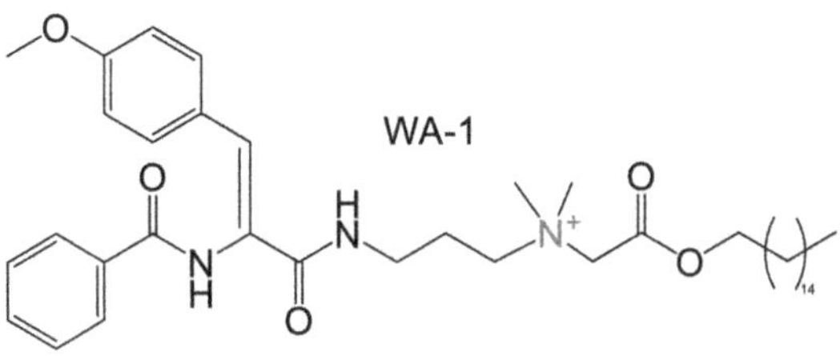	50	CS/0.5 M HCl	96.6%	33
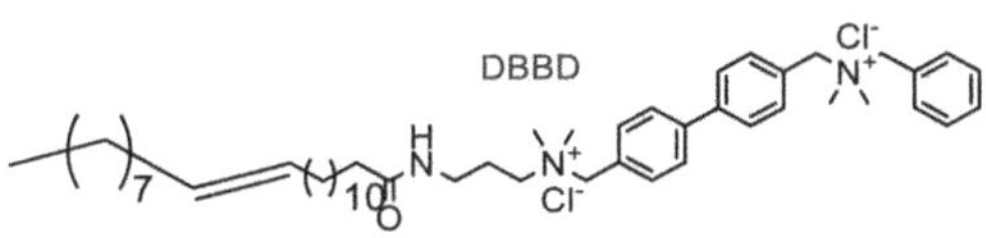	20 mM	MS/15% HCl	96.6%	34
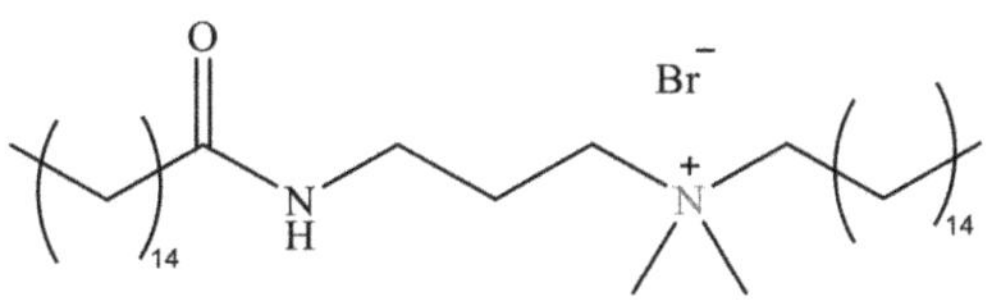	1×10^{-3} M	N80 steel/1 M HCl	98.5%	36

Structure and abbreviation	Opt. Conc.	Medium	%IE (EIS)	Reference
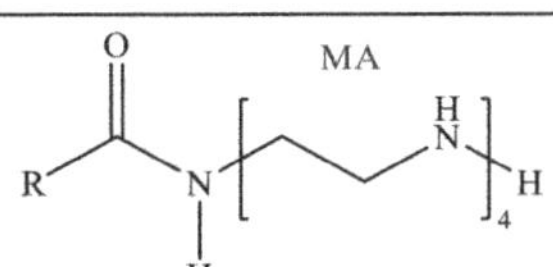	50 ppm	CS/2 M HCl	74.5%	37
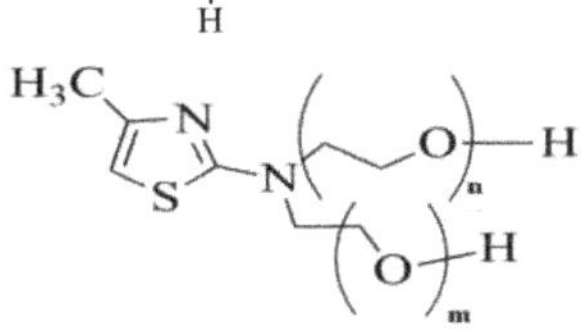	1×10^{-3} M	X65 S/15% HCl	90.5%	39

(*continued*)

Table 11.2 (*continued*)

Structure and abbreviation	Opt. Conc.	Medium	%IE (EIS)	Reference
Tween-80 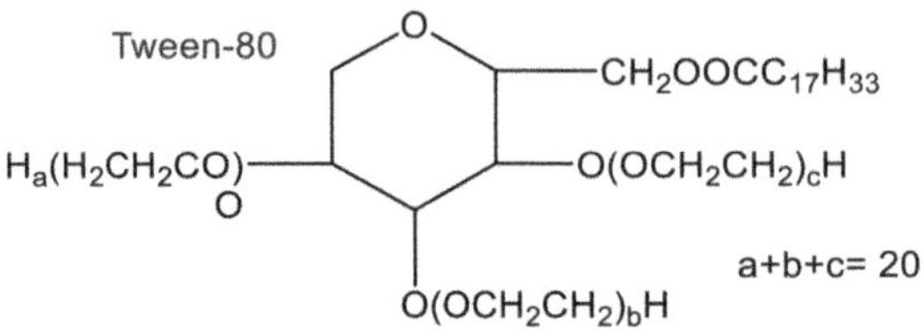	10 ppm	MS/1 M HCl	45.79%	40
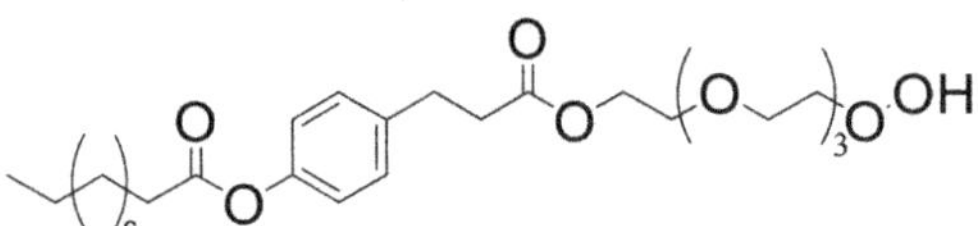 HTOPD	35 ppm	CS/0.5 M HCl	91.7%	42
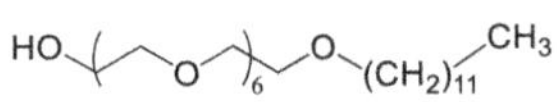 AEO7	40 μmol L^{-1}	CS/0.5 M HCl	87.4%	44
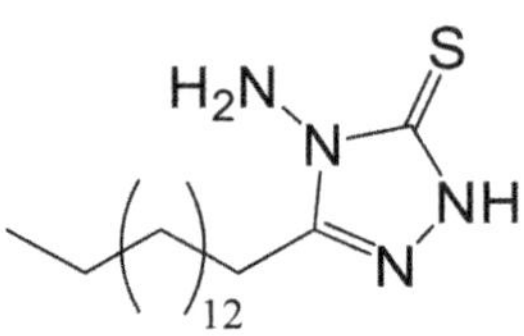	5×10^{-4} M	CS/1 M HCl	96.6%	46
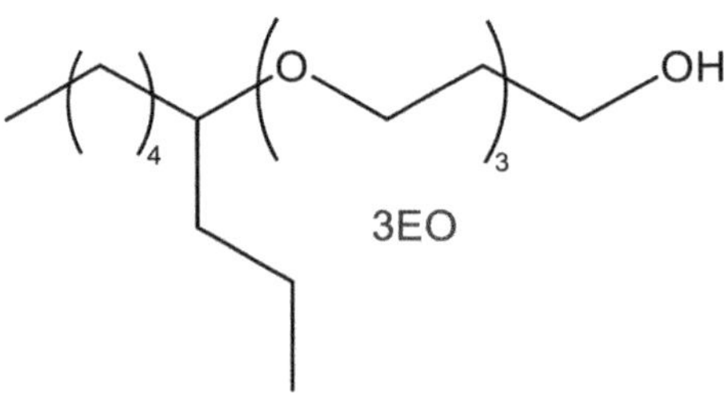 3EO	150 ppm	Steel/3.5% NaCl	76.6%	48
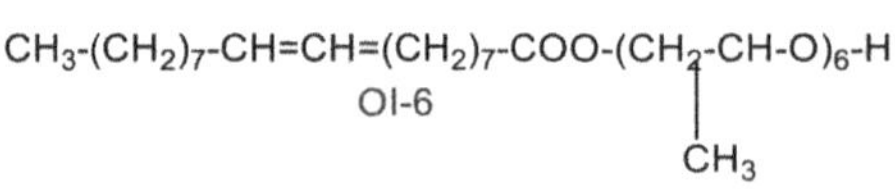 OI-6	2000 ppm	Al (6061)/2 M HCl	—	50

Structure and abbreviation	Opt. Conc.	Medium	%IE (EIS)	Reference
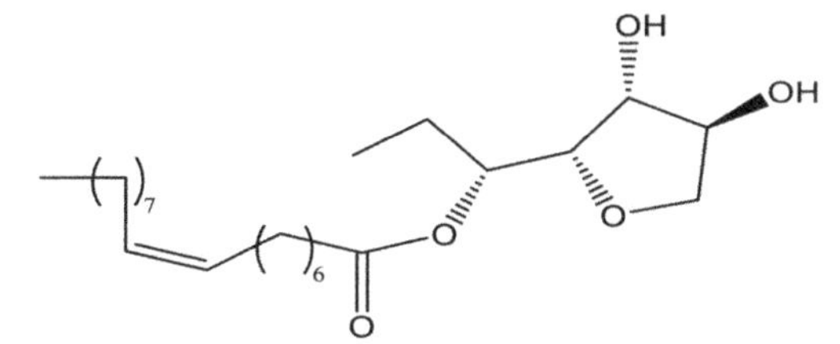	125 ppm	MS/1 M HCl	92.8%	41
Tween 60 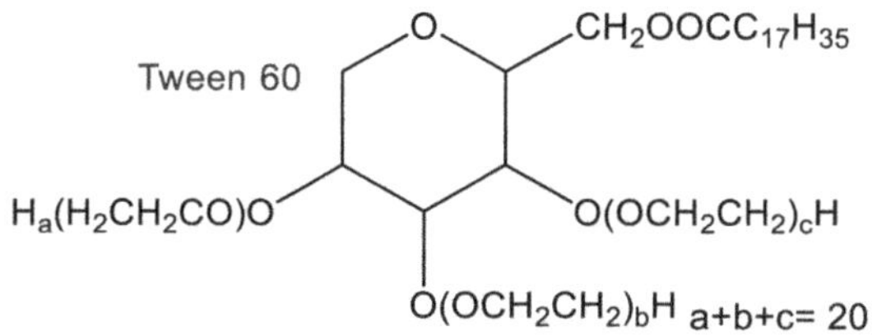	300 ppm	CS/0.5 M H_2SO_4	—	43
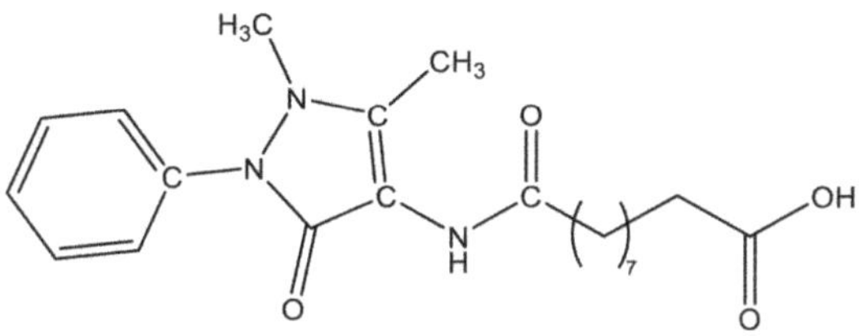	800 ppm	CS/1 M HCl	—	45
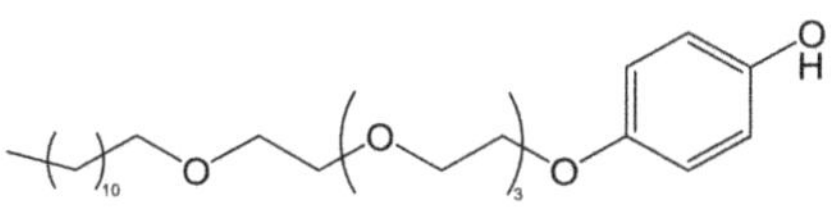	500 ppm	CS/1 M HCl	79.7%	47
Triton X100 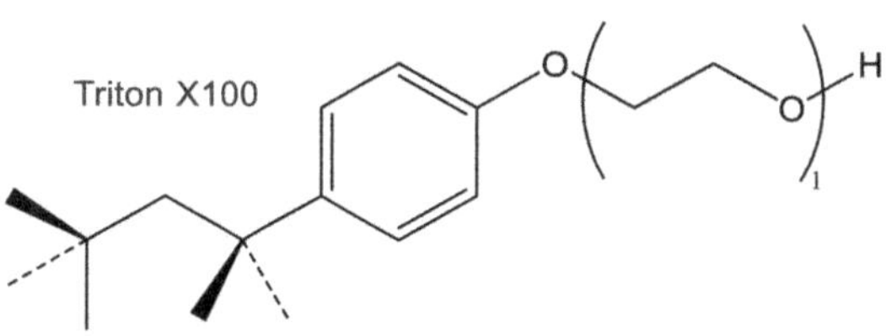	0.01 mol L^{-1}	Al/0.01 M NaOH	—	49
nonoxynol-9 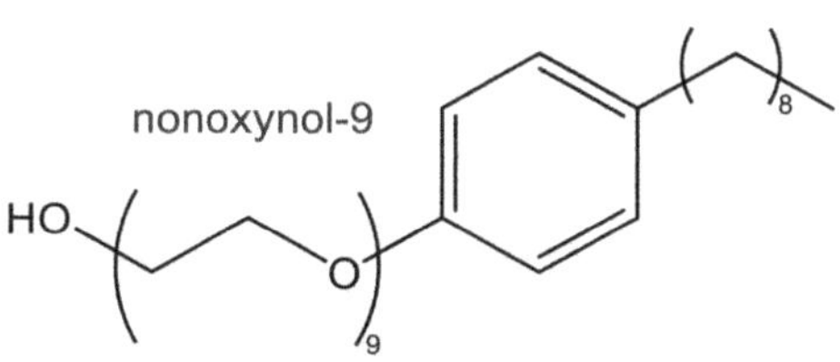	2000 ppm	Al/4.0 NaOH	92.8%	51

Table 11.3 Summary of some major reports on gemini surfactants as corrosion inhibitors.

Structure and abbreviation	Opt. Conc.	Medium	%IE (EIS)	Reference
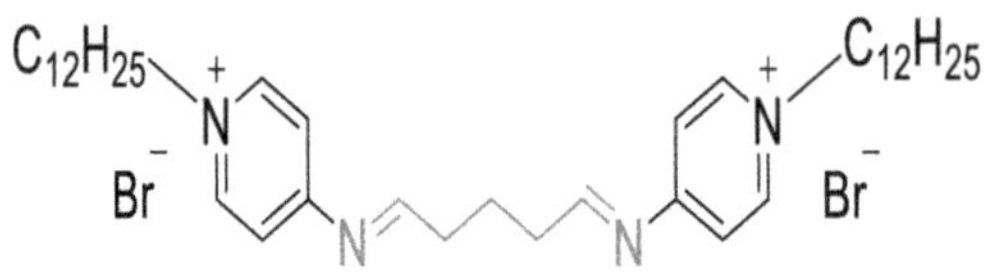	5 mM	1018 steel/1 M HCl	95.6%	20
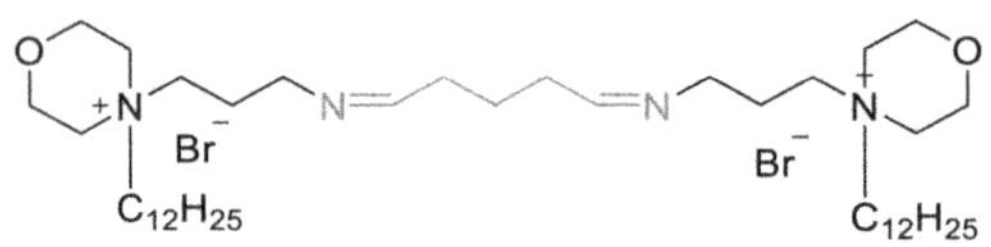	5×10^{-3} M	CS/1 M HCl	94.8%	53
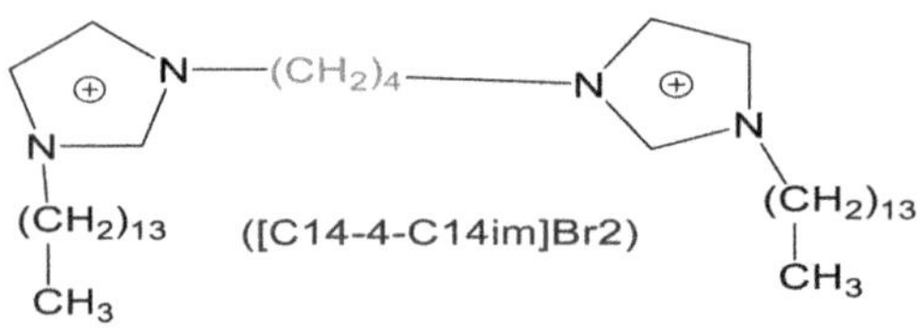 ([C14-4-C14im]Br2)	100×10^{-6} M	A3CS/1 M HCl	97.1%	55
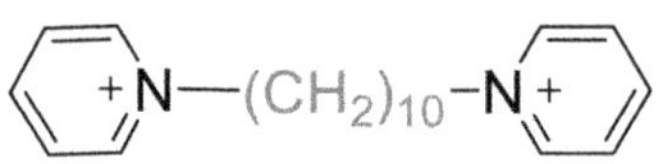 DBP	300 ppm	P110 steel/1 M HCl	91.7%%	57
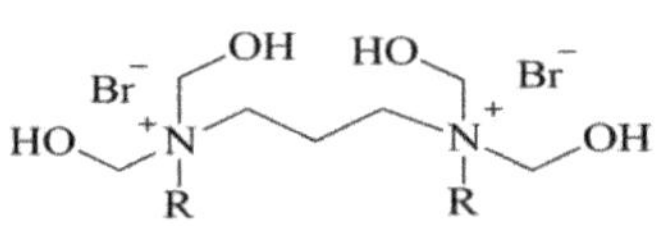 R = $C_{12}H_{25}$	3×10^{-5} M	MS/1 M HCl	88.6%	58
PDHDB 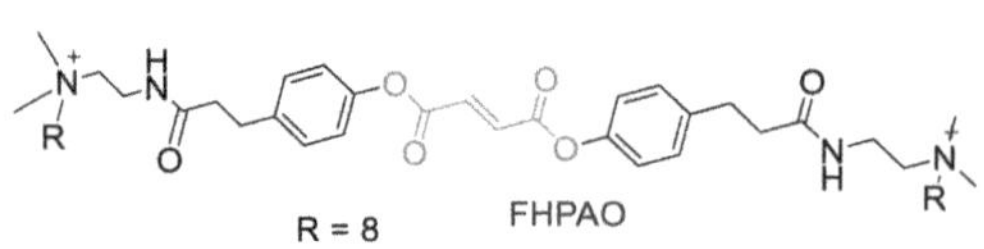 R = 8 FHPAO	5 mM	CS/pH-2	—	60

Structure and abbreviation	Opt. Conc.	Medium	%IE (EIS)	Reference
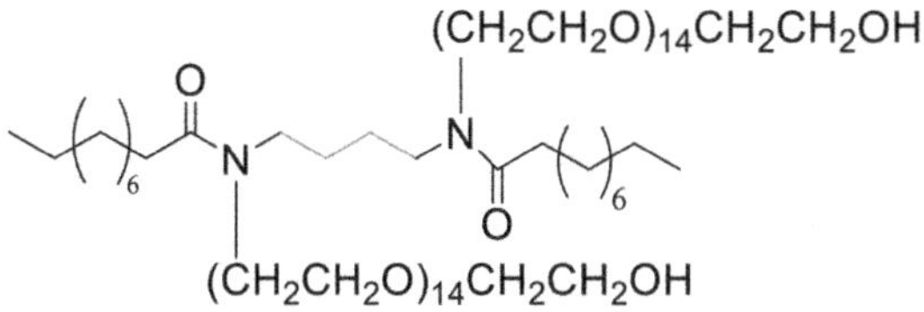	1×10^{-3} M	CS/1 M HCl	93.7%	52
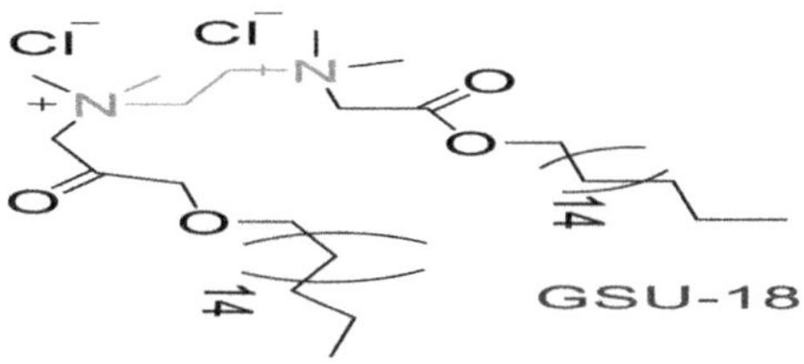 GSU-18	1×10^{-3} M	N80 steel/1.0 M H_2SO_4	95.0%	54
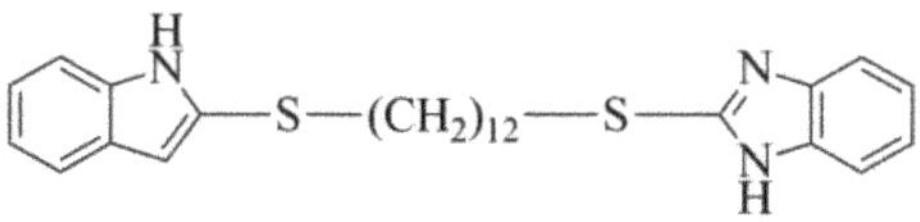 BITD	5×10^{-4} M	MS/1 M HCl	97.0%	56
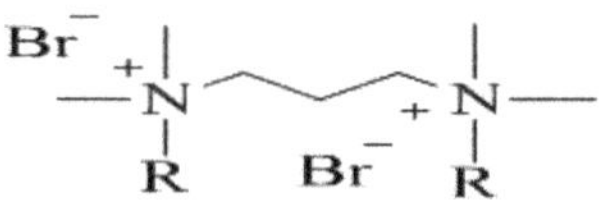 $R = C_{12}H_{25}$ PDDB	3×10^{-5} M	MS/1 M HCl	88.2%	58
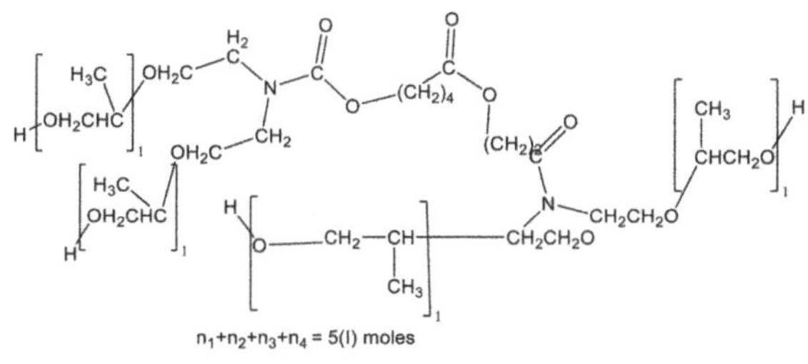	10^{-3} M	CS/0.5 M HCl	—	59
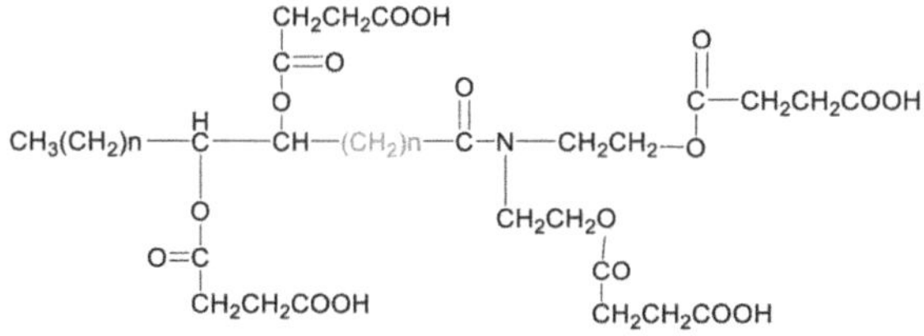	160 mol L^{-1}	MS/0.5 M HCl	—	61

(*continued*)

Table 11.3 *(continued)*

Structure and abbreviation	Opt. Conc.	Medium	%IE (EIS)	Reference
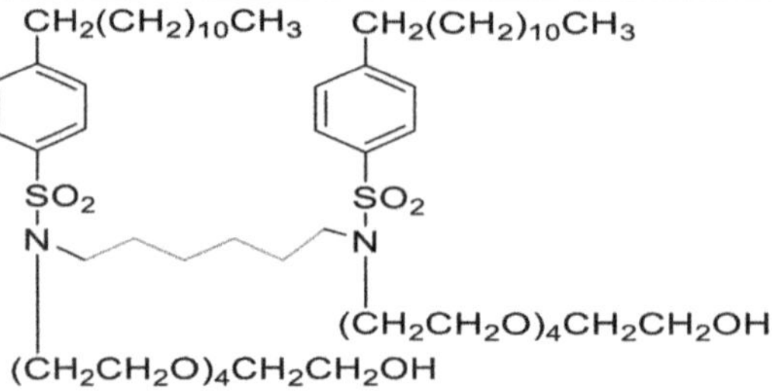	1×10^{-3} M	CS/0.5 M H_2SO_4	94.6%	62
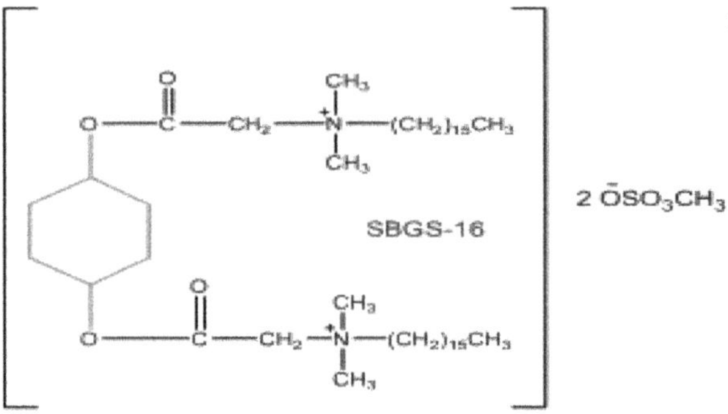	1 mM	MS/1.0 M H_2SO_4	96.4%	64
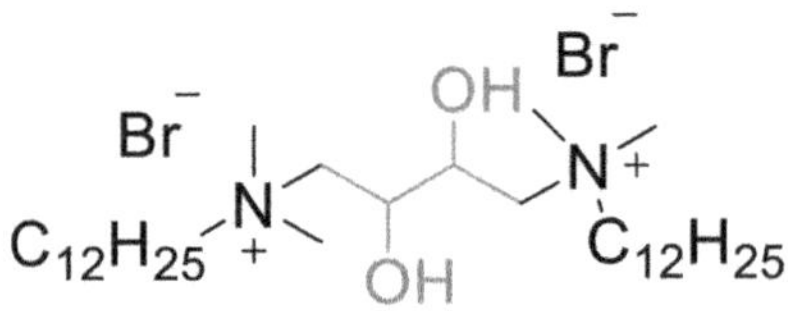 12-4-(2OH)-12	10 μM	MS/1.0 M HCl	82.5%	66
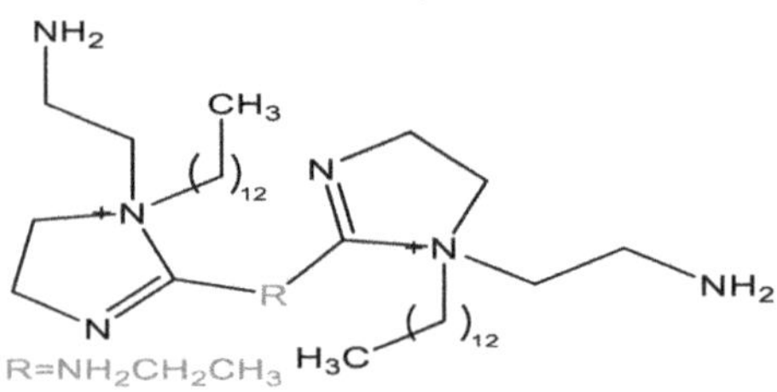			80.9%	68
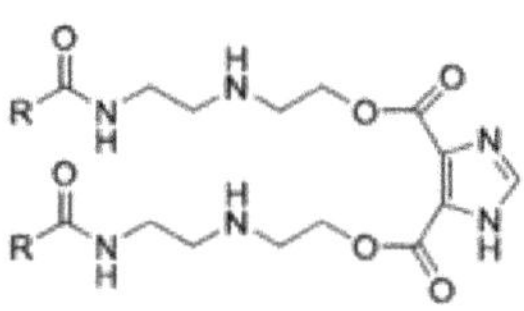	100 ppm	UNS S41425 SS/NACE TM-0177 solution	99.4%	70
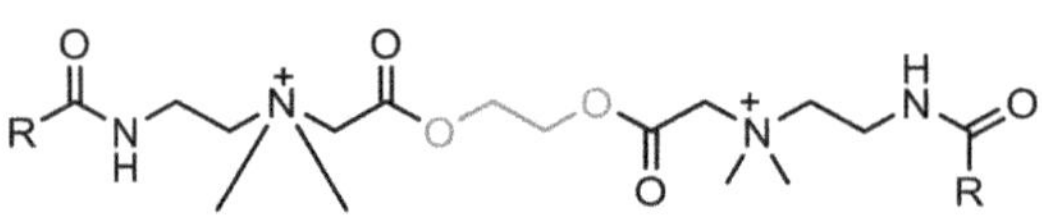 R=$C_{15}H_{31}$ CGSES16	2.5×10^{-3} M	CS/1 M HCl	96.3%	72

Structure and abbreviation	Opt. Conc.	Medium	%IE (EIS)	Reference
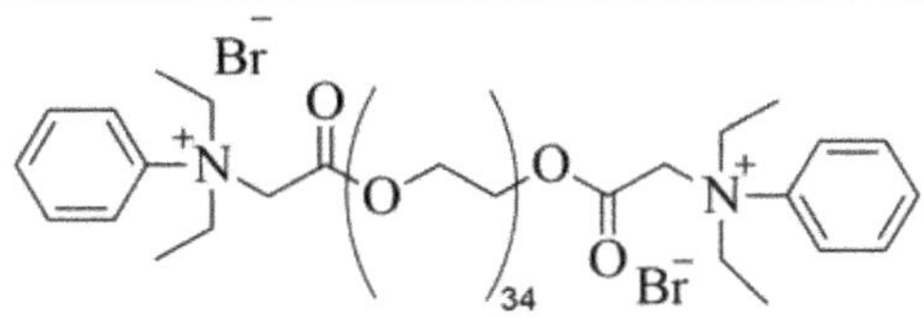 G1500Br	1×10^{-2} M	MS/1 M HCl	89.2%	63
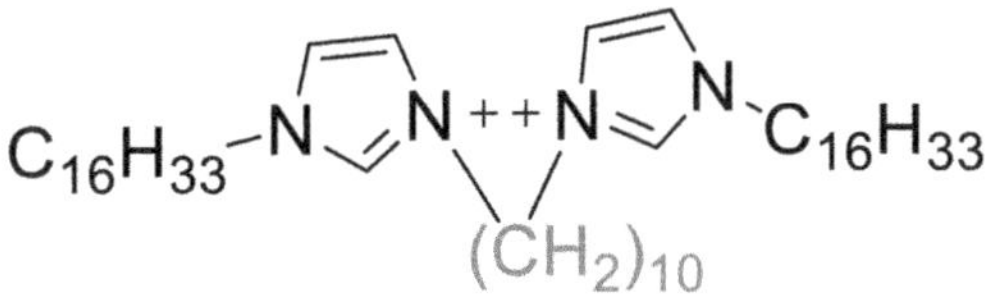	50 ppm	CS/0.5 M H_2SO_4	85.2%	65
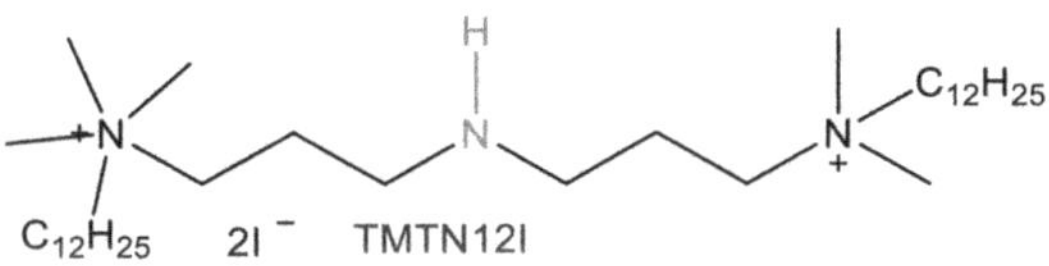	1.5 mM	AISI 304 SS/3 M HCl	96	67
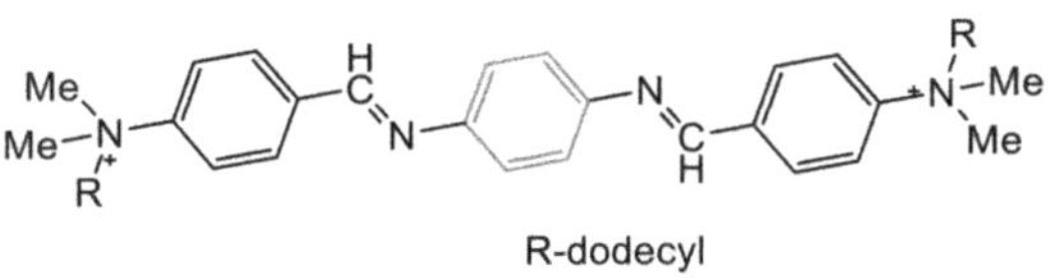	200 ppm	API 5L X52 CS/1 M HCl	94.8%	69
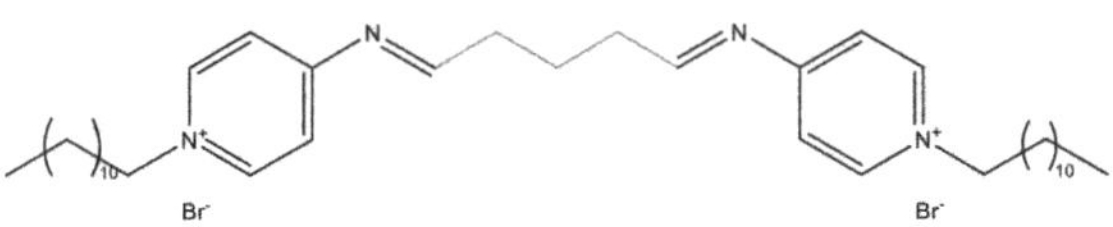	5 mM	MS/1 M HCl	95.6%	71
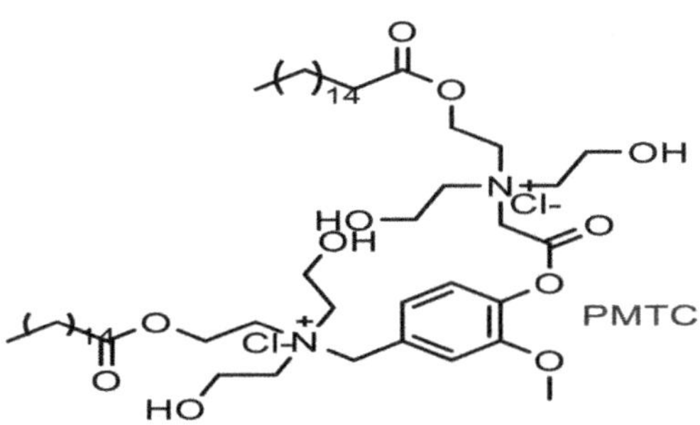	1000 μM	CS/1 M HCl	86.2%	73

Table 11.4 Summary of some major reports on ionic liquid-based surfactants as corrosion inhibitors.

Structure and abbreviation	Opt. Conc.	Medium	%IE (EIS)	Reference
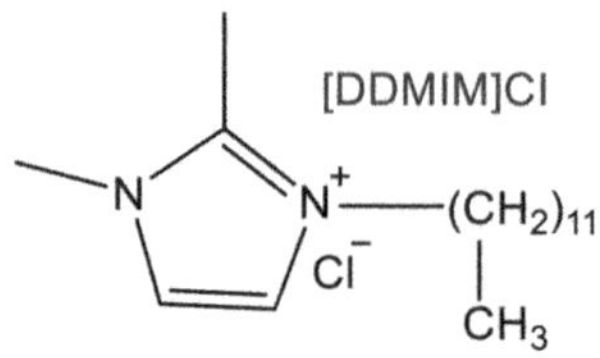 [DDMIM]Cl	MS/15% HCl	1000 μM	96.5%	74
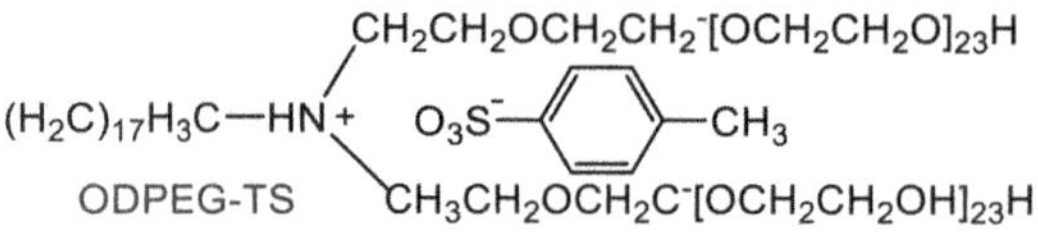 ODPEG-TS	Steel/1 M HCl	0.098 M	96.7%	76
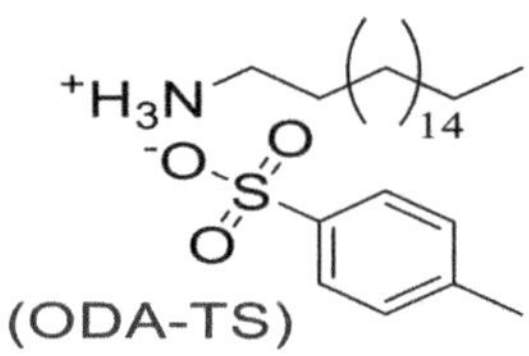 (ODA-TS)	Steel/1 M HCl	0.341 mM	98.0%	78
 ([BMMB]+Br?)	CS/1 M HCl	6.84×10^{-4} M	97.9%	80
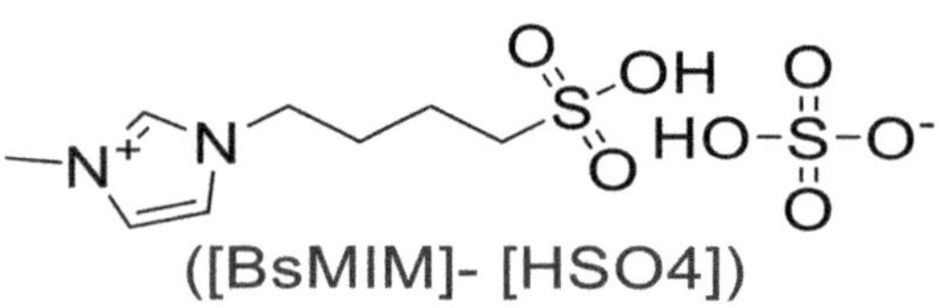 ([BsMIM]- [HSO4])	304 SS/1 M H_2SO_4	1×10^{-2} M	—	82
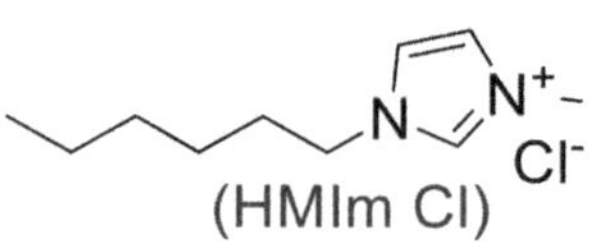 (HMIm Cl)	MS/2 M HCl	1%	88.0%	84

Structure and abbreviation	Opt. Conc.	Medium	%IE (EIS)	Reference
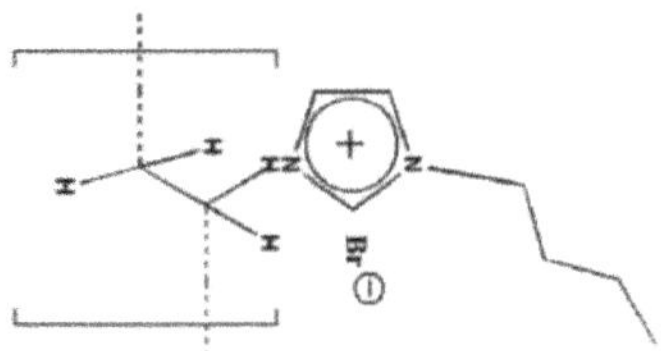	MS/1 M HCl	400 ppm	92.0%	75
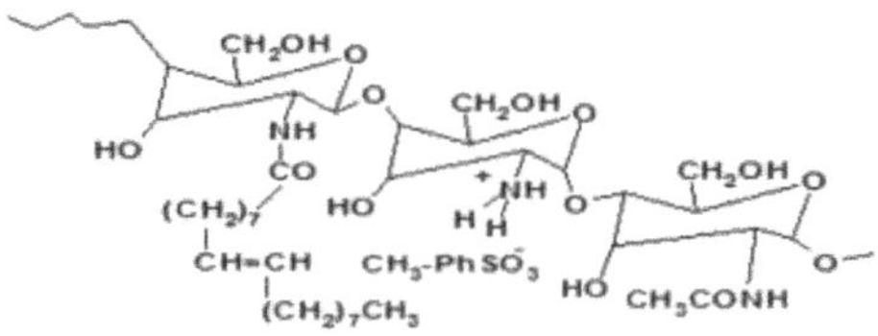 (CSPTA-OA)	Steel/HCl	250 ppm	95.6%	77
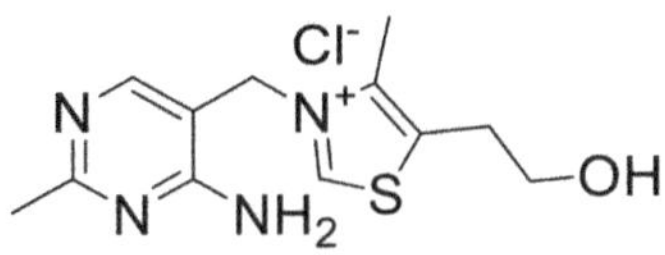 (AMPMHMC)	CS/1 M HCl	40 ppm	90.3%	79
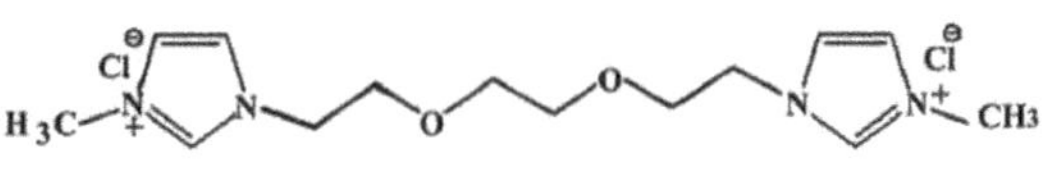 (TSIL)	MS/1 M HCl	100 ppm	78.9%	81
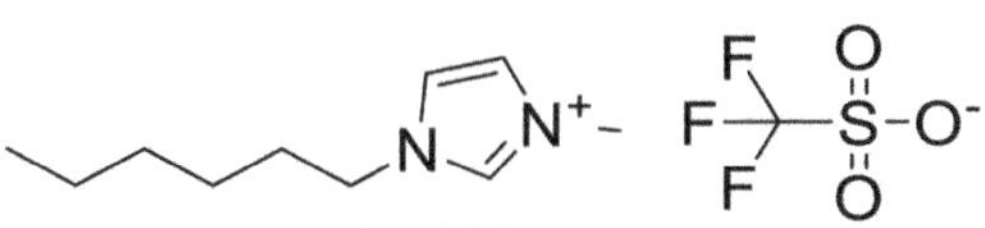 [HMIM][TfO]	MS/1 M HCl	500 ppm	79.9%	83
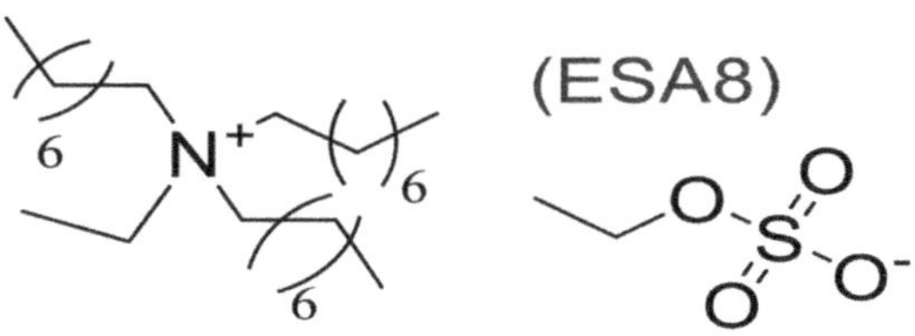 (ESA8)	X60 steel/1 M H_2SO_4	100 ppm	—	85

(continued)

Table 11.4 (*continued*)

Structure and abbreviation	Opt. Conc.	Medium	%IE (EIS)	Reference
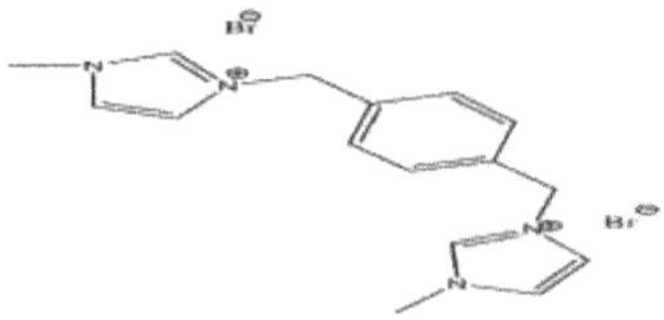 [m- 2HEA][OI]	MS/0.1 M HCl	2.5 mM	—	86
2HEA][OI] 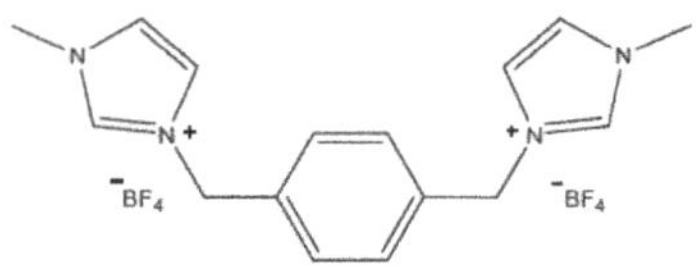	MS/1 M H_2SO_4	600 ppm	94.0%	88
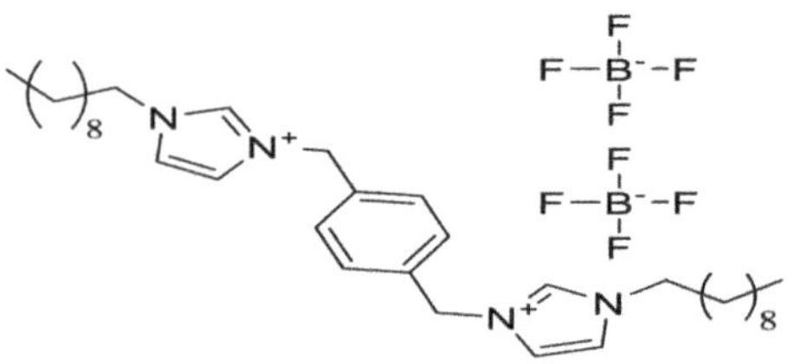	304 SS/0.5 M H_2SO_4	100 ppm	98.7%	90

attributed to their environmental friendly behaviour. Because of their natural and biological origin, biosurfactants can be considered eco-friendly alternatives to be used in place of traditional surfactant molecules. The increasing demands for green chemistry and sustainable developments do not permit the use of traditional toxic surfactants of synthetic origin as corrosion inhibitors. Their synthesis and purification require the use and discharge of huge amounts of toxic organic solvents. Generally, biosurfactants are derived as secondary metabolites by the living organism, especially by microbes such as fungi and bacteria, *etc.* The molecular weight of the biosurfactants varies between 1000 kDa and 500 Da. Some of the main classes of biosurfactants include phospholipids, neutral lipids (lipopeptides), glycolipids, glycolipids and polymers. In general, biosurfactants consist of various molecules, including fatty acids, amides, carboxylic acids, lactones, lipids and sugar, *etc.* Biosurfactants possess numerous industrial and biological applications other than corrosion inhibition. Similar to the traditional surfactants, biosurfactants become effectual by adsorbing on the metal-electrolyte interfaces.

Structure and abbreviation	Opt. Conc.	Medium	%IE (EIS)	Reference
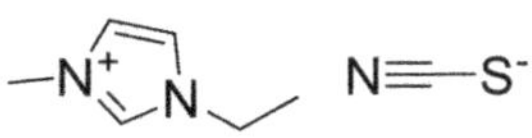 (EMIM)+(SCN)-	API 5L X52/0.5 M H_2SO_4	75 ppm	88.0%	87
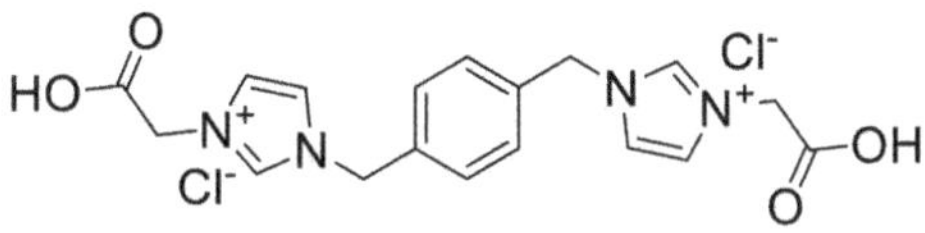	CS/0.5 M HCl	0.1 mM	94.8%	89
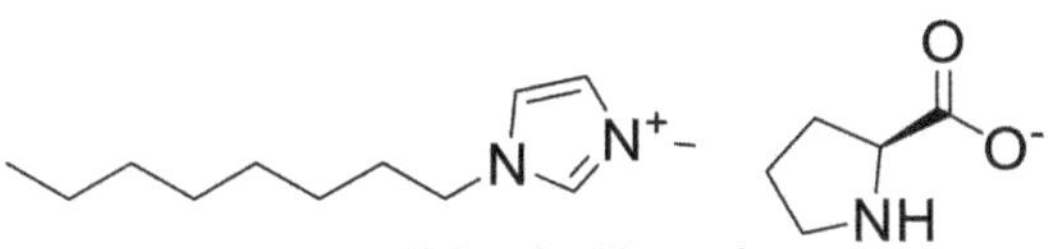 ([Omim]Lpro)	MS/0.5 M H_2SO_4	10 mM	82.2%	91

11.2 Summary

Surface active molecules are generally called surfactants and from the ongoing discussion it has been established that surfactants act as effective corrosion inhibitors for different metal/electrolyte systems. These molecules are composed of a polar hydrophilic head and a non-polar hydrophobic hydrocarbon chain. Generally, these molecules become effective by their adsorption on the metallic surface in such as way that the polar heads orient toward the metal surface and non-polar tails orient towards the electrolyte.[93–95] Surfactants can be classified as cationic-, anion-, zwitterionic- and non-ionic-type depending upon the charge(s) present on them. They can also be classified as bolaform and gemini surfactants. Through electrochemical investigations it is established that surfactants and biosurfactants behave as mixed- and interface-type corrosion inhibitors. Obviously, they adsorb at the interface of metal and electrolyte, thereby imposing a barrier for the charge transfer process. This type of adsorption also blocks the active sites presented over the metallic surface responsible for corrosion, which ultimately decreases

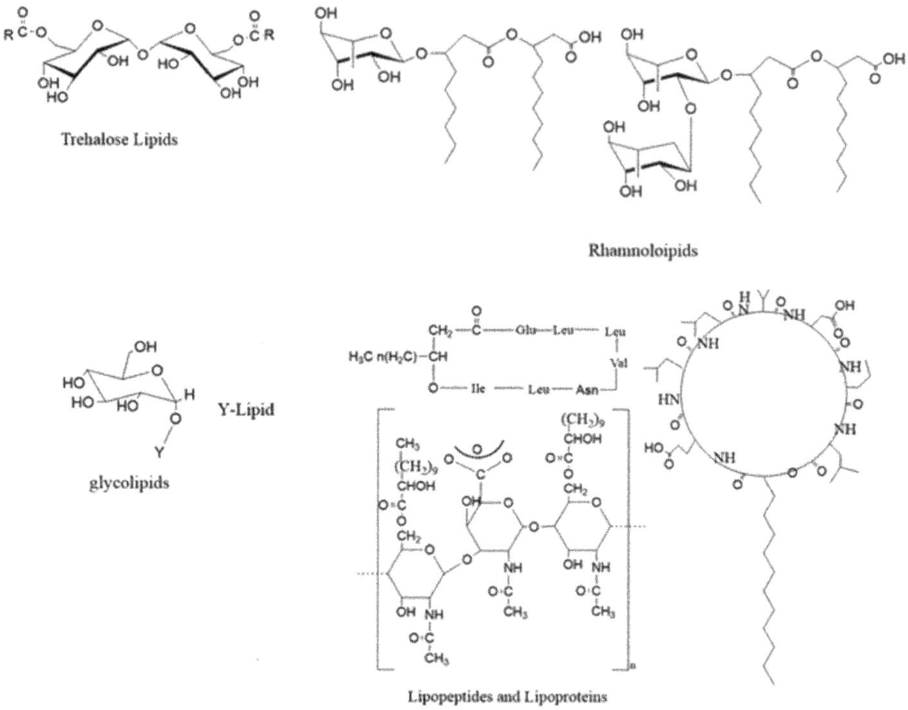

Figure 11.4 Some common examples of biosurfactants.

the value of the corrosion current density. Adsorption of the surfactants and biosurfactants mostly followed the Langmuir adsorption isotherm model. It has also been observed that increases in the hydrophobic hydrocarbon chain cause subsequent increases in the inhibition performance of surfactants and biosurfactants. Numerous surface investigations such as SEM, AFM, EDX, XRD and XPS *etc.* are widely used to describe the adsorption behaviour of corrosion inhibition. Computational based studies, especially DFT, MDS and MCS, are also used to describe the mode and nature of interactions between surfactants and biosurfactants and metallic surfaces.

References

1. M. J. Rosen and J. T. Kunjappu, *Surfactants and Interfacial Phenomena*, John Wiley & Sons, 2012.
2. D. G. Cooper, Biosurfactants, *Microbiol. Sci.*, 1986, **3**, 145–149.
3. J. Zajic, W. Seffens and C. Panchal, Biosurfactants, *Crit. Rev. Biotechnol.*, 1983, **1**, 87–107.
4. V. P. Torchilin, Structure and design of polymeric surfactant-based drug delivery systems, *J. Control. Release*, 2001, **73**, 137–172.

5. H. Wang, Dispersing carbon nanotubes using surfactants, *Curr. Opin. Colloid Interface Sci.*, 2009, **14**, 364–371.
6. W. G. Gorman and G. D. Hall, Use of dielectric constants in the classification of surfactants, *J. Pharm. Sci.*, 1963, **52**, 442–446.
7. T. F. Tadros, *Applied Surfactants: Principles and Applications*, John Wiley & Sons, 2006.
8. G. Bleys and P. Joos, Adsorption kinetics of bolaform surfactants at the air/water interface, *J. Phys. Chem.*, 1985, **89**, 1027–1032.
9. S. Goh, S. Chin, L. Lee, M. A. Yarmo and N. Nik Yusof, *Synthesis and Characterization of New Bolaform Surfactant from Palm Oil Derivative, Advanced Materials Research*, Trans Tech Publ, 2012, pp. 95–99.
10. F. M. Menger and J. S. Keiper, Gemini surfactants, *Angew. Chem., Int. Ed.*, 2000, **39**, 1906–1920.
11. Y. Zhu, M. L. Free, R. Woollam and W. Durnie, A review of surfactants as corrosion inhibitors and associated modeling, *Prog. Mater. Sci.*, 2017, **90**, 159–223.
12. N. Negm, A. Al Sabagh, M. Migahed, H. A. Bary and H. El Din, Effectiveness of some diquaternary ammonium surfactants as corrosion inhibitors for carbon steel in 0.5 M HCl solution, *Corros. Sci.*, 2010, **52**, 2122–2132.
13. M. Migahed and A. Al-Sabagh, Beneficial role of surfactants as corrosion inhibitors in petroleum industry: a review article, *Chem. Eng. Commun.*, 2009, **196**, 1054–1075.
14. R. Aslam, M. Mobin, J. Aslam, A. Aslam, S. Zehra and S. Masroor, Application of surfactants as anticorrosive materials: A comprehensive review, *Adv. Colloid Interface Sci.*, 2021, **295**, 102481.
15. E. A. Badr, Inhibition effect of synthesized cationic surfactant on the corrosion of carbon steel in 1 M HCl, *J. Ind. Eng. Chem.*, 2014, **20**, 3361–3366.
16. S. M. Shaban, I. Aiad, A. H. Moustafa and O. H. Aljoboury, Some alginates polymeric cationic surfactants; surface study and their evaluation as biocide and corrosion inhibitors, *J. Mol. Liq.*, 2019, **273**, 164–176.
17. Z. Wei, X. Chen, J. Duan, G. Zhan, Y. Wei and A. Zhang, Branched chain *versus* straight chain fluorinated surfactant: a comparative study of their anticorrosion performance on carbon steel, *J. Mol. Liq.*, 2019, **280**, 327–333.
18. A. Fouda, M. Elmorsi, T. Fayed, S. M. Shaban and O. Azazy, Corrosion inhibition of novel prepared cationic surfactants for API N80 carbon steel pipelines in oil industries, *Surf. Eng. Appl. Electrochem.*, 2018, **54**, 180–193.
19. A. Fouda, S. Rashwan, S. M. Shaban, H. E. Ibrahim and M. Elbhrawy, Evaluation of a novel cationic surfactant based on 2-(2 (dimethylamino) ethoxy) ethanol as a corrosion inhibitor for carbon steel 1018 in 1.0 M HCl solution, *Egypt. J. Pet.*, 2018, **27**, 295–306.
20. J. Li, Y. Xie, W. Li, R. Zhu and W. Wu, Synthesis and inhibition behavior of the quaternary ammonium salt-type tetrameric surfactant for corrosion of N80 steel in HCl medium, *Prot. Met. Phys. Chem. Surf.*, 2019, **55**, 789–794.

21. A. Khamis, M. M. Saleh, M. I. Awad and B. El-Anadouli, Enhancing the inhibition action of cationic surfactant with sodium halides for mild steel in 0.5 M H_2SO_4, *Corros. Sci.*, 2013, **74**, 83–91.
22. K. Shalabi, A. Helmy, A. El-Askalany and M. Shahba, New pyridinium bromide mono-cationic surfactant as corrosion inhibitor for carbon steel during chemical cleaning: Experimental and theoretical studies, *J. Mol. Liq.*, 2019, **293**, 111480.
23. A. Yousefi, S. Javadian and J. Neshati, A new approach to study the synergistic inhibition effect of cationic and anionic surfactants on the corrosion of mild steel in HCl solution, *Ind. Eng. Chem. Res.*, 2014, **53**, 5475–5489.
24. I. Aiad, M. M. El-Sukkary, E. Soliman, M. Y. El-Awady and S. M. Shaban, Inhibition of mild steel corrosion in acidic medium by some cationic surfactants, *J. Ind. Eng. Chem.*, 2014, **20**, 3524–3535.
25. M. Hegazy, A. Badawi, S. Abd El Rehim and W. Kamel, Corrosion inhibition of carbon steel using novel *N*-(2-(2-mercaptoacetoxy)ethyl)-*N*, *N*-dimethyl dodecan-1-aminium bromide during acid pickling, *Corros. Sci.*, 2013, **69**, 110–122.
26. A. Fouda, M. Migahed, A. Atia and I. Mousa, Corrosion inhibition and adsorption behavior of some cationic surfactants on carbon steel in hydrochloric acid solution, *J. Bio- Tribo-Corros.*, 2016, **2**, 1–12.
27. R. H. Tammam, A. M. Fekry and M. M. Saleh, Understanding different inhibition actions of surfactants for mild steel corrosion in acid solution, *Int. J. Electrochem. Sci.*, 2016, **11**, 1310–1326.
28. H. M. Abd El-Lateef, A. H. Tantawy and A. A. Abdelhamid, Novel quaternary ammonium-based cationic surfactants: Synthesis, surface activity and evaluation as corrosion inhibitors for C1018 carbon steel in acidic chloride solution, *J. Surfactants Deterg.*, 2017, **20**, 735–753.
29. S. Masroor, M. Mobin, M. J. Alam and S. Ahmad, The novel iminium surfactant p-benzylidene benzyldodecyl iminium chloride as a corrosion inhibitor for plain carbon steel in 1 M HCl: electrochemical and DFT evaluation, *RSC Adv.*, 2017, **7**, 23182–23196.
30. P. Han, C. Chen, W. Li, H. Yu, Y. Xu, L. Ma and Y. Zheng, Synergistic effect of mixing cationic and nonionic surfactants on corrosion inhibition of mild steel in HCl: experimental and theoretical investigations, *J. Colloid Interface Sci.*, 2018, **516**, 398–406.
31. M. Saleh, M. G. Mahmoud and H. M. Abd El-Lateef, Comparative study of synergistic inhibition of mild steel and pure iron by 1-hexadecylpyridinium chloride and bromide ions, *Corros. Sci.*, 2019, **154**, 70–79.
32. Y. El Kacimi, H. El Bakri, K. Alaoui, R. Touir, M. Galai, M. E. Touhami and M. Doubi, Synergistic effect study of Cetyl Trimethyl Ammonium Bromide with iodide ions at low concentration for mild steel corrosion in 5.0 M HCl medium, *Chem. Data Collect.*, 2020, **30**, 100558.
33. W. S. Abdrabo, B. Elgendy, K. A. Soliman, H. M. Abd El-Lateef and A. H. Tantawy, Synthesis, assessment and corrosion protection investigations of some novel peptidomimetic cationic surfactants: Empirical and theoretical insights, *J. Mol. Liq.*, 2020, **315**, 113672.

34. D. Wang, Y. Li, B. Chen and L. Zhang, Novel surfactants as green corrosion inhibitors for mild steel in 15% HCl: Experimental and theoretical studies, *Chem. Eng. J.*, 2020, **402**, 126219.
35. X. Li, S. Deng and X. Xie, Inhibition effect of tetradecylpyridinium bromide on the corrosion of cold rolled steel in 7.0 M H_3PO_4, *Arabian J. Chem.*, 2017, **10**, S3715–S3724.
36. A. Fouda, M. Elmorsi, S. M. Shaban, T. Fayed and O. Azazy, Evaluation of *N*-(3-(dimethyl hexadecyl ammonio) propyl) palmitamide bromide as cationic surfactant corrosion inhibitor for API N80 steel in acidic environment, *Egypt. J. Pet.*, 2018, **27**, 683–694.
37. A. A. Ali and W. I. El-Dougdoug, Preparation and evaluation of amido poly amine surfactant based on *Melia azedarach* seeds oil as corrosion inhibitor of C-steel in 2.0 M HCl pickling medium, *Green Chem. Lett. Rev.*, 2017, **10**, 346–358.
38. A. Fouda, S. Etaiw, A. T. El-Habab and A. Wahba, Synthesis, characterization, and application of new nonionic surfactant as a corrosion inhibitor for carbon steel in 1 M hydrochloric acid solution, *J. Bio- Tribo-Corros.*, 2020, **6**, 1–9.
39. N. El Basiony, A. Elgendy, A. E. El-Tabey, A. Al-Sabagh, G. M. Abd El-Hafez, M. Abd El-raouf and M. Migahed, Synthesis, characterization, experimental and theoretical calculations (DFT and MC) of ethoxylated aminothiazole as inhibitor for X65 steel corrosion in highly aggressive acidic media, *J. Mol. Liq.*, 2020, **297**, 111940.
40. W. Zhang, H.-J. Li, C. Wang, L.-J. Wang, G. Li, H. Ma, Q. Pan and Y.-C. Wu, Synergistic effect of 1-(2, 5-dioxoimidazolidin-4-yl) urea and Tween-80 towards the corrosion mitigation of mild steel in HCl, *New J. Chem.*, 2019, **43**, 13899–13910.
41. A. El-Maksoud, F. El-Dossoki, M. Migahed, M. Gouda and E.-S. R. El-Gharkawy, New imidazol-1-ium bromide derivative surfactants as corrosion inhibitors for carbon steel in 1 M HCl solutions: Experimental and theoretical studies, *J. Bio- Tribo-Corros.*, 2021, **7**, 1–15.
42. A. A. Abd-Elaal, N. Elbasiony, S. M. Shaban and E. Zaki, Studying the corrosion inhibition of some prepared nonionic surfactants based on 3-(4-hydroxyphenyl) propanoic acid and estimating the influence of silver nanoparticles on the surface parameters, *J. Mol. Liq.*, 2018, **249**, 304–317.
43. F. Branzoi and V. Branzoi, Investigation of some nonionic surfactants as corrosion inhibitors for carbon steel in sulfuric acid medium, *Int. J. Electrochem. Sci.*, 2017, **12**, 7638–7658.
44. M. H. Sliem, M. Afifi, A. B. Radwan, E. M. Fayyad, M. F. Shibl, F. E.-T. Heakal and A. M. Abdullah, AEO7 surfactant as an eco-friendly corrosion inhibitor for carbon steel in HCl solution, *Sci. Rep.*, 2019, **9**, 1–16.
45. N. A. Negm, M. A. El Hashash, A. Abd-Elaal, S. M. Tawfik and A. Gharieb, Amide type nonionic surfactants: synthesis and corrosion inhibition evaluation against carbon steel corrosion in acidic medium, *J. Mol. Liq.*, 2018, **256**, 574–580.

46. M. Sobhi, R. El-Sayed and M. Abdallah, The effect of non ionic surfactants containing triazole, thiadiazole and oxadiazole as inhibitors of the corrosion of carbon steel in 1 M hydrochloric acid, *J. Surfactants Deterg.*, 2013, **16**, 937–946.
47. M. Abdallah, H. Altass, R. El-Sayed, A. S. Al Gorair, B. A. Jahdaly and M. Sobhi, Synthesis and estimation of some surface-active compounds derived from fused pyridine as corrosion inhibitors for aluminum in hydrochloric acid solutions, *Prot. Met. Phys. Chem. Surf.*, 2021, **57**, 811–819.
48. K. Shalabi and A. A. Nazeer, Ethoxylates nonionic surfactants as promising environmentally safe inhibitors for corrosion protection of reinforcing steel in 3.5% NaCl saturated with $Ca(OH)_2$ solution, *J. Mol. Struct.*, 2019, **1195**, 863–876.
49. Z. Liu, P. Hedayati, E. J. Sudhölter, R. Haaring, A. R. Shaik and N. Kumar, Adsorption behavior of anionic surfactants to silica surfaces in the presence of calcium ion and polystyrene sulfonate, *Colloids Surf., A*, 2020, **602**, 125074.
50. N. A. Negm, A. F. El-Farargy, E. A. Abdel Halim, S. El-lboudy and A. E. S. I. Ahmed, Novel biobased nonionic surfactants: synthesis, surface activity and corrosion inhibition efficiency against aluminum alloy dissolution in acidic media, *J. Surfactants Deterg.*, 2014, **17**, 1203–1211.
51. M. Deyab, Effect of nonionic surfactant as an electrolyte additive on the performance of aluminum-air battery, *J. Power Sources*, 2019, **412**, 520–526.
52. A. El-Tabei and M. Hegazy, A corrosion inhibition study of a novel synthesized gemini nonionic surfactant for carbon steel in 1 M HCl solution, *J. Surfactants Deterg.*, 2013, **16**, 757–766.
53. M. A. Hegazy and A. S. El-Tabei, Synthesis, surface properties, synergism parameter and inhibitive performance of novel cationic gemini surfactant on carbon steel corrosion in 1 M HCl solution, *J. Surfactants Deterg.*, 2013, **16**, 221–232.
54. M. M. Khalaf, A. H. Tantawy, K. A. Soliman and H. M. Abd El-Lateef, Cationic gemini-surfactants based on waste cooking oil as new 'green' inhibitors for N80-steel corrosion in sulphuric acid: a combined empirical and theoretical approaches, *J. Mol. Struct.*, 2020, **1203**, 127442.
55. T. Zhou, J. Yuan, Z. Zhang, X. Xin and G. Xu, The comparison of imidazolium Gemini surfactant $[C_{14}\text{-}4\text{-}C_{14}im]Br_2$ and its corresponding monomer as corrosion inhibitors for A3 carbon steel in hydrochloric acid solutions: Experimental and quantum chemical studies, *Colloids Surf., A*, 2019, **575**, 57–65.
56. M. Chafiq, A. Chaouiki, M. Damej, H. Lgaz, R. Salghi, I. H. Ali, M. Benmessaoud, S. Masroor and I.-M. Chung, Bolaamphiphile-class surfactants as corrosion inhibitor model compounds against acid corrosion of mild steel, *J. Mol. Liq.*, 2020, **309**, 113070.
57. T. Han, J. Guo, Q. Zhao, Y. Wu and Y. Zhang, Enhanced corrosion inhibition of carbon steel by pyridyl gemini surfactants with different alkyl chains, *Mater. Chem. Phys.*, 2020, **240**, 122156.

58. C. Yin, M. Kong, J. Zhang, Y. Wang, Q. Ma, Q. Chen and H. Liu, Influence of hydroxyl groups on the inhibitive corrosion of Gemini surfactant for carbon steel, *ACS Omega*, 2020, **5**, 2620–2629.
59. A. Mona, S. Abdelhamed and M. D. Allah, Synthesis of novel Gemini surfactants based on succinic acid and their application as inhibitors for carbon steel corrosion, *J. Bio- Tribo-Corros.*, 2020, **6**, 1–12.
60. S. M. Shaban, I. Aiad, F. A. Yassin and A. Mosalam, The tail effect of some prepared cationic surfactants on silver nanoparticle preparation and their surface, thermodynamic parameters, and antimicrobial activity, *J. Surfactants Deterg.*, 2019, **22**, 1445–1460.
61. A. Adewuyi, A. Göpfert and T. Wolff, Succinyl amide gemini surfactant from Adenopus breviflorus seed oil: A potential corrosion inhibitor of mild steel in acidic medium, *Ind. Crops Prod.*, 2014, **52**, 439–449.
62. A. El-Tabei and M. Hegazy, Synthesis and characterization of a novel nonionic gemini surfactant as corrosion inhibitor for carbon steel in acidic solution, *Chem. Eng. Commun.*, 2015, **202**, 851–863.
63. A. A. Abd-Elaal, S. M. Shaban and S. M. Tawfik, Three Gemini cationic surfactants based on polyethylene glycol as effective corrosion inhibitor for mild steel in acidic environment, *J. Assoc. Arab Univ. Basic Appl. Sci.*, 2017, **24**, 54–65.
64. H. M. Abd El-Lateef, K. A. Soliman and A. H. Tantawy, Novel synthesized Schiff base-based cationic gemini surfactants: electrochemical investigation, theoretical modeling and applicability as biodegradable inhibitors for mild steel against acidic corrosion, *J. Mol. Liq.*, 2017, **232**, 478–498.
65. A. E. Elkholy and F. E.-T. Heakal, Electrochemical measurements and semi-empirical calculations for understanding adsorption of novel cationic Gemini surfactant on carbon steel in H_2SO_4 solution, *J. Mol. Struct.*, 2018, **1156**, 473–482.
66. M. Mahdavian, A. R. Tehrani-Bagha, E. Alibakhshi, S. Ashhari, M. J. Palimi, S. Farashi, S. Javadian and F. Ektefa, Corrosion of mild steel in hydrochloric acid solution in the presence of two cationic gemini surfactants with and without hydroxyl substituted spacers, *Corros. Sci.*, 2018, **137**, 62–75.
67. M. Pakiet, I. H. Kowalczyk, R. L. Garcia, R. Akid and B. E. Brycki, Influence of different counterions on gemini surfactants with polyamine platform as corrosion inhibitors for stainless steel AISI 304 in 3 M HCl, *J. Mol. Liq.*, 2018, **268**, 824–831.
68. M. Migahed, M. El-Rabiei, H. Nady and E. Zaki, Novel Gemini cationic surfactants as anti-corrosion for X-65 steel dissolution in oilfield produced water under sweet conditions: Combined experimental and computational investigations, *J. Mol. Struct.*, 2018, **1159**, 10–22.
69. I. Hamed, M. M. Osman, O. H. Abdelraheem and M. I. Nessim, Inhibition of API 5L X52 pipeline steel corrosion in acidic medium by gemini surfactants: Electrochemical evaluation and computational study, *Int. J. Corros.*, 2019, **2019**, 4857181.

70. A. C. Hernandez, E. V. Velez, J. Uruchurtu and J. G. G. Rodriguez, A Corrosion inhibition study of a novel nonionic gemini surfactant derived from palm oil for supermartensitic stainless steel in sour solution, *ECS Trans.*, 2019, **94**, 41.
71. M. Hegazy, R. Samy, A. Labena, M. A. Wadaan and W. N. Hozzein, 4, 4′-(((1*E*, 5*E*)-pentane-1, 5-diylidene) bis (azanylylidene)) bis (1-dodecylpyridin-1-ium) bromide as a novel corrosion inhibitor in an acidic solution (part I), *Mater. Sci. Eng., C*, 2020, **110**, 110673.
72. N. S. Bin-Hudayb, E. E. Badr and M. Hegazy, Adsorption and corrosion performance of new cationic Gemini surfactants derivatives of fatty amido ethyl aminium chloride with ester spacer for mild steel in acidic solutions, *Materials*, 2020, **13**, 2790.
73. S. M. Shaban, S. Abd Elsamad, S. M. Tawfik, A.-H. Adel and I. Aiad, Studying surface and thermodynamic behavior of a new multi-hydroxyl Gemini cationic surfactant and investigating their performance as corrosion inhibitor and biocide, *J. Mol. Liq.*, 2020, **316**, 113881.
74. K. Qiao and Y. Zeng, Comparative study on two imidazolium-based ionic liquid surfactants as corrosion inhibitors for N80 steel in 15% hydrochloric acid solution, *Mater. Corros.*, 2020, **71**, 1913–1926.
75. E. K. Ardakani, E. Kowsari and A. Ehsani, Imidazolium-derived polymeric ionic liquid as a green inhibitor for corrosion inhibition of mild steel in 1.0 M HCl: Experimental and computational study, *Colloids Surf., A*, 2020, **586**, 124195.
76. A. M. Atta, G. A. El-Mahdy, H. A. Allohedan and M. Abdullah, Adsorption characteristics and corrosion inhibition efficiency of ethoxylated octadecylamine ionic liquid in aqueous acid solution, *Int. J. Electrochem. Sci.*, 2016, **11**, 882–898.
77. G. A. El-Mahdy, A. M. Atta, H. A. Al-Lohedan and A. O. Ezzat, Influence of green corrosion inhibitor based on chitosan ionic liquid on the steel corrodibility in chloride solution, *Int. J. Electrochem. Sci.*, 2015, **10**, 5812–5826.
78. A. M. Atta, G. A. El-Mahdy, H. A. Al-Lohedan and A. R. O. Ezzat, A new green ionic liquid-based corrosion inhibitor for steel in acidic environments, *Molecules*, 2015, **20**, 11131–11153.
79. A. Farag, Postanal minimally invasive surgery “PAMIS” assisted extralevator abdominoperineal excision “ELAPE” for cancer: A novel approach in supine position, *Arab J. Gastroenterol.*, 2019, **20**, 53–55.
80. P. Kannan, J. Karthikeyan, P. Murugan, T. S. Rao and N. Rajendran, Corrosion inhibition effect of novel methyl benzimidazolium ionic liquid for carbon steel in HCl medium, *J. Mol. Liq.*, 2016, **221**, 368–380.
81. E. Kowsari, M. Payami, R. Amini, B. Ramezanzadeh and M. Javanbakht, Task-specific ionic liquid as a new green inhibitor of mild steel corrosion, *Appl. Surf. Sci.*, 2014, **289**, 478–486.
82. Y. Ma, F. Han, Z. Li and C. Xia, Acidic-functionalized ionic liquid as corrosion inhibitor for 304 stainless steel in aqueous sulfuric acid, *ACS Sustainable Chem. Eng.*, 2016, **4**, 5046–5052.

83. M. E. Mashuga, L. O. Olasunkanmi, A. S. Adekunle, S. Yesudass, M. M. Kabanda and E. E. Ebenso, Adsorption, thermodynamic and quantum chemical studies of 1-hexyl-3-methylimidazolium based ionic liquids as corrosion inhibitors for mild steel in HCl, *Materials*, 2015, **8**, 3607–3632.
84. A. Yousefi, S. Javadian, N. Dalir, J. Kakemam and J. Akbari, Imidazolium-based ionic liquids as modulators of corrosion inhibition of SDS on mild steel in hydrochloric acid solutions: experimental and theoretical studies, *RSC Adv.*, 2015, **5**, 11697–11713.
85. N. V. Likhanova, P. Arellanes-Lozada, O. Olivares-Xometl, H. Hernández-Cocoletzi, I. V. Lijanova, J. Arriola-Morales and J. Castellanos-Aguila, Effect of organic anions on ionic liquids as corrosion inhibitors of steel in sulfuric acid solution, *J. Mol. Liq.*, 2019, **279**, 267–278.
86. T. E. Schmitzhaus, M. R. Ortega Vega, R. Schroeder, I. L. Muller, S. Mattedi and C. d. F. Malfatti, *N*-methyl-2-hydroxyethylammonium oleate ionic liquid performance as corrosion inhibitor for mild steel in hydrochloric acid medium, *Mater. Corros.*, 2020, **71**, 1885–1902.
87. M. C. Luna, T. Le Manh, R. C. Sierra, J. M. Flores, L. L. Rojas and E. A. Estrada, Study of corrosion behavior of API 5L X52 steel in sulfuric acid in the presence of ionic liquid 1-ethyl 3-methylimidazolium thiocyanate as corrosion inhibitor, *J. Mol. Liq.*, 2019, **289**, 111106.
88. A. P. Hanza, R. Naderi, E. Kowsari and M. Sayebani, Corrosion behavior of mild steel in H_2SO_4 solution with 1, 4-di [1′-methylene-3′-methyl imidazolium bromide]-benzene as an ionic liquid, *Corros. Sci.*, 2016, **107**, 96–106.
89. S. Cao, D. Liu, H. Ding, J. Wang, H. Lu and J. Gui, Corrosion inhibition effects of a novel ionic liquid with and without potassium iodide for carbon steel in 0.5 M HCl solution: An experimental study and theoretical calculation, *J. Mol. Liq.*, 2019, **275**, 729–740.
90. M. Zaky, M. Nessim and M. Deyab, Synthesis of new ionic liquids based on dicationic imidazolium and their anti-corrosion performances, *J. Mol. Liq.*, 2019, **290**, 111230.
91. X. Zheng, S. Zhang, M. Gong and W. Li, Experimental and theoretical study on the corrosion inhibition of mild steel by 1-octyl-3-methylimidazolium L-prolinate in sulfuric acid solution, *Ind. Eng. Chem. Res.*, 2014, **53**, 16349–16358.
92. C. N. Mulligan, Environmental applications for biosurfactants, *Environ. Pollut.*, 2005, **133**, 183–198.
93. P. Parthipan, D. Sabarinathan, S. Angaiah and A. Rajasekar, Glycolipid biosurfactant as an eco-friendly microbial inhibitor for the corrosion of carbon steel in vulnerable corrosive bacterial strains, *J. Mol. Liq.*, 2018, **261**, 473–479.
94. G. Płaza and V. Achal, Biosurfactants: Eco-friendly and innovative biocides against biocorrosion, *Int. J. Mol. Sci.*, 2020, **21**, 2152.
95. N. Kosaric, Biosurfactants and their application for soil bioremediation, *Food Technol. Biotechnol.*, 2001, **39**, 295–304.

Subject Index